S0-EXP-530

# Plant Systematics and Evolution

Supplement 8

P. K. Endress and E. M. Friis (eds.)

# Early Evolution of Flowers

Springer-Verlag Wien New York

Prof. Dr. Peter K. Endress
Institut für Systematische Botanik der Universität,
Zürich, Schweiz

Prof. Dr. Else Marie Friis
Naturhistorika Riksmuseet
Sektionen för Paleobotanik, Stockholm, Sweden

QK
653
.E2
1994

This work is subject to copyright.
All rights are reserved, whether the whole or part of the material is concerned, specifically those of translation, reprinting, re-use of illustrations, broadcasting, reproduction by photocopying machines or similar means, and storage in data banks.
© 1994 Springer-Verlag/Wien
Typesetting: Thomson Press (India) Ltd., New Delhi, 110001
Printed in Austria by A. Holzhausens Nfg., A-1070 Wien
Binding: Hermann Scheibe Ges. m.b.H., A-2351 Wr. Neudorf

Printed on acid-free and chlorine-free bleached paper

With 130 Figures

Library of Congress Cataloging-in-Publication Data

Early evolution of flowers/P. K. Endress, E. M. Friis (eds.).
p. cm. – (Plant systematics and evolution. Supplement; 8)
ISBN 3-211-82599-1 (Wien: alk. paper). – ISBN 0-387-82599-1 (New York: alk. paper)
1. Flowers. 2. Angiosperms – Evolution. 3. Flowers – Morphology.
I. Endress, Peter K. II. Friis, Else Marie. III. Series: Plant systematics and evolution. Supplementum; 8.
QK653.E2 1994
582.13′04463 – dc20

ISSN 0172-6668
ISBN 3-211-82599-1 Springer-Verlag Wien New York
ISBN 0-387-82599-1 Springer-Verlag New York Wien

cc46641

# Contents



*Listed in Current Contents*

Pl. Syst. Evol. [Suppl.] 8: 1–6 (1994)

# Introduction – Major trends in the study of early flower evolution

Our knowledge of early angiosperm phylogeny and of flower evolution is rapidly expanding. New evidence comes from different directions, and it seems important to constantly synthesize new results from these fields. (1) Palaeobotany brought two major developments, first comparative structural and stratigraphic pollen studies (since BRENNER 1963 and DOYLE 1969), followed by the discovery of excellently preserved Cretaceous flowers (since FRIIS & SKARBY 1981). (2) Development and biology of extant flowers and structure of modern pollen was studied in many living conservative groups during the same period. Application of the SEM for comparative floral development was especially helpful (since ENDRESS 1972). The comparative study of pollen with combined SEM and TEM techniques provided a powerful approach to link extant and fossil material (since WALKER 1976). Comparative pollination biology of primitive flowers in relation to floral structure brought another new dimension (since GOTTSBERGER 1974). (3) The application of cladistic techniques facilitated the rigorous discussion on the origin of the angiosperms from other seed plants and on early angiosperm diversification (since CRANE 1985, DOYLE & DONOGHUE 1986). (4) More recently comparative molecular studies with DNA and RNA became possible due to new techniques, applied for macrosystematics of angiosperms (since JANSEN & PALMER 1987). (5) The latest development is the direct analysis of floral development with molecular genetic techniques, which, however, as yet has been done only for few model species of evolutionarily more advanced angiosperm groups (COEN & MEYEROWITZ 1991).

### Results from the fossil record

The fossil record of pollen and leaves has provided a rough framework for understanding patterns in early angiosperm evolution (BRENNER 1963, 1976; DOYLE & HICKEY 1976; HUGHES 1976). Based on these organs DOYLE & HICKEY (1976) demonstrated a progressive complexity and diversity in angiosperm morphology in the Early to mid-Cretaceous Potomac group sequence from eastern North America. Recent discoveries of well-preserved angiosperm floral structures from Cretaceous strata in Europe and North America indicate that a parallel increase in complexity in floral organs took place during the same time interval (e.g., FRIIS & CREPET 1987, FRIIS & ENDRESS 1990). Although the fossil floral structures are much more rare in the Cretaceous strata than are leaves and pollen, they provide a broader and more profound basis for comparison with modern angiosperms and the establishment of systematic affinities of early angiosperms. Results obtained so far from the study of Cretaceous flowers support the generally accepted view that the magnoliids contain the most primitive angiosperms, and that the hamamelidids diversified from the magnoliids early in the history of angiosperms. Early Cretaceous flowers are typically minute, unisexual or bisexual with few floral parts, undifferentiated perianth, massive stamens typically with valvate dehiscence, and uniovulate carpels, indicating that flowers of general lauralean or piperalean organization were well-established in the Early Cretaceous (e.g., FRIIS & ENDRESS 1990, CRANE & al., contribution 3 in this volume).

Larger floral structures related to the woody *Magnoliales* are rare in the Cretaceous fossil record (DILCHER & CRANE 1985) and appeared later than the smaller flowers. The abundance of small simple flowers early in the history of angiosperms may reflect a particular rapid diversification of lauralean taxa rather than reflecting the basal floral organization in angiosperms, and rooting of the angiosperms is still much debated (contribution by DOYLE in this volume). Based on leaf architecture and sedimentological data DOYLE & HICKEY (1976) suggested that early angiosperms were riparian weeds of magnoliid and monocot affinity (paleoherbs). A herbaceous growth form was also suggested by TAYLOR & HICKEY (1990) based on the leaf architecture of a small compression/impression fossil from the Aptian of Australia. Little is known about growth form in early angiosperms, but the scarcity of larger pieces of angiosperm wood in the Cretaceous, particularly in the Early and mid-Cretaceous, could be an indication for the absence of larger angiosperm trees in the early phases of angiosperm diversification (UPCHURCH & WOLFE 1987, HERENDEEN 1991).

While the fossil record clearly indicates that the first major radiation of angiosperms took place in the Early Cretaceous (e.g., DOYLE & HICKEY 1976, CRANE & LIDGARD 1989) phylogenetic analysis indicates that angiosperms may have originated much earlier, possibly in the Triassic (DOYLE & DONOGHUE 1993, CRANE 1993). An early radiation of angiosperms has also been claimed by CORNET (e.g., 1989a, b) based on dispersed pollen and the enigmatic *Sanmiguelia* plant. However, the nature of these fossils is still poorly understood and their angiosperm affinity remains to be established. Although several of the dispersed pollen grains show angiospermous features such as reticulate and tectate exine (e.g., CORNET 1989b), details in wall ultrastructure and aperture configuration indicate that they are more related to non-angiospermous anthophytes (see discussion in FRIIS & PEDERSEN 1995).

## Results from the study of the structure and biology of extant flowers

The extant *Magnoliidae* contain many families with a single or only few species. In the classification of CRONQUIST (1988) of the 39 families 14 contain a single genus. Some of the families with more than one genus have a pronouncedly scattered relictual distribution (such as *Chloranthaceae*, *Winteraceae*, *Lardizabalaceae*), and some are known to have occurred in a much wider area in the Cretaceous or early Tertiary (e.g., *Chloranthaceae*, *Winteraceae*) and have been more diverse than today (KRUTZSCH 1989).

Spiral phyllotaxis, which is mostly considered to be the basal pattern of arrangement of floral organs, is less generally common in the extant *Magnoliidae* than was earlier believed (TUCKER 1984, ENDRESS 1986). Irregular and whorled patterns are equally common. All these patterns may co-occur at low systematic levels, even within single species. Variability in floral organ number may be equally extensive (ENDRESS 1986). There are many families with predominantly unicarpellate flowers and many with uniovulate carpels. There are several families without a distinct perianth, and many with a simple perianth that is not differentiated into calyx and corolla. The delimitation of a flower towards the bract region outside the flower is not always clear-cut. Valvate dehiscence of anthers is widespread (ENDRESS & HUFFORD 1989). In some groups, especially in *Ceratophyllales*, *Piperales* and in some *Laurales*, the flowers are small, sometimes exceedingly so. Large, beetle-pollinated flowers show peculiar specializations. Their organization does seem less primitive than that of some smaller, less functionally differentiated flowers (GOTTSBERGER 1974, 1988; GOTTSBERGER & al. 1980). Flies or small moths may contain pollinators as primitive as beetles (THIEN & al. 1985, PELLMYR 1992, KATO & INOUE 1994). Floral secretions, such as nectar and stigmatic fluid, in addition to pollen, may constitute important rewards for pollinators in magnoliid flowers. Protogyny is the dominant kind of dichogamy in bisexual flowers (ENDRESS 1990), which could also indicate that secretions, and not pollen, were a primitive reward. Optical and olfactorial attraction of pollinators is by stamens and carpels in many groups, and not by the perianth. The advent of anatropous ovules may have preceded that of closed carpels (TOMLINSON 1991, LLOYD & WELLS 1992). However, some paleoherbs, such as *Ceratophyllaceae* and *Piperales*, and also *Chloranthaceae*, have orthotropous ovules, which may be viewed as apomorphies due to particular ovary architectures.

A consequence of these various new developments is that classification of major angiosperm groups is under extensive revision. The application of some informal names for the most problematical groups, such as, e.g., "paleoherbs", seems wise before a new classification can be worked out on a relatively consolidated new basis. As a working basis we here still use the classification by CRONQUIST (1988) in addition to the informal names.

As its appears now, earlier assumptions on primitive angiosperms were not completely wrong. What remains valid in particular is that the *Magnoliidae* are the most basal group. However, it has become uncertain whether the woody *Magnoliales* are at the base. There are equally good arguments to place herbaceous groups (paleoherbs) at the base of the *Magnoliidae* (e.g., DOYLE & HICKEY 1976, ZIMMER & al. 1989, LES & al. 1991, TAYLOR & HICKEY 1992, CHASE & al. 1993, DOYLE & DONOGHUE 1993, QIU & al. 1993, and discussion in the contribution by DOYLE in this volume).

This volume includes new advances in our understanding of early angiosperm and flower evolution and brings together contributions from palaeobotanical as well as neobotanical studies. The contributions were presented at two symposia held at the XV International Botanical Congress in Yokohama on September 1, 1993, and organized by PETER K. ENDRESS, ELSE MARIE FRIIS, HARUFUMI NISHIDA, and MICHIO TAMURA.

The first contribution discusses various hypotheses on the evolution of flowers and floral organs based on recent phylogenetic-cladistic analyses (DOYLE). The following three contributions give a survey of the vast diversity of Cretaceous magnoliid floral organs discovered during the past few years, and two chapters deal with higher dicots. Early Cretaceous assemblages from Portugal (Friis & al., contribution 2) and from eastern North America (Crane & al., contribution 3) exhibit a wide range of magnoliid fossils, and particularly taxa related to the *Laurales* appear to be well differentiated, but it is characteristic that none of the fossils can be assigned with certainty to a modern family. Magnoliid fossils continue to be important also in Late Cretaceous (Turonian) floras of eastern North America and include several taxa that are comparable to extant taxa at the family level (Crepet & Nixon, contribution 4). The earliest diversification of the *Hamamelididae* started in the Early Cretaceous; and a large variety of taxa are present by the Late Cretaceous (Drinnan & al., contribution 5). By that time most higher dicotyledon subclasses were established, and the new genus *Elsemaria* described from the rich Coniacian to Santonian flora of Hokkaido, Japan, exemplifies a permineralized capsular fruit of possible dilleniid affinity (Nishida, contribution 6). For age and sequence of stratigraphic names used in the palaeobotanical contributions readers may consult Fig. 4 in contribution 5 (Drinnan & al.). The neobotanical papers deal with floral development and biological aspects of previously poorly known magnoliids: *Schisandra* of *Schisandraceae* (Tucker & Bourland, contribution 7), *Barclaya* of *Nymphaeaceae* (Williamson & Schneider, contribution 8), *Ceratophyllum* of *Ceratophyllaceae* (Endress, contribution 9, not read at the Symposium). Among ranunculids petal evolution is traced in *Ranunculaceae* (Kosuge, contribution 10). Potential relationships of magnoliid flowers concerning outstanding features to those of other groups are discussed for monocots (Erbar & Leins, contribution 11) and for dilleniids and caryophyllids (Leins & Erbar, contribution 12).

## References


Brenner, G. J., 1963: The spores and pollen of the Potomac Group of Maryland. – Bull. Maryland Dept. Geol. Mines Water Resources **27**: 1–215.

– 1976: Middle Cretaceous floral provinces and early migration of angiosperms. – In Beck, C. B., (Ed.): Origin and early evolution of angiosperms, pp. 23–47. – New York: Columbia University Press.

Chase, M. W., Soltis, D. E., Olmstead, R. G., Morgan, D., Les, D. H., Mishler, B. D., Duvall, M. R., Price, R. A., Hills, H. G., Qiu, Y.-L., Kron, K. A., Rettig, J. H., Conti, E., Palmer, J. H., Manhart, J. R., Sytsma, K. J., Michaels, H. J., Kress, W. J., Karol, K. G., Clark, W. D., Hedrén, M., Gaut, B. S., Jansen, R. K., Kim, K.-J., Wimpee, C. F., Smith, J. F., Furnier, G. R., Strauss, S. H., Xiang, Q.-Y., Plunckett, G. M., Soltis, P. S., Swensen, S. M., Williams, S. E., Gadek, P. A., Quinn, C. J., Eguiarte, L. E., Golenberg, E., Learn, G. H., Jr., Graham, S. W., Barrett, S. C. H., Dayanandan, S., Albert, V. A., 1993: Phylogenetics of seed plants: an analysis of nucleotide sequences from the plastid gene *rbc*L. – Ann. Missouri Bot. Gard. **80**: 528–580.

Coen, E. S., Meyerowitz, E. M., 1991: The war of the whorls: genetic interactions controlling flower development. – Nature **353**: 31–37.

Cornet, B., 1989a: The reproductive morphology and biology of *Sanmiguelia lewisii*, and its bearing on angiosperm evolution in the Late Triassic. – Evol. Trends Pl. **3**: 25–51.

– 1989b: Late Triassic angiosperm-like pollen from the Richmond Rift Basin of Virginia, U.S.A. – Palaeontograph. **213B**: 37–87.
CRANE, P. R., 1985: Phylogenetic relationships in seed plants. – Cladistics **1**: 329–348.
– 1993: Time for the angiosperms. – Nature **366**: 631–632.
– LIDGARD, S., 1989: Angiosperm diversification and paleolatitudinal gradients in Cretaceous floristic diversity. – Science **246**: 675–678.
CRONQUIST, A., 1988: The evolution and classification of flowering plants. 2nd edn. – Bronx: New York Botanical Garden.
DILCHER, D. L., CRANE, P. R., 1985: *Archaeanthus*: an early angiosperm from the Cenomanian of the western interior of North America. – Ann. Missouri Bot. Gard. **71**: 351–383.
DOYLE, J. A., 1969: Cretaceous angiosperm pollen of the Atlantic coastal plain and its evolutionary significance. – J. Arnold Arbor. **50**: 1–35.
– DONOGHUE, M. J., 1986: Seed plant phylogeny and the origin of angiosperms: an experimental cladistic approach. – Bot. Rev. **52**: 321–431.
– – 1993: Phylogenies and angiosperm diversification. – Paleobiol. **19**: 141–167.
– HICKEY, L. J., 1976: Pollen and leaves from the mid-Cretaceous Potomac Group and their bearing on early angiosperm evolution. – In BECK, C. B., (Ed.): Origin and early evolution of angiosperms, pp. 139–206. – New York: Columbia University Press.
ENDRESS, P. K., 1972: Zur vergleichenden Entwicklungsmorphologie, Embryologie und Systematik bei *Laurales*. – Bot. Jahrb. Syst. **92**: 331–428.
– 1986: Reproductive structures and phylogenetic significance of extant primitive angiosperms. – Pl. Syst. Evol. **152**: 1–28.
– 1990: Evolution of reproductive structures and functions in primitive angiosperms (*Magnoliidae*). – Mem. New York Bot. Gard. **55**: 5–34.
– HUFFORD, L. D., 1989: The diversity of stamen structures and dehiscence patterns among *Magnoliidae*. – Bot. J. Linn. Soc. **100**: 45–85.
FRIIS, E. M., CREPET, W. L., 1987: Time of appearance of floral features. – In FRIIS, E. M., CHALONER, W. G., CRANE, P. R., (Eds): The origins of angiosperms and their biological consequences, pp. 145–179. – Cambridge: Cambridge University Press.
– ENDRESS, P. K., 1990: Origin and evolution of angiosperm flowers. – Adv. Bot. Res. **17**: 99–162.
– PEDERSEN, K. R., 1995: Angiosperm pollen in situ in Cretaceous reproductive organs. – In JANSONIUS, J., MCGREGOR, D. C., (Eds): Palynology: principles and applications. – Amer. Assoc. Stratigraphic Palynol. Foundation **1** (in press).
– SKARBY, A., 1981: Structurally preserved angiosperm flowers from the Upper Cretaceous of southern Sweden. – Nature **291**: 485–486.
GOTTSBERGER, G., 1974: The structure and function of the primitive angiosperm flower – a discussion. – Acta Bot. Neerl. **23**: 461–471.
– 1988: The reproductive biology of primitive angiosperms. – Taxon **37**: 630–643.
– SILBERBAUER-GOTTSBERGER, I., EHRENDORFER, F., 1980: Reproductive biology in the primitive relic angiosperm *Drimys brasiliensis* (*Winteraceae*). – Pl. Syst. Evol. **135**: 11–39.
HERENDEEN, P. S., 1991: Charcoalified angiosperm wood from the Cretaceous of eastern North America and Europe. – Rev. Palaeobot. Palynol. **70**: 225–239.
HUGHES, N. F., 1976: Palaeobiology of angiosperm origins. – Cambridge: Cambridge University Press.
JANSEN, R. K., PALMER, J. D., 1987: A chloroplast DNA inversion marks an ancient evolutionary split in the sunflower family (*Asteraceae*). – Proc. Natl. Acad. Sci. USA **84**: 5818–5822.
KATO, M., INOUE, T., 1994: Origin of insect pollination. – Nature **368**: 195.
KRUTZSCH, W., 1989: Paleogeography and historical phytogeography (paleochorology) in the Neophyticum. – Pl. Syst. Evol. **162**: 5–61.

LES, D. H., GAVIN, D. K., WIMPEE, C. F., 1991: Molecular evolutionary history of ancient aquatic angiosperms. – Proc.. Natl. Acad. Sci. USA **88**: 10119–10123.

LLOYD, D. G., WELLS, M. S., 1992: Reproductive biology of a primitive angiosperm. *Pseudowintera colorata* (*Winteraceae*), and the evolution of pollination systems in the *Anthophyta*. – Pl. Syst. Evol. **181**: 77–95.

PELLMYR, O., 1992: Evolution of insect pollination and angiosperm diversification. – Trends Ecol. Evol. **7**: 46–49.

QIU, Y.-L., CHASE, M. W., LES, D. H., PARKS, C. R., 1993: Molecular phylogenetics of the *Magnoliidae*: cladistic analyses of nucleotide sequences of the plastid gene *rbc*L. – Ann. Missouri Bot. Gard. **80**: 587–606.

TAYLOR, D. W., HICKEY, L. J., 1990: An Aptian plant with attached leaves and flowers: implications for angiosperm origin. – Science **247**: 702–704.

– – 1992: Phylogenetic evidence for the herbaceous origin of angiosperms. – Pl. Syst. Evol. **180**: 137–156.

THIEN, L. B., BERNHARDT, P., GIBBS, G. W., PELLMYR, O., BERGSTRÖM, G., GROTH, I., MCPHERSON, G., 1985: The pollination of *Zygogynum* (*Winteraceae*) by a moth, *Sabatinca* (*Micropterigidae*): an ancient association? – Science **227**: 540–543.

TOMLINSON, P. B., 1991: Pollen scavenging. – Natl. Geogr. Res. Explor. **7**: 188–195.

TUCKER, S. C., 1984: Origin of symmetry in flowers. – In WHITE, R. A., DICKISON, W. C., (Eds): Contemporary problems in plant anatomy, pp. 351–395. – Orlando: Academic Press.

UPCHURCH, G. R., WOLFE, J. A., 1987: Mid-Cretaceous to Early Tertiary vegetation and climate: evidence from fossil leaves and wood. – In FRIIS, E. M., CHALONER, W. G., CRANE, P. R., (Eds): The origins of angiosperms and their biological consequences, pp. 75–105. – Cambridge: Cambridge University Press.

WALKER, J. W., 1976: Comparative pollen morphology and phylogeny of the ranalean complex. – In BECK, C. B., (Ed.): Origin and early evolution of angiosperms, pp. 241–299. – New York: Columbia University Press.

ZIMMER, E. A. HAMBY, R. K., ARNOLD, M. L., LEBLANC, D. A., THERIOT, E. C., 1989: Ribosomal RNA phylogenies and flowering plant evolution. – In FERNHOLM, B., BREMER, K., JÖRNVALL, H., (Eds): The hierarchy of life, pp. 205–214. – Amsterdam: Elsevier.

PETER K. ENDRESS
ELSE MARIE FRIIS

Pl. Syst. Evol. [Suppl.] 8: 7–29 (1994)

# Origin of the angiosperm flower: a phylogenetic perspective

JAMES A. DOYLE

Received February 21, 1994

**Key words:** Angiosperms, anthophytes, *Gnetales*, *Bennettitales*. – Flower, carpel, phylogeny, cladistics.

**Abstract:** Morphological and molecular analyses agree that angiosperms are monophyletic and somehow related to *Gnetales*, but uncertainties on rooting (among woody magnoliids or paleoherbs) and the position of fossils permit varied scenarios for origin of the flower. Trees linking angiosperms with *Bennettitales*, *Pentoxylon*, and *Gnetales* and this "anthophyte" clade with *Caytonia* imply that flowers arose in the common ancestor of anthophytes and carpels are cupule-bearing sporophylls. However, trees linking angiosperms with *Caytonia* and/or glossopterids imply that flowers originated more than once, as may certain fossil anthophytes. Trees linking anthophytes with coniferopsids suggest that flowers evolved by aggregation of fertile shoots into pseudanthia. New data on fossils or the control of floral morphogenesis in angiosperms and *Gnetales* might distinguish among these hypotheses.

Phylogenetic analyses of fossil and modern seed plants based on several morphological and molecular data sets have strengthened the hypothesis that angiosperms are monophyletic and revived the concept that their closest living relatives are the *Gnetales*. However, these analyses have given inconsistent results concerning relationships of angiosperms and *Gnetales* with fossil groups and basal relationships within angiosperms, and these lead to very different scenarios for the origin of the angiosperm flower. These conflicts are due partly to different assumptions on homologies and partly to the inclusion of different taxa, but experimental analyses show that many of the same alternatives are almost equally parsimonious even with one data set. Rather than defend one preferred scheme, it may therefore be more profitable to explore several of the most parsimonious alternatives and consider what sorts of new data might help to decide among them. To do so, I will concentrate on a data set designed to overcome weaknesses of previous studies, which dealt separately with relationships within angiosperms and between angiosperms and other seed plants, and to explore apparent conflicts between analyses based on morphological and rRNA data (DOYLE & al. 1994).

**Previous euanthial scenarios**

The first comprehensive phylogenetic analyses of seed plants by CRANE (1985) and DOYLE & DONOGHUE (1986, 1992) yielded generally consistent trees. Angiosperms were linked with Mesozoic *Bennettitales*, *Pentoxylon*, and living *Gnetales*, together called anthophytes because they all have flower-like reproductive structures, and this group was linked with so-called Mesozoic seed ferns. This result implies that the flower, in the sense of a short, specialized axis bearing closely aggregated sporophylls, rather than an unspecialized axis or an elongate strobilus, originated before the carpel. In CRANE (1985) anthophytes were linked with Triassic corystosperms and angiosperms were nested within anthophytes, related directly to *Gnetales*, whereas in DOYLE & DONOGHUE (1986, 1992) anthophytes were linked with the primarily Jurassic genus *Caytonia* and angiosperms were the sister group of other anthophytes (Fig. 1).

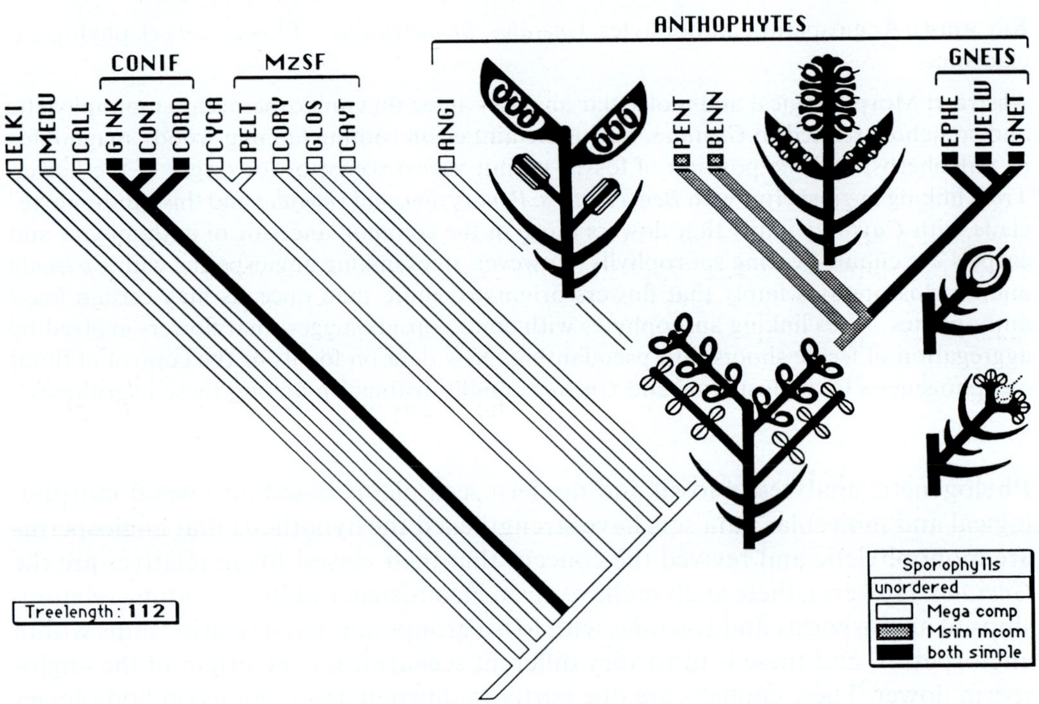


Fig. 1. Representative most parsimonious seed plant tree of DOYLE & DONOGHUE (1992), showing distribution of the sporophyll character (both micro- and megasporophylls pinnately organized, microsporophylls pinnately organized but megasporophylls simple, both micro- and megasporophylls simple). Sketches summarize the morphology of the reproductive structures in angiosperms, *Bennettitales*, *Gnetales*, and the hypothetical common ancestor of anthophytes. ELKI *Elkinsia* (=Devonian "seed fern" in DOYLE & DONOGHUE 1992), MEDU *Medullosaceae*, CALL *Callistophyton*, GINK *Ginkgoales*, CONI *Coniferales*, CORD *Cordaitales*, CYCA *Cycadales*, PELT *Peltaspermum*, CORY *Corystospermaceae*, GLOS *Glossopteridales*, CAYT *Caytoniaceae*, ANGI angiosperms, PENT *Pentoxylon*, BENN *Bennettitales*, EPHE *Ephedra*, WELW *Welwitschia*, GNET *Gnetum*, CONIF coniferopsids, MzSF Mesozoic "seed ferns," GNETS *Gnetales*. Tree rooted by specifying Devonian "progymnosperms" (not shown) as outgroup

If the original angiosperm carpel contained several anatropous, bitegmic ovules, as is widely assumed, it has no obvious prototype in other anthophytes. *Bennettitales* had a terminal ovuliferous receptacle covered with numerous orthotropous ovules and interseminal scales, whereas female flowers of *Gnetales* have a single orthotropous ovule surrounded by a second integument that apparently corresponds to the perianth of the male flower (CRANE 1985). A possible carpel prototype occurs in *Caytonia*, which had ovulate structures consisting of a rachis bearing two rows of anatropous cupules, each containing several ovules. The cupules themselves would be derived from leaflets with ovules on their adaxial surface (HARRIS 1940, REYMANÓWNA 1974). If the number of ovules per cupule was reduced to one, each cupule would be comparable to an anatropous, bitegmic ovule; the carpel might correspond to the rachis, expanded and folded to enclose the cupules (GAUSSEN 1946, STEBBINS 1974, DOYLE 1978). In *Bennettitales*, CRANE (1985) and DOYLE & DONOGHUE (1986) proposed that *Caytonia*-like sporophylls were reduced to single orthotropous ovules (which are at least sometimes bitegmic, suggesting that they are also cupules) and sterilized to produce the interseminal scales. In *Gnetales*, the number of ovules per flower would be reduced to only one, and the cupule wall would be lost (a possible weakness of this scheme, discussed further below). Microsporophylls would be reduced in both angiosperms (to stamens with two lateral pairs of microsporangia, considered synangia) and *Gnetales*, where *Welwitschia* probably shows the basic condition, with a cup-like androecium representing either a whorl of six microsporophylls (CRANE 1985, 1988; DOYLE & DONOGHUE 1986, 1992) or more likely two opposite microsporophylls with three synangia each (NIXON & al. 1994). Assuming that the cupule-bearing structures of *Caytonia* are sporophylls, this scenario is a classical euanthial theory (where the flower is an axis bearing sporophylls), reminiscent of ARBER & PARKIN (1907, 1908) but with stronger character support and more specific details on the relatives of anthophytes and the homologies of floral parts.

Two related alternatives may be mentioned at this point. First, DOYLE & DONOGHUE (1986: 363) and CRANE (1988: 253) discussed the possibility that the ovuliferous receptacle of *Bennettitales* is not an axis but rather a megasporophyll, shifted to a terminal position and modified into a radial structure. Such a structure might be analogous to the developmentally terminal single carpel of *Tasmannia (Drimys) lanceolata* (*Winteraceae*) (TUCKER & GIFFORD 1966). Second, MEYEN (1988) suggested that angiosperms were derived from *Bennettitales* by gamoheterotopy (transfer of characters from one sex to another). Although multiovulate angiosperm carpels are very different from the structures on a bennettitalean ovuliferous receptacle, such carpels are more like bennettitalean microsporophylls, which were often flat and bore two rows of adaxial synangia. A regulatory mutation that led to remodeling of female structures on the microsporophyll plan, with bitegmic ovules in place of microsynangia, could provide a prototype for the angiosperm flower.

In a subsequent analysis of angiosperms, concentrating on "magnoliid" groups and representing more advanced taxa with reconstructed ancestors or relatively primitive placeholders, DONOGHUE & DOYLE (1989) used the outgroups found in the seed plant analysis to polarize characters within angiosperms. The resulting trees were "rooted" in or next to *Magnoliales* in a restricted sense (those families

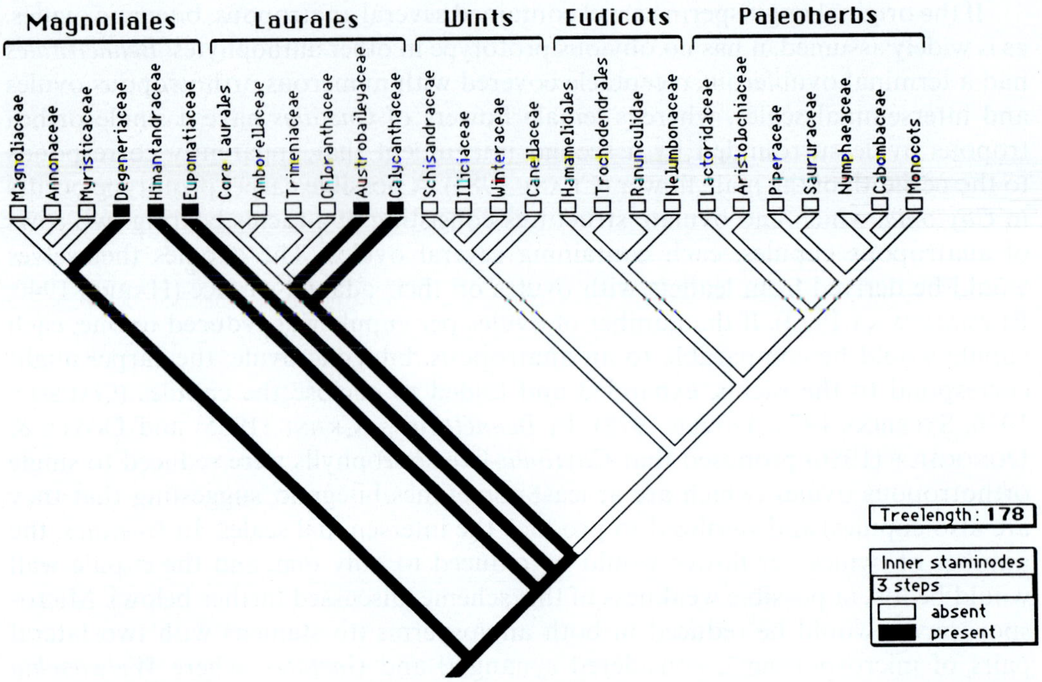


Fig. 2. Representative most parsimonious angiosperm tree of DONOGHUE & DOYLE (1989), showing distribution of the inner staminode character. Core Laurales *Hortonia*, *Monimiaceae*, *Atherospermataceae*, *Siparunaceae*, *Gomortega*, *Hernandiaceae*, *Lauraceae*; Wints winteroids. In other most parsimonious trees, *Magnoliales* are a monophyletic group and include *Canellaceae*, but in this tree *Magnoliales* are paraphyletic and *Canellaceae* belong to the winteroid clade

with granular exine structure), which formed either a basal clade or a paraphyletic basal grade (e.g., Fig. 2). The remaining angiosperms included two other woody magnoliid lines (*Laurales* and winteroids, including *Winteraceae*, *Illiciales*, and possibly *Canellaceae*, which are basal *Magnoliales* in other most parsimonious trees), in uncertain order, then a clade consisting of groups with tricolpate and derived pollen, called eudicots by DOYLE & HOTTON (1991), linked with a clade consisting of herbaceous magnoliids (*Aristolochiaceae*, *Lactoridaceae*, *Nymphaeales*, and *Piperales*, not including *Chloranthaceae*, which belong to *Laurales*) and monocots, called paleoherbs. These results can be related to three models for the primitive flower discussed by ENDRESS (1986): *Chloranthaceae*, with extremely simple flowers; *Winteraceae*, with variable numbers of irregularly arranged parts; and *Austrobaileyaceae*, *Himantandraceae*, and *Eupomatiaceae*, with inner staminodes. The DONOGHUE & DOYLE trees are most consistent with the third model, since as shown in Fig. 2 the taxa with inner staminodes are at or near the base of *Magnoliales* and *Laurales*, and least consistent with the chloranthaceous model; trees with *Chloranthaceae* basal are seven steps longer. The third model is also favored by the analysis of LOCONTE & STEVENSON (1991), which placed *Calycanthales* (*Calycanthaceae* plus *Idiospermaceae*), with inner staminodes, at the base of the angiosperms.

DOYLE & DONOGHUE (1986) and DONOGHUE & DOYLE (1989) recognized that these results were uncertain, since in both studies they had found very different trees that were almost equally parsimonious. They also knew that the results could be incorrect due to circular reasoning. In the seed plant analysis, angiosperms were scored based on the assumption that they were originally like *Magnoliales* or *Winteraceae*. However, if this assumption was incorrect, it may have led to incorrect placement of angiosperms in the seed plant trees, and this would have led to an inappropriate choice of outgroups in the angiosperm analysis. Thus it is possible that the basal position of *Magnoliales* in the angiosperm trees was a consequence of the earlier assumption that the first angiosperms were like this group. If different outgroups had been used, another group might have been basal in angiosperms, and the inferred primitive flower might have been different.

These doubts were increased by analyses of rRNA sequence data (HAMBY & ZIMMER 1992, DOYLE & al. 1994), which indicate that the angiosperms are not rooted among *Magnoliales* and other woody magnoliids, but rather among paleoherbs, with *Nymphaeales* basal. Suggestively, a similar rooting was only one step less parsimonious than the magnolialean rooting in terms of the morphological data set of DONOGHUE & DOYLE (1989). In trees based on *rbc*L (cpDNA) sequences (CHASE & al. 1993, QIU & al. 1993), the angiosperms are rooted between the aquatic genus *Ceratophyllum* and all the other groups, which then split into eudicots and a monosulcate clade (monocots and magnoliids).

Fossil data also raise doubts concerning the magnolialean rooting, since monocots, other probable paleoherbs, and uniovulate carpels resembling those of *Chloranthaceae* are conspicuous in the early record of angiosperms (HICKEY & DOYLE 1977, TAYLOR & HICKEY 1990, CRANE & al. 1994, FRIIS & al. 1994). However, these observations do not clearly favor any alternative rooting, since the oldest angiosperm-containing floras (Barremian-Aptian) also provide evidence of clades whose modern representatives are woody and usually have multiovulate carpels (*Magnoliales*, winteroids: DOYLE & HOTTON 1991, DOYLE & DONOGHUE 1993), and the abundance of uniovulate carpels could be largely a function of the early radiation of *Chloranthaceae* and related *Laurales*.

To address these problems, DOYLE & DONOGHUE (1990) and DOYLE & al. (1994) conducted a new seed plant analysis with a greater sample of angiosperm diversity, specifically nine potentially basal angiosperm groups, included both individually and together. When *Magnoliales* were substituted for angiosperms as a whole, results of the previous sort were obtained, as might be expected, but substituting other groups for angiosperms or including all nine angiosperm taxa changed the inferred relationships among angiosperms and other seed plants. However, in the nine-angiosperm analyses, trees like those found by DOYLE & DONOGHUE (1986) and several others were only one step less parsimonious than the shortest trees. While these results do not strongly refute the previously proposed euanthial scenario, they do imply that there are several other very different pathways of floral evolution that must also be considered.

## Challenges to the anthophyte concept

Trees found by DOYLE & al. (1994) and recent observations on Mesozoic seed plants raise the possibility that flower-like structures evolved independently in angiosperms and other anthophyte groups. Some of the fossil data may be consistent with a single origin of the flower, but they raise questions concerning the homologies

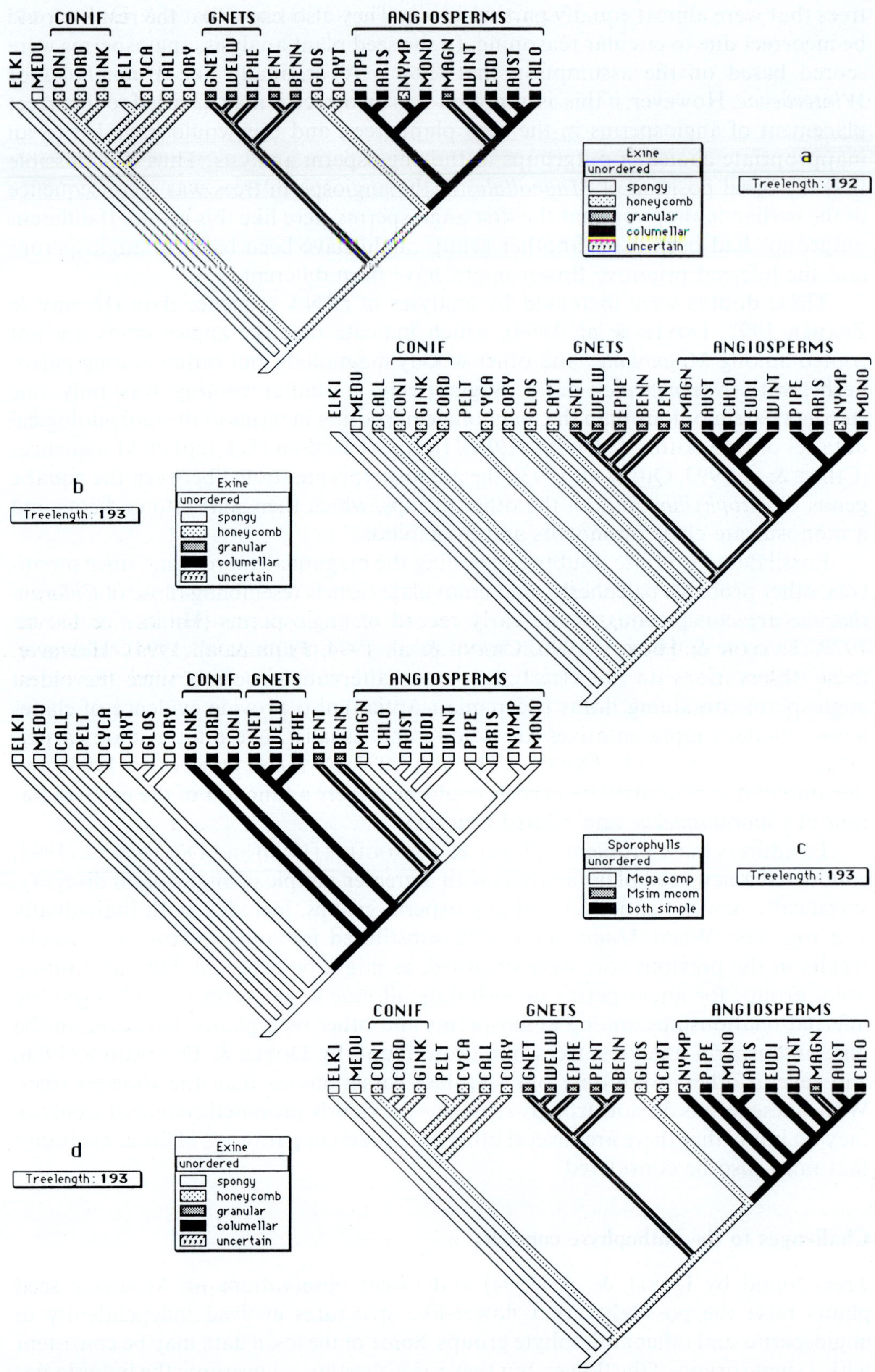
CONIF
GNETS
ANGIOSPERMS
ELKI MEDU CONI CORD GINK PELT CYCA CALL CORY GNET WELW EPHE PENT BENN GLOS CAYT PIPE ARIS NYMP MONO MAGN WINT EUDI AUST CHLO
Exine
unordered
spongy
honeycomb
granular
columellar
uncertain
a
Treelength: 192
CONIF
GNETS
ANGIOSPERMS
ELKI MEDU CALL CONI GINK CORD PELT CYCA CORY GLOS CAYT GNET WELW EPHE BENN PENT MAGN AUST CHLO EUDI WINT PIPE ARIS NYMP MONO
b
Treelength: 193
Exine
unordered
spongy
honeycomb
granular
columellar
uncertain
CONIF
GNETS
ANGIOSPERMS
ELKI MEDU CALL PELT CYCA CAYT GLOS CORY GINK CORD CONI GNET WELW EPHE PENT BENN MAGN CHLO AUST EUDI WINT PIPE ARIS NYMP MONO
Sporophylls
unordered
Mega comp
Msim mcom
both simple
c
Treelength: 193
CONIF
GNETS
ANGIOSPERMS
ELKI MEDU CONI CORD GINK PELT CYCA CALL CORY GNET WELW EPHE PENT BENN GLOS CAYT NYMP PIPE MONO ARIS EUDI WINT MAGN AUST CHLO
d
Treelength: 193
Exine
unordered
spongy
honeycomb
granular
columellar
uncertain

of ovule-bearing structures postulated by CRANE (1985) and DOYLE & DONOGHUE (1986) (Fig. 1).

Flowers would clearly arise more than once in alternative trees in which anthophytes and derived polyphyletically from Mesozoic seed ferns. Trees of this sort, with angiosperms linked directly with *Caytonia* and the resulting clade linked with *Bennettitales*, *Pentoxylon*, and *Gnetales*, were one step less parsimonious than the shortest trees in terms of the data sets of DOYLE & DONOGHUE (1986, 1992). However, this situation was reversed in the nine-angiosperm analysis of DOYLE & al. (1994): in the most parsimonious trees, angiosperms were linked with *Caytonia* and glossopterids (Fig. 3a), and trees of the previous kind were one step longer (Fig. 3b). Similar trees (with various relationships of *Caytonia* and glossopterids) were found in several of the analyses in which individual groups were substituted for angiosperms (*Nymphaeales*, monocots, *Aristolochiaceae*, tricolpate eudicots, *Winteraceae*, *Austrobaileya*). These results take on special relevance in light of rRNA analyses, which place one of these groups, *Nymphaeales*, at the base of the angiosperms (HAMBY & ZIMMER 1992, DOYLE & al. 1994).

The change in relative status of these trees reflects the ambiguity of character support for the anthophyte clade. Although DOYLE & DONOGHUE (1986: 353) showed 10 characters uniting the anthophytes, four of these could have arisen lower on the tree, because their state is unknown in fossil outgroups (tunica layer in the apical meristem, scalariform pitting in the secondary xylem, lignin showing Mäule reaction, siphonogamy). Two are equivocal as synapomorphies because of variation in the outgroups or within anthophytes: simply pinnate leaves could have arisen below glossopterids, with a reversal to pinnately compound leaves in *Caytonia*; syndetocheilic (paracytic) stomata could have arisen at the base of anthophytes and been lost in *Pentoxylon* and *Ephedra*, or evolved independently in angiosperms, *Bennettitales*, and the *Welwitschia-Gnetum* clade. The four unequivocal synapomorphies were once-pinnate microsporophylls, one ovule per cupule (bitegmic ovule), loss of air sacs on the pollen, and granular or columellar exine structure, but the first three of these depend on outgroup relationships (e.g., whether anthophytes are nested among groups with saccate pollen, like *Caytonia* and glossopterids). Furthermore, there are competing characters that potentially unite angiosperms with *Caytonia* (flat guard cells) or *Caytonia* and glossopterids (reticulate venation, anatropous cupules).

◀

Fig. 3. Most parsimonious and one step less parsimonious seed plant trees found in the nine-angiosperm analyses of DOYLE & al. (1994). PIPE *Piperales* (*Piperaceae*, *Saururaceae*); ARIS *Aristolochiaceae*; NYMP *Nymphaeales* (*Nymphaeaceae*, *Cabombaceae*); MONO monocots; MAGN "core" *Magnoliales* (*Magnoliaceae*, *Degeneriaceae*, *Myristicaceae*, *Annonaceae*); WINT *Winteraceae*; EUDI eudicots (groups with tricolpate and derived pollen); AUST *Austrobaileyaceae*; CHLO *Chloranthaceae*; other abbreviations as in Fig. 1. *a* Representative most parsimonious tree, with angiosperms linked directly with *Caytonia* and glossopterids; *b* tree with previously most parsimonious arrangement of non-angiosperm groups, as in Fig. 1, with anthophytes linked with *Caytonia* and *Magnoliales* basal in angiosperms; *c* neo-englerian tree, with anthophytes linked with coniferopsids and *Gnetales* basal in anthophytes; *d* angiosperms rooted among paleoherbs, with basal groups arranged as implied by rRNA data (DOYLE & al. 1994)

Factors favoring the shift to the tree in Fig. 3a include elimination of the distinction between once-pinnate and pinnately compound microsporophylls (Doyle & Donoghue 1992) and changes in two pollen characters that were needed in order to remove biases in the previous analyses. Granular and columellar infratectal structures were combined as one state in the seed plant analysis (Doyle & Donoghue 1986) but split in the angiosperm analysis (Donoghue & Doyle 1989), with granular designated as ancestral. This implicitly treated honeycomb alveolar, granular, and columellar as an ordered series, which would bias against direct derivation of columellar from alveolar structure, as in trees where angiosperms are linked with alveolar outgroups like *Caytonia* and glossopterids, and columellar taxa like most paleoherbs are basal in angiosperms. Similarly, Donoghue & Doyle (1989) recognized two endexine states in the angiosperm analysis, with absence of extra-apertural endexine (as in *Magnoliales*) designated as ancestral, and hence closer to the laminated endexine condition of gymnosperms. Since these biases are hard to justify, Doyle & al. (1994) redefined these characters as unordered, contributing to the new result. Another factor was the inclusion of angiosperm taxa with anomocytic as well as paracytic stomata (paleoherbs, eudicots). Doyle & al. (1994) added one new character shared by angiosperms and *Gnetales*, double fertilization (fusion of both sperm with megagametophyte nuclei, with or without endosperm formation, as documented in *Ephedra* by Friedman 1990, 1992), but this is equivocal as an anthophyte synapomorphy because its state is unknown in fossils.

Since *Bennettitales*, *Pentoxylon*, and *Gnetales* are still the next-closest relatives of angiosperms, trees like Fig. 3a could be thought of as implying that *Caytonia* and/or glossopterids are anthophytes, rather than as breaking up the anthophytes. However, because there is no evidence that the relatively large, complex sporophylls of *Caytonia* and glossopterids were aggregated into typical flowers, the flower would have evolved independently in angiosperms and the bennettitalean-gnetalean line. An important point is that these trees implicitly assume that *Caytonia* and glossopterids had anthophyte states in characters that are unknown in fossils (tunica, Mäule reaction, siphonogamy, double fertilization). The lack of independent evidence for this assumption could be regarded as a weakness of this scheme, although it is not wholly implausible in view of the small size and anthophyte-like anatomical features of *Caytonia* seeds (reduced megaspore wall, thick nucellar cuticle: Harris 1954, Reymanówna 1974, 1974; Crane 1985).

Even if anthophytes as a group are monophyletic, there are growing indications that flowers may have arisen more than once among them, or at least evolved from a rather different prototype than the one postulated by Doyle & Donoghue (1986). In other words, anthophytes may be a valid clade, but their name may be inappropriate.

In *Bennettitales*, Crane (1988) showed that reports of bisexual flowers in the oldest (Late Triassic) representatives are questionable, and his preliminary phylogenetic analysis indicated that bisexual flowers were derived within the order. Several forms apparently lacked a perianth. Furthermore, the Triassic fossil *Westersheimia* consists of several ovuliferous receptacles rather than a single terminal one, with no sign of other floral parts. It is not clear whether this structure was a branch bearing several flowers or a pinnately compound appendage, but the latter interpretation would support the concept that the ovuliferous receptacle of other *Bennettitales* was a secondarily terminal, radial megasporophyll, rather than an axis with reduced, uniovulate (unicupulate) megasporophylls. Triassic microsporophylls called *Leguminanthus* had broad, sheathing bases, implying that they are not basally fused like the microsporophylls of most *Bennettitales*, and probably not whorled. If this condition was basic in *Bennettitales*, it would negate one of the apparent synapomorphies (whorled microsporophylls) that link *Bennettitales* and *Gnetales* in the analyses of DOYLE & DONOGHUE (1986, 1992).

The reproductive structures of *Pentoxylon* have never seemed very flower-like, although this could be a result of loss of parts: they consist of branched microsporophylls and pedunculate "heads" of ovules (*Carnoconites*) that were borne on separate short shoots, with no associated perianth. CRANE (1985) and DOYLE & DONOGHUE (1986) interpreted *Carnoconites* as an axis with reduced uniovulate megasporophylls, like the ovuliferous receptacle of *Bennettitales*. However, although this interpretation is consistent with the morphology of *C. cranwelliae*, which had single heads on unbranched peduncles, CRANE (1988: 253) pointed out that it is questionable for *C. compactus*, which had branched peduncles with several heads. This suggests rather a compound sporophyll with ovule-bearing pinnae, possibly comparable to *Westersheimia*. Finally, ROTHWELL & SERBET (1994) have found that heads of *Carnoconites* have bilateral internal anatomy, also implying that they are foliar structures rather than axes bearing reduced sporophylls.

In *Gnetales*, the most significant new data come from observations by KIRCHNER (1992) and VAN KONIJNENBURG-VAN CITTERT (1992) on *Piroconites kuespertii*, an Early Jurassic fossil in which VAN KONIJNENBURG-VAN CITTERT found typically gnetalean striate ephedroid pollen, and associated organs, here referred to collectively as *Piroconites*. *P. kuespertii* is a scale-like structure with numerous trilocular synangia on its upper surface, which is sometimes found attached to the adaxial surface of a lanceolate, parallel-veined bract (*Chlamydolepis lautneri*). These bracts are somewhat similar to the putatively associated leaves (*Desmiophyllum gothanii*). The individual synangia are reminiscent of the trilocular microsynangia of *Welwitschia*, but *Welwitschia* has only three synangia per microsporophyll (assuming that the androecium consists of two opposite sporophylls: NIXON & al. 1994). The female structures (*Bernettia inopinata*) appear to be constructed on a similar plan, but the ovule-bearing portion has numerous ovules in the place of microsynangia and is always associated with the bract (KIRCHNER 1992, VAN KONIJNENBURG-VAN CITTERT, pers. comm.; and my own observations on material at Utrecht). How these organs were borne is unknown, but *Piroconites* specimens are often preserved next to each other, suggesting that they were associated on the plant (van KONIJNENBURG-van CITTERT 1992). They may thus have formed flower-like structures, but these would have been much larger and probably less tightly integrated than flowers of modern *Gnetales*.

The Late Triassic genus *Dechellyia* (ASH 1972) also seems related to *Gnetales*, based on its ephedroid pollen and opposite, linear leaves with two or four parallel veins, but it differs from the modern genera in having superficially conifer-like male strobili and conspicuously winged seeds borne oppositely at the tips of branches. These structures are not obviously comparable to those of *Piroconites*, although the winged seeds might be derived from a *Bernettia*-like prototype by

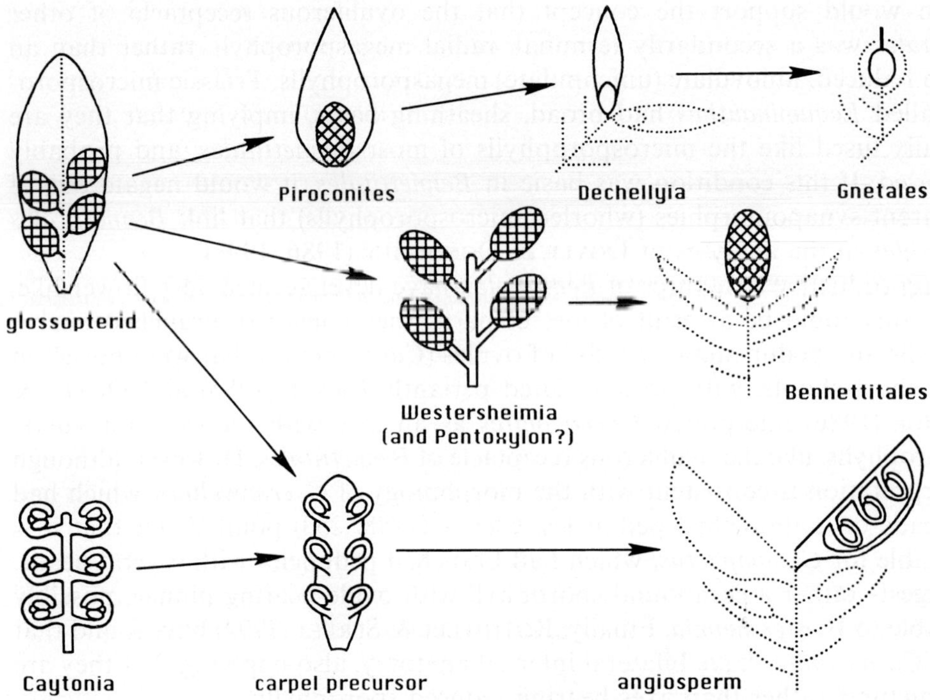


Fig. 4. Alternative scenarios for evolution of anthophyte ovulate structures suggested by the Jurassic gnetalean relative *Piroconites* (*Bernettia*) (KIRCHNER 1992, VAN KONIJNENBURG-VAN CITTERT, 1992), the Triassic gnetalean relative *Dechellyia* (ASH 1972), and the compound ovuliferous receptacles of the Triassic bennettitalean *Westersheimia* (CRANE 1988)

reduction of the number of ovules to one and modification of the subtending bract into the wing.

Considering potential outgroups of the anthophytes, the ovule-bearing structures of *Piroconites* (*Bernettia*) are unexpectedly similar to those of Permian glossopterids, suggesting the comparisons presented in Fig. 4. Glossopterid fructifications consist of one or more "cupules" with densely packed ovules on their adaxial surface (confirmed by orientation of the vascular bundles: TAYLOR & TAYLOR 1992), which are themselves attached to the midrib of a subtending leaf. As recognized by RETALLACK & DILCHER (1981), the morphological relationship of the cupule to the subtending leaf is problematical: it could be an appendage on a reduced branch in the axil of the leaf (cf. MEEUSE 1976), a sporophyll adnate to the leaf (analogous to fusion between stamens and petals in angiosperms), or part of the leaf itself (analogous to the adaxial fertile segment of the leaf in the fern order *Ophioglossales*, and possibly also represented in the Cretaceous "seed fern" *Ktalenia* of TAYLOR & ARCHANGELSKY 1985, cf. DOYLE & DONOGHUE 1986: 362). Any of these interpretations might also apply to the bract-sporophyll complex of *Piroconites*; better evidence on the morphological relationship of parts in the two groups could strengthen or refute their potential homology. Whatever its homologies, a whole

*Bernettia*-like structure would have to be reduced to a single terminal ovule between *Piroconites* and modern *Gnetales*. *Dechellyia* might represent an intermediate stage in this process, with the subtending bract still present as the wing. The morphological gap between these forms and modern *Gnetales* parallels the stratigraphic gap between Triassic-Early Jurassic ephedroid pollen (apparently representing stem-relatives of the group) and its better-documented Early Cretaceous radiation (which apparently involved crown-group *Gnetales*: DOYLE & DONOGHUE 1993).

This concept might also be extended to *Bennettitales* and *Pentoxylon*, although less easily. A glossopterid cupule or *Bernettia*, with densely packed ovules, might be compared with an ovuliferous receptacle of *Bennettitales* or a head of *Pentoxylon*, under the hypothesis discussed above that these structures are sporophylls shifted to a terminal position rather than axes bearing numerous reduced sporophylls. This might involve "spread" of the ovules to both sides of the cupule and thickening of the intervening tissue. As noted by VAN KONIJNENBURG-VAN CITTERT (1992), *Piroconites* is also reminiscent of *Bennettistemon ovatum*, a bennettitalean micro-spororophyll with numerous densely packed microsporangia. However, unlike *Piroconites*, *Bennettitales* and *Pentoxylon* have no obvious homolog of the subtending leaf of glossopterids, which would have to be lost or otherwise dissociated from the ovuliferous structure. A still greater obstacle is the presence of a "cupule" around the ovule of some *Bennettitales* and possibly *Pentoxylon* (HARRIS 1954, CRANE 1985, 1988). In the schemes of CRANE (1985) and DOYLE & DONOGHUE (1986, 1992; Fig. 1), this layer is homologous with the cupule of *Caytonia*, glossopterids, and corystosperms. Under this interpretation, the ovuliferous receptacle could correspond to a glossopterid structure bearing several cupules, but not to a single glossopterid cupule or ovule-bearing structure of *Piroconites*.

How angiosperms might relate to this scheme is unclear (Fig. 4). A key requirement is to explain both the carpel and the presence of two ovule integuments (this is a weakness of the anthocorm theory of MEEUSE 1976, which homologizes the cupule of Mesozoic seed ferns with the carpel, leaving no homolog for the second integument). One possibility is that anthophytes are monophyletic but more closely related to glossopterids than to *Caytonia*. The angiosperm carpel might then be derived by conversion of glossopterid cupules into bitegmic ovules (by reduction to one ovule per cupule) and enclosure by folding of the subtending leaf (STEBBINS 1974, RETALLACK & DILCHER 1981). Another possibility is that anthophytes are diphyletic, with *Bennettitales*, *Pentoxylon*, and *Gnetales* related to glossopterids and angiosperms related to *Caytonia*. The carpel could then be derived from a *Caytonia*-like prototype by expansion of the rachis and reduction of the cupules to bitegmic ovules (GAUSSEN 1946, DOYLE 1978, CRANE 1985, DOYLE & DONOGHUE 1986). However, unless glossopterids and *Caytonia* had more aggregation of fertile parts than seems likely, both schemes would imply that flower-like structures evolved independently in angiosperms and other anthophytes.

None of these schemes would rule out some sort of relationship between glossopterids and *Caytonia*, since these groups are more or less closely related in previous analyses, based on simple-reticulate leaf venation, seed characters (small size, platyspermy, thick nucellar cuticle), and saccate pollen, and their cupules might be ultimately homologous. The main problem is the fact that the ovulate structures of glossopterids are attached to a subtending leaf, but there is no evidence for such an association in *Caytonia* (Doyle & Donoghue 1986: 362–363). One possibility is that the glossopterid condition was ancestral, and the leaf was lost or reduced to form the sporophyll rachis in the *Caytonia* line. Another is that the ancestral structure was a cupule-bearing structure more like that of *Caytonia*, which became associated with the leaf in the glossopterid line. Better evidence on the morphological nature of structures in glossopterids and *Caytonia* and/or discovery of more primitive relatives of these groups would be most welcome.

## Neo-englerian scenarios

A very different alternative is that anthophytes are related not to Mesozoic seed ferns but rather to coniferopsids. Trees of this sort, called neo-englerian because they recall the early twentieth century German-Austrian view that angiosperms are derived from conifers via *Gnetales* (e.g., Wettstein 1907), were only two steps less parsimonious than the shortest trees in terms of the data set of Doyle & Donoghue (1986), and basically similar trees have been found by Nixon & al. (1994) and Rothwell & Serbet (1994). Coniferopsids have reduced, simple leaves and sporophylls relative to the first seed plants (seed ferns). In most (though not all) neo-englerian trees, the basal group in anthophytes is not angiosperms but *Gnetales*, which resemble coniferopsids in having linear leaves (except *Gnetum*, which is apparently derived) and reduced sporophylls, and Mesozoic seed ferns form a clade located elsewhere on the tree. In terms of the nine-angiosperm data set of Doyle & al. (1994), some neo-englerian trees were only one step less parsimonious than the shortest trees (Fig. 3c). Neo-englerian trees were also found when *Chloranthaceae* were substituted for angiosperms, as might be expected based on the presence of several "gnetalean" features in this family, such as opposite leaves, two-trace nodes, spicate inflorescences of simple flowers, and carpels with one orthotropous ovule (cf. Doyle & Donoghue 1986: 386). In the trees of Nixon & al. (1994), angiosperms are actually nested within *Gnetales* and therefore derived from a gnetalean ground plan.

Neo-englerian trees seem more plausible if groups with simple flowers and orthotropous ovules (such as *Chloranthaceae*, *Piperales*, or *Ceratophyllum*) are assumed to be basal in angiosperms, which is the case in the trees of Taylor & Hickey (1992), Chase & al. (1993), and Nixon & al. (1994). However, this was not the sort of arrangement seen in neo-englerian trees derived from the nine-angiosperm data set of Doyle & al. (1994): instead, the basal angiosperm group was *Magnoliales* (Fig. 3c). This counterintuitive result seems to follow from the fact that *Magnoliales* are actually more like *Gnetales* than *Chloranthaceae* are in many characters, such as absence of chloranthoid marginal teeth on the leaves and boat-shaped pollen with a continuous tectum and granular exine structure. Nine-angiosperm trees with *Chloranthaceae* basal are much less parsimonious, by five steps.

Neo-englerian trees have very different implications for homologies of angiosperm reproductive structures (Fig. 5). The original organization would presumably be that seen in cordaites and Paleozoic conifers: a compound strobilus consisting of an axis with bracts and axillary fertile short shoots bearing scale leaves and simple sporophylls, with either a few microsporangia or an ovule with one integument. Subsequently, the ovuliferous axillary shoots were transformed into the cone scales

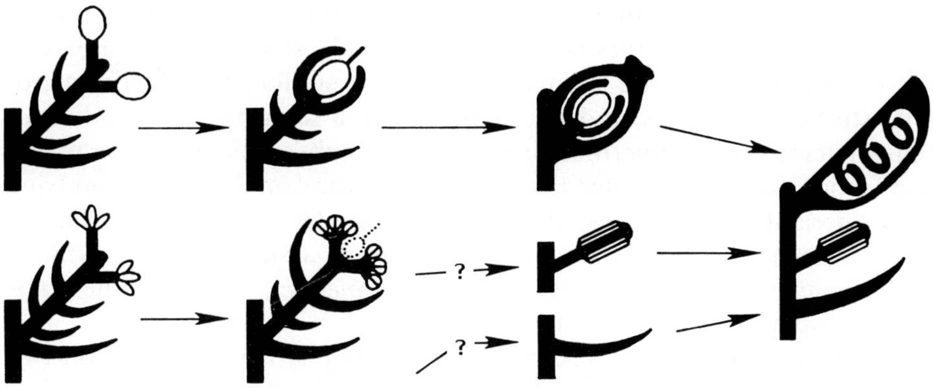


Fig. 5. Scenario for evolution of angiosperm reproductive structures implied by neo-englerian trees (Fig. 3c), with anthophytes linked with coniferopsids and angiosperms derived from a gnetalean prototype by aggregation of bracts and axillary fertile shoots. Position of the abortive ovule in the male flower of *Welwitschia* indicated by dotted lines

of modern conifers (FLORIN 1951). Within anthophytes, the basic condition would be one ovule per short shoot, as in *Gnetales*. Within *Gnetales*, Fig. 5 assumes that the basic morphology was most like that of *Welwitschia*, as argued by CRANE (1985, 1988), based on the detailed observations of MARTENS (1971): male flower with two pairs of perianth parts and two microsporophylls; female flower with an ovule with a micropylar tube, surrounded by an outer integument corresponding in position to the inner perianth pair in male flower, plus an outer perianth pair. The male flower of *Welwitschia* has an abortive terminal ovule, but because this could be either a vestige from a bisexual ancestor (ARBER & PARKIN 1908) or an autapomorphy, it is indicated with dotted lines (this ovule is unitegmic, which is an argument for homology of the outer integument in the female flower with the perianth). In general, conditions in other *Gnetales* can be ascribed to reduction. *Ephedra* has one pair of perianth parts in the male flower and no appendages other than the outer integument in the female flower (this develops from two lateral primordia and may therefore correspond to the outer rather than the inner perianth pair of *Welwitschia*: TAKASO 1985). *Gnetum* has a simple perianth in the male flower but two outer integuments in the female flower, perhaps corresponding to both perianth pairs in *Welwitschia*.

Not all of these homologies are clearcut, so other interpretations should be kept in mind. One problem is the presence of two small nubs on either side of the terminal ovule in the male flower of *Welwitschia* (MARTENS 1971). In the context of their trees, DOYLE & DONOGHUE (1986: 416) surmised that these nubs might be vestiges of the *Caytonia*-anthophyte cupule or of bennettitalean interseminal scales (another possible homology, suggested by the comparisons made in the previous section, is with the leaf or wing associated with the ovulate structures of *Piroconites* and *Dechellyia*: Fig. 4). In contrast, FRIIS (pers. comm.) suggests that they are reduced bracts (perianth parts, in the present terminology), like those making up the outer integument in the female flower. However, this would imply that the male flower is an axis bearing both microsporophylls and a structure corresponding to a whole female flower, or else a single axis bearing sterile appendages, microsporophylls, sterile appendages, and a terminal ovule, a sequence with no known parallels in coniferopsids. Furthermore, since the two nubs are lateral, they would correspond in position to the outer perianth pair of the female flower, implying that the inner perianth pair was lost. A weakness of the CRANE (1985) and DOYLE & DONOGHUE (1986) schemes is the fact that they must assume that the original cupule of *Caytonia* and *Bennettitales* was lost in *Gnetales*; however, the FRIIS scenario must assume equally major losses.

Assuming the gnetalean prototype postulated in Fig. 5, I would envision two very different scenarios for evolution of a typical bisexual angiosperm flower.

One scenario would involve transformation of a coniferopsid-gnetalean axillary fertile short shoot into an angiosperm flower by elaboration of its component parts. Like the schemes discussed in the first section, this is basically a euanthial interpretation. Floral parts are borne in the same order in a bisexual angiosperm flower and the male flower of *Welwitschia*, but they are much simpler in the latter. It is not too difficult to imagine transformation of scale-like microsporophylls with a few free microsporangia into stamens with two lateral synangia; the microsporophylls of *Welwitschia*, with three synangia, might represent an intermediate step. However, carpels enclosing several bitegmic ovules would have to be derived from single unitegmic ovules, which are borne on stalk-like sporophylls in some cordaites but completely sessile in modern conifers and *Gnetales*. Both the carpel and the outer integument of the ovules would presumably have to arise de novo. The need to assume so much de novo origin of structures is in itself reason to question this scenario.

The other scenario would involve aggregation of several bracts and short shoots into a superficially flower-like but actually multiaxial structure. This is a pseudanthial interpretation, as proposed by WETTSTEIN (1907) and reformulated by MEEUSE (1972, 1976) in terms of "advanced cycadopsid" groups (especially Mesozoic seed ferns and *Gnetales*, interpreted as having axes with bracts and axillary fertile structures – a reasonable interpretation for *Gnetales* but questionable for *Caytonia*, in which the fertile structures appear to be pinnately organized sporophylls: HARRIS 1940, REYMANÓWNA 1974, DOYLE 1978, CRANE 1985). However, an updated version would not rely on WETTSTEIN's assumption that "*Amentiferae*" are primitive angiosperms, since these are now generally recognized to be advanced eudicots. A pseudanthial scenario seems easiest to envision if the original angiosperm carpel had one orthotropous ovule, as in *Chloranthaceae*, *Piperaceae* (although here the unilocular ovary appears to consist of three or four fused carpels: TUCKER & al. 1993), or *Ceratophyllum*. The outer integument might be homologous with the outer integument of *Gnetales*, and hence with the perianth of the male flower and the scale leaves on the axillary short shoot of cordaites and Paleozoic conifers. The carpel wall might be the subtending bract, expanded and folded around the axillary unit. As recognized by TAYLOR (1991) in postulating a similar carpel prototype, subsequent elaboration would be required to produce other kinds of carpel: ovules would have to be multiplied along the ventral margins of the carpel (bract) and reflexed into an anatropous orientation.

The position and morphology of the stamens in a bisexual angiosperm flower pose additional problems for a pseudanthial interpretation. This is because microsporophylls do not occupy the appropriate position in *Gnetales* if the flower is homologized with a whole compound strobilus. In the male flower of *Welwitschia*, the microsporophylls are located between the abortive unitegmic ovule and the perianth, which corresponds to the outer integument in the female flower. The same arrangement occurs in occasional bisexual flowers of *Ephedra* (MEHRA 1950, CRANE 1985). Under the scheme in Fig. 5, the corresponding position for the stamens in an angiosperm would be between the inner and outer integuments of the ovule, not below the carpels on the floral axis (the primary axis of the compound strobilus). One interpretation that would place stamens and carpels on the same axis is that each stamen represents a whole gnetalean male flower (cf. WETTSTEIN 1907). However, this would require transformation of a stem-like radial structure with a perianth and sporophylls into a single bilateral stamen with a microsynangium on either side. The subtending bract would also have to be accounted for, whether by loss, fusion with the axillary shoot to become part of the stamen, or transformation into a perianth part. The last hypothesis might be plausible for angiosperms that have stamens opposite the petals, but it would require dissociation of the two structures in other groups. Another interpretation is that the angiosperm stamen corresponds to a single gnetalean microsporophyll that was shifted from one order of branching to another (from the axillary short shoot to the primary axis of the compound strobilus) – an example of heterotopy.

Given the association of angiosperms, *Gnetales*, and *Bennettitales* in all recent analyses, any interpretation of the angiosperm flower must also be consistent with origin of the bennettitalean flower. The similar order of parts in male flowers of *Welwitschia* and bisexual flowers of *Bennettitales* would be consistent with an elaboration scenario, but this would require transformation of simple microsporophylls into complex structures with pinnately arranged synangia (i.e., even more elaboration than in angiosperms), de novo origin of the cupule, and multiplication of the ovules, plus origin of the interseminal scales. A pseudanthial scenario would pose a different set of problems, particularly in explaining why the organization of the so-called flowers of *Gnetales* and *Bennettitales* should seem so similar if the bennettitalean flower is actually an aggregation of gnetalean bracts and short shoots. FRIIS (pers. comm.) suggests that the cupule of *Bennettitales* is homologous with the outer integument (perianth) of *Gnetales*, which would imply that the ovuliferous receptacle is an axis bearing fertile shoots (although without clearly identifiable subtending bracts, unless these are represented by the interseminal scales). This would fit her view that the ovule in the male flower of *Welwitschia* is reduced, but it implies a serial juxtaposition of leaf-like pinnate microsporophylls and secondary ovuliferous shoots on the same axis. This poses essentially the same problems discussed in attempting to explain the bisexual flower of angiosperms as a pseudanthium, although if angiosperms and *Bennettitales* are related, the necessary reorganization might have occurred only once, in their common ancestor.

A general reason to be suspicious of neo-englerian scenarios is that there may be an intrinsic bias in favor of them at the stage of character analysis (DOYLE & DONOGHUE 1986: 373). Whereas trees linking anthophytes with Mesozoic seed ferns assume that the more complex, pinnately organized leaves and sporophylls of angiosperms, *Bennettitales*, and Mesozoic seed ferns are homologous and the simpler leaves and sporophylls of coniferopsids and *Gnetales* are convergent, neo-englerian trees assume that the simple organs are homologous and the complex organs are convergent. However, simple structures look intrinsically more similar than complex structures, even if they are derived (reduced) from very different precursors. In other words, it is easier to go wrong in homologizing reduced structures than complex ones, and therefore to err in favor of trees of the neo-englerian type. Nevertheless, whether the convergences are between coniferopsids and *Gnetales* or between other anthophytes and Mesozoic seed ferns, the repeated appearance of trees linking anthophytes with Mesozoic seed ferns and with coniferopsids implies that both sorts of tree should be taken seriously.

## Paleobotanical and molecular tests

In searching for data that might provide a test of these alternatives, it should be noted that all of the trees considered imply that the closest living relatives of angiosperms are *Gnetales*. This conclusion is confirmed by analyses of rRNA and *rbc*L (cpDNA) sequences (Hamby & Zimmer 1992, Chase & al. 1993, Doyle & al. 1994) and strengthened by the embryological studies of Friedman (1990, 1992, 1994), which show that *Ephedra* resembles angiosperms in having not only double fertilization but also similar embryogeny: after two divisions of each diploid fusion nucleus (probably an autapomorphy), embryo development follows a typically angiosperm-like cellular pattern. Most of the radical differences in scenarios for floral evolution are a function of how fossil taxa (*Caytonia*, glossopterids, *Piroconites*, *Dechellyia*, *Bennettitales*, *Pentoxylon*) fit in around living angiosperms, *Gnetales*, and coniferopsids. On the one hand, this reaffirms the importance of fossils; on the other, it suggests what sorts of paleobotanical data we should look for.

Critical data on these problems could come from more complete material of known fossil taxa or from discovery of anthophytes with previously unknown character combinations. For example, determination that *Caytonia* had anthophyte states in characters that are now unknown would strengthen its relationship with the anthophytes and argue against neo-englerian trees, although it would not say whether anthophytes are monophyletic or diphyletic. Discovery that *Caytonia* did not have anthophyte states in these characters would strengthen the hypothesis that anthophytes are monophyletic, but whether they are related to Mesozoic seed ferns or to coniferopsids would remain unresolved. Among fossils that are less well known, *Piroconites* may pose problems for the euanthial interpretation shown in Fig. 1, but it seems even less consistent with neo-englerian trees, since its reproductive structures are much less coniferopsid-like than those of modern *Gnetales*. On the other hand, documentation of additional coniferopsid-like features in *Dechellyia* could have an opposite effect. It would be useful to know if *Piroconites* and *Dechellyia* had a tubular micropyle, one of the synapomorphies linking *Gnetales* and *Bennettitales* in the analyses of Doyle & Donoghue (1986, 1992). Further evidence on the plant that produced *Eucommiidites* pollen, another probable anthophyte (Pedersen & al. 1989), would also be of interest.

Discovery of any fossil that could be confidently placed on the stem-lineage leading to angiosperms could also resolve many questions. Thus the Late Triassic

*Crinopolles* pollen group (CORNET 1989), which has monocot-like graded reticulate sculpture, or the Middle Jurassic leaf *Phyllites* (SEWARD 1904), which has paleoherb-like palmate venation, could constitute evidence that the angiosperms are rooted near paleoherbs rather than *Magnoliales*, providing that these fossils are on the angiosperm stem-lineage, as proposed for *Crinopolles* based on their gymnosperm-like endexine (DOYLE & HOTTON 1991, DOYLE & DONOGHUE 1993). However, more evidence concerning other organs is needed before it can be assumed that these fossils are really related to angiosperms.

Although there has been great progress in understanding the diversity of Early Cretaceous angiosperm reproductive structures (cf. CRANE & al. 1994, FRIIS & al. 1994), it seems less likely that Cretaceous fossils will provide decisive evidence on these questions. This is because so far all Cretaceous angiosperm relatives appear to belong to the angiosperm crown-group, rather than being on the angiosperm stem-lineage (DOYLE & DONOGHUE 1993). They may therefore have a great deal to say about relationships within angiosperms, but less about outgroup relationships or ancestral states. However, this could change if it is found that taxa on the angiosperm stem-lineage persisted into the Cretaceous, or if sampling becomes exhaustive enough to establish that the very earliest Cretaceous angiosperms all had one type of floral morphology.

Even though the most obvious tests are paleobotanical, certain data from extant plants could help. For example, demonstration that the androecium of *Welwitschia* (and the androecial column of *Ephedra* and *Gnetum*) consists of two basically pinnate microsporophylls with three synangia each (NIXON & al. 1994), rather than a whorl of six simple microsporophylls, would remove one character (simple microsporophylls) supporting neo-englerian trees in the data sets of DOYLE & DONOGHUE (1986, 1992). However, the most promising new kinds of neontological data are molecular.

First, although molecular data may not show how fossils fit into seed plant phylogeny (barring recovery of DNA sequences from fossils), they may say something about rooting of the angiosperms and morphology of the primitive flower. For example, in the most recent analyses of rRNA sequences (DOYLE & al. 1994), angiosperms are rooted among paleoherbs, with *Nymphaeales* basal, then *Piperales* (not including *Chloranthaceae*), *Aristolochiaceae* plus monocots, and a clade consisting of woody magnoliids and eudicots, within which relationships are poorly resolved. This does not represent a severe conflict with morphological data: with the nine-angiosperm morphological data set of DOYLE & al. (1994), trees with this arrangement of basal groups are only one step less parsimonious than the shortest trees (Fig. 3d). DOYLE & al. (1994) did experiments to evaluate the relative strength of these results, using a simplified rRNA data set with 11 angiosperm taxa and bootstrap analysis (FELSENSTEIN 1985), which asks how often a given clade is seen in analyses based on characters sampled randomly from the original matrix. The strongest results are that both angiosperms and *Gnetales* are monophyletic, at extremely high bootstrap levels of 100% and 99%, and the two groups are related, at the 88% level. This contradicts trees of NIXON & al. (1994), in which angiosperms are nested within *Gnetales* (which are therefore paraphyletic). Within angiosperms, *Nymphaeales* are basal in only 54% of the bootstrap replicates, but some combination of paleoherbs is almost always basal. *Magnoliales* are almost never basal, except at chance levels (less than 1%), and the same is true of *Chloranthaceae*.

These results call into question the use of *Magnolia* as a model for the primitive flower, but they do not support a radically different prototype (DOYLE & al. 1994). *Piperales*, which have orthotropous ovules and no perianth and figure prominently in recent pseudanthial theories (MEEUSE 1972, 1976), are low on the tree, but not basal. *Chloranthaceae* are still higher in the tree, supporting the view that their simple flowers are reduced. Instead, the arrangement of taxa implies that the primitive flower was more like that of *Cabomba*, *Lactoris*, *Saruma*, and monocots: trimerous and bisexual, with a perianth, stamens differentiated into anther and filament, and carpels with several anatropous ovules. However, position of the ovules might have been laminar rather than marginal (the most common condition in angiosperms), since TAYLOR (1991) characterizes placentation in *Nymphaeales* as lateral (along the radial walls of the carpel), chaotic, or even exmedial (on the carpel midrib). The seeds might have contained both perisperm and endosperm, an unusual feature of both *Nymphaeales* and *Piperales*. Looking outside the angiosperms, the inference that carpels originally had several anatropous ovules would conflict with neo-englerian pseudanthial schemes of the sort shown in Fig. 5, where the original carpel had a single orthotropous ovule. On the other hand, the possibility that placentation was originally laminar or exmedial rather than marginal might make it easier to derive the carpel from a glossopterid prototype (Fig. 4), since in glossopterids the cupules corresponding to the bitegmic ovules of angiosperms were attached to the middle, not the margins, of the subtending leaf.

Molecular data could also impinge upon neo-englerian scenarios by indicating whether or not conifers are related to angiosperms and *Gnetales*, but this would not be decisive, since the most parsimonious trees of DOYLE & DONOGHUE (1992) included some in which conifers are the closest living relatives of anthophytes but Mesozoic seed ferns are interpolated between the two groups. The rRNA trees do not support a neo-englerian arrangement, since *Ginkgo*, cycads, and conifers form a clade, within which conifers are linked with cycads (HAMBY & ZIMMER 1992, DOYLE & al. 1994). However, bootstrap analysis shows that the support for these relationships is weak.

In all these cases, present rRNA data must be considered inconclusive, since other molecular data sets have given different results, such as the basal position of *Ceratophyllum* in trees based on *rbc*L (CHASE & al. 1993). A general problem is the fact that "long branches" on which there has been a large amount of molecular evolution tend to attract each other due to spurious convergences (FELSENSTEIN 1978). It is possible that the basal position of *Ceratophyllum* in the *rbc*L analyses is due to this effect, since the position of *Ceratophyllum* in unrooted angiosperm trees is unstable (QIU & al. 1993), suggesting a high level of homoplasy on the line leading to this genus. However, these conflicts may be resolved by study of additional sequences, recognition of rarer and thus potentially more reliable genome re-arrangements (cf. RAUBESON & JANSEN 1992), or development of methods to correct for long branch attraction. For example, the fact that *Nymphaeales* remain basal in rRNA trees constructed using neighbor joining (HAMBY & ZIMMER 1992), a distance method that compensates to some extent for the problem of unequal branch lengths (SWOFFORD & OLSEN 1990, HUELSENBECK & HILLIS 1993), is an argument against the view that the nymphaealean rooting found in the other rRNA analyses is due to long branch attraction.

A very different molecular approach might use genes involved in floral development as a key to morphological homologies in angiosperms and *Gnetales* (DOYLE 1993). In the scheme of COEN & MEYEROWITZ (1991) and BOWMAN & al. (1991) for control of the identity of floral whorls in *Arabidopsis*, based on developmental mutants, the "A" genes specify perianth, while the "C" genes specify fertile parts. Activity of the "B" genes plus the "A" genes specifies petals as opposed to sepals, while activity of the "B" genes plus the "C" genes specifies stamens as opposed to carpels. A homologous system has been found in *Antirrhinum*, suggesting that it is general for eudicots, although evidence from magnoliids is needed before it can

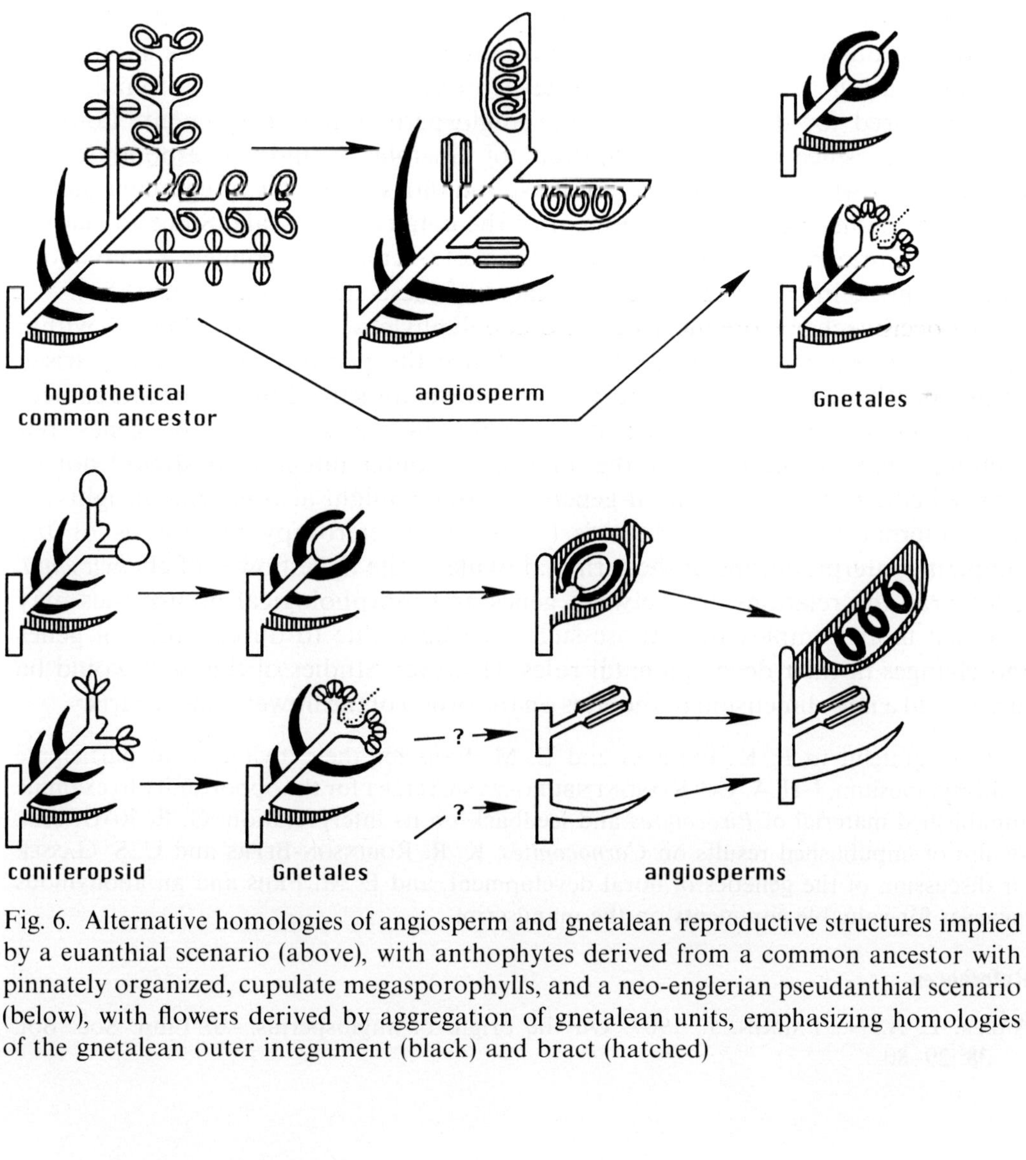


Fig. 6. Alternative homologies of angiosperm and gnetalean reproductive structures implied by a euanthial scenario (above), with anthophytes derived from a common ancestor with pinnately organized, cupulate megasporophylls, and a neo-englerian pseudanthial scenario (below), with flowers derived by aggregation of gnetalean units, emphasizing homologies of the gnetalean outer integument (black) and bract (hatched)

be assumed to be basic for angiosperms as a whole. In addition, ROBINSON-BEERS & al. (1992) have identified genes necessary for development of the integuments in *Arabidopsis*. To use such genes as evidence for the homology of structures in different groups, it will be necessary not only to identify the homologs of particular genes, but also to show that they are expressed in the putatively homologous structures.

It is easiest to see how this approach might distinguish between two extreme hypotheses (Fig. 6), the euanthial scenario of DOYLE & DONOGHUE (1986) and a neo-englerian pseudanthial scenario. The contrasting predictions may be seen by focusing on the outer integument of *Gnetales*. If the euanthial scenario is correct, the common ancestor of angiosperms and *Gnetales* had flowers with the standard order of parts, and it may be hypothesized that the system of floral genes and their present roles had arisen by this point. Therefore the outer integument of *Gnetales* would correspond to the perianth of angiosperms, and homologs of the "A" genes, such as *apetala 2* (COEN & MEYEROWITZ 1991), should be active during its development. In contrast, if the pseudanthial scenario is correct, the outer integument of *Gnetales* does not correspond to the angiosperm perianth, but rather to the angiosperm outer integument. Therefore homologs of the "A" genes should not be active during development of this layer in Gnetales, but rather homologs of genes required for development of the angiosperm outer integument. Looking further back, since the female "flower" of *Gnetales* would correspond to the ovuliferous cone scale of modern conifers, homologs of "outer integument genes" might also be involved in development of the conifer cone scale. Since the carpel wall would correspond to the subtending bract of *Gnetales* and conifers, homologous genes might be involved in the development of these structures.

How this scheme might relate to a polyphyletic origin of flowers within anthophytes (Fig. 4) is more ambiguous. Since the perianth would have arisen independently in angiosperms and *Gnetales*, development of the outer integument in *Gnetales* would not be expected to involve homologs of the "A" genes, but homologs of genes that specify the angiosperm outer integument should not be involved either. A developmental-genetic approach might also provide insights on other alternatives, such as MEYEN's (1988) gamoheterotopy hypothesis or the competing interpretations of the terminal ovule in the male flower of *Welwitschia*. In general, the relationships between genes and morphological homologies may be much more complex than those suggested here, due to duplications of genes and changes in their developmental roles. However, studies of this sort would be sure to add a new dimension to theories on the origin of the flower and its parts.

I am grateful to P. K. ENDRESS and E. M. FRIIS for their invitation to participate in this symposium, J. H. A. VAN KONIJNENBURG-VAN CITTERT for the opportunity to examine unpublished material of *Piroconites* and feedback on its interpretation, G. R. ROTHWELL for use of unpublished results on *Carnoconites*, K. R. ROBINSON-BEERS and C. S. GASSER for discussion of the genetics of floral development, and E. M. FRIIS and an anonymous reviewer for valuable comments on the manuscript.

## References

ARBER, E. A. N., PARKIN, J., 1907: On the origin of angiosperms. – J. Linn. Soc. Bot. **38**: 29–80.

– – 1908: Studies on the evolution of the angiosperms. The relationship of the angiosperms to the Gnetales. – Ann. Bot. **22**: 489–515.

ASH, S. R., 1972: Late Triassic plants from the Chinle Formation in northeastern Arizona. – Palaeontology **15**: 598–618.

BOWMAN, J. L., SMYTH, D. R., MEYEROWITZ, E. M., 1991: Genetic interactions among floral homeotic genes of *Arabidopsis*. – Development **112**: 1–20.

CHASE, M. W. & 41 others, 1993: Phylogenetics of seed plants: an analysis of nucleotide sequences from the plastid gene *rbc*L. – Ann. Missouri Bot. Gard. **80**: 526–580.

COEN, E. S., MEYEROWITZ, E. M., 1991: The war of the whorls: genetic interactions controlling flower development. – Nature **353**: 31–37.

CORNET, B., 1989: Late Triassic angiosperm-like pollen from the Richmond rift basin of Virginia, U.S.A. – Palaeontographica, Abt. B, **213**: 37–87.

CRANE, P. R., 1985: Phylogenetic analysis of seed plants and the origin of angiosperms. – Ann. Missouri Bot. Gard. **72**: 716–793.

– 1988: Major clades and relationships in the "higher" gymnosperms. – In BECK, C. B., (Ed.): Origin and evolution of gymnosperms, pp. 218–272. – New York: Columbia University Press.

– FRIIS, E. M., PEDERSEN, K. R., 1994: Palaeobotanical evidence on the early radiation of magnoliid angiosperms. – Pl. Syst. Evol. [Suppl.] **8**: 51–72.

DONOGHUE, M. J., DOYLE, J. A., 1989: Phylogenetic analysis of angiosperms and the relationships of *Hamamelidae*. – In CRANE, P. R., BLACKMORE, S., (Eds): Evolution, systematics, and fossil history of the *Hamamelidae*, 1, pp. 17–45. – Oxford: Clarendon Press.

DOYLE, J. A., 1978: Origin of angiosperms. – Annual Rev. Ecol. Syst. **9**: 365–392.

– 1993: Cladistic and paleobotanical perspectives on the origin of angiosperm organs. – J. Cell. Biochem. [Suppl.] **17B**: 8.

– DONOGHUE, M. J., 1986: Seed plant phylogeny and the origin of angiosperms: an experimental cladistic approach. – Bot. Rev. **52**: 321–431.

– – 1990: Reexamination of seed plant and angiosperm relationships. – Amer. J. Bot. **77** [6, Suppl.]: 84.

– – 1992: Fossils and seed plant phylogeny reanalyzed. – Brittonia **44**: 89–106.

– – 1993: Phylogenies and angiosperm diversification. – Paleobiology **19**: 141–167.

– HOTTON, C. L., 1991: Diversification of early angiosperm pollen in a cladistic context. – In BLACKMORE, S., BARNES, S. H., (Eds): Pollen and spores: patterns of diversification, pp. 169–195. – Oxford: Clarendon Press.

– DONOGHUE, M. J., ZIMMER, E. A., 1994: Integration of morphological and ribosomal RNA data on the origin of angiosperms. – Ann. Missouri Bot. Gard. **81**: 419–450.

ENDRESS, P. K., 1986: Reproductive structures and phylogenetic significance of extant primitive angiosperms. – Pl. Syst. Evol. **152**: 1–28.

FELSENSTEIN, J., 1978: Cases in which parsimony or compatibility methods will be positively misleading. – Syst. Zool. **27**: 401–410.

– 1985: Confidence limits on phylogenies: an approach using the bootstrap. – Evolution **39**: 783–791.

FRIIS, E. M., PEDERSEN, K. R., CRANE, P. R., 1994: Angiosperm floral structures from the Early Cretaceous of Portugal. – Pl. Syst. Evol. [Suppl.] **8**: 31–49.

FLORIN, R., 1951: Evolution in cordaites and conifers. – Acta Horti Berg. **15**: 285–388.

FRIEDMAN, W. E., 1990: Sexual reproduction in *Ephedra nevadensis* (*Ephedraceae*): further evidence of double fertilization in a nonflowering seed plant. – Amer. J. Bot. **77**: 1582–1598.

– 1992: Evidence of a pre-angiosperm origin of endosperm: implications for the evolution of flowering plants. – Science **255**: 336–339.

– 1994: The evolution of embryogeny in seed plants and the developmental origin and early history of endosperm. – Amer. J. Bot. **81**: 1468–1487.

GAUSSEN, H., 1946: Les Gymnospermes, actuelles et fossiles. – Trav. Lab. Forest. Toulouse, tome II, sect. 1, vol. 1, fasc. 3, ch. 5: 1–26.

HAMBY, R. K., ZIMMER, E. A., 1992: Ribosomal RNA as a phylogenetic tool in plant systematics. – In SOLTIS, P. S., SOLTIS, D. E., DOYLE, J. J., (Eds): Molecular systematics of plants, pp. 50–91. – New York: Chapman and Hall.

HARRIS, T. M., 1940: *Caytonia*. – Ann. Bot., n.s. **4**: 713–734.

– 1954: Mesozoic seed cuticles. – Svensk Bot. Tidskr. **48**: 281–291.

HICKEY, L. J., DOYLE, J. A., 1977: Early Cretaceous fossil evidence for angiosperm evolution. – Bot. Rev. **43**: 1–104.

HUELSENBECK, J. P., HILLIS, D. M., 1993: Success of phylogenetic methods in the four-taxon case. – Syst. Biol. **42**: 247–264.

KIRCHNER, M., 1992: Untersuchungen an einigen Gymnospermen der fränkischen Rhät-Lias-Grenzschichten. – Palaeontographica, Abt. B, **224**: 17–61.

LOCONTE, H., STEVENSON, D. W., 1991: Cladistics of the *Magnoliidae*. – Cladistics **7**: 267–296.

MARTENS, P., 1971: Les Gnétophytes. Encyclopedia of plant anatomy **12**(2). – Stuttgart: Borntraeger.

MEEUSE, A. D. J., 1972: Sixty-five years of theories of the multiaxial flower. – Acta Biotheor. **21**: 167–202.

– 1976: Floral evolution and emended anthocorm theory. – Int. Biosci. Monogr. (Hissar, Madras) **1**: 1–188.

MEHRA, P. N., 1950: Occurrence of hermaphrodite flowers and the development of female gametophyte in *Ephedra intermedia* SHRENK et MEY. – Ann. Bot., n.s. **14**: 165–180.

MEYEN, S. V., 1988: Origin of the angiosperm gynoecium by gamoheterotopy. – Bot. J. Linn. Soc. **97**: 171–178.

NIXON, K. C., CREPET, W. L., STEVENSON, D. W., FRIIS, E. M., 1994: Phylogenetic relationships of seed plants. – Ann. Missouri Bot. Gard. **81**: 484–533.

PEDERSEN, K. R., CRANE, P. R., FRIIS, E. M., 1989: Pollen organs and seeds with *Eucommiidites* pollen. – Grana **28**: 279–294.

QIU, Y.-L., CHASE, M. W., LES, D. H., PARKS, C. R., 1993: Molecular phylogenetics of the *Magnoliidae*: cladistic analyses of nucleotide sequences of the plastid gene *rbc*L. – Ann. Missouri Bot. Gard. **80**: 587–606.

RAUBESON, L. A., JANSEN, R. K., 1992: Chloroplast DNA evidence on the ancient evolutionary split in vascular land plants. – Science **255**: 1697–1699.

RETALLACK, G., DILCHER, D. L., 1981: Arguments for a glossopterid ancestry of angiosperms. – Paleobiology **7**: 54–67.

REYMANÓWNA, M., 1973: The Jurassic flora from Grojec near Kraków in Poland. Part II. *Caytoniales* and anatomy of *Caytonia*. – Acta Palaeobot. **14**: 45–87.

– 1974: On anatomy and morphology of *Caytonia*. – Birbal Sahni Inst. Palaeobot. Spec. Publ. **2**: 50–57.

ROBINSON-BEERS, K. R., PRUITT, E., GASSER, C. S., 1992: Ovule development in wild type *Arabidopsis* and two female-sterile mutants. – Pl. Cell. **4**: 1237–1249.

ROTHWELL, G. R., SERBET, R., 1994: Lignophyte phylogeny and the evolution of spermatophytes: a numerical cladistic analysis. – Syst. Bot. **19**: 443–482.

SEWARD, A. C., 1904: Catalogue of Mesozoic plants in the British Museum. The Jurassic flora, part 1. – London: British Museum (Natural History).

STEBBINS, G. L., 1974: Flowering plants: evolution above the species level. – Cambridge, Mass.: Harvard University Press.

SWOFFORD, D. L., OLSEN, G. J., 1990: Phylogeny reconstruction. – In HILLIS, D. M., MORITZ, C., (Eds): Molecular systematics, pp. 411–501. – Sunderland, Mass.: Sinauer.

TAKASO, T., 1985: A developmental study of the integument in gymnosperms 3. *Ephedra distachya* L. and *E. equisetina* BGE. – Acta Bot. Neerl. **34**: 33–48.

TAYLOR, D. W., 1991: Angiosperm ovules and carpels: their characters and polarities, distribution in basal clades, and structural evolution. – Postilla **208**: 1–40.

– HICKEY, L. J., 1990: An Aptian plant with attached leaves and flowers: implications for angiosperm origin. – Science **247**: 702–704.

– 1992: Phylogenetic evidence for the herbaceous origin of angiosperms. – Pl. Syst. Evol. **180**: 137–156.

TAYLOR, E. L., TAYLOR, T. N., 1992: Reproductive biology of the Permian *Glossopteridales* and their suggested relationship to flowering plants. – Proc. Natl. Acad. Sci. USA **89**: 11495–11497.

TAYLOR, T. N., ARCHANGELSKY, S., 1985: The Cretaceous pteridosperms *Ruflorinia* and *Ktalenia* and implications on cupule and carpel evolution. – Amer. J. Bot. **72**: 1842–1853.

TUCKER, S. C., GIFFORD, E. M., Jr., 1966: Organogenesis in the carpellate flower of *Drimys lanceolata*. – Amer. J. Bot. **53**: 433–442.

– DOUGLAS, A. W., LIANG, H.-X., 1993: Utility of ontogenetic and conventional characters in determining phylogenetic relationships of *Saururaceae* and *Piperaceae* (*Piperales*). – Syst. Bot. **18**: 614–641.

VAN KONIJNENBURG-VAN CITTERT, J. H. A., 1992: An enigmatic Liassic microsporophyll, yielding *Ephedripites* pollen. Rev. Palaeobot. Palynol. **71**: 239–254.

WETTSTEIN, R. R. von, 1907: Handbuch der systematischen Botanik, II. – Leipzig, Wien: Deuticke.

Address of the author: JAMES A. DOYLE, Section of Evolution and Ecology, University of California, Davis, CA 95616, USA.

Accepted March 25, 1994 by E. M. FRIIS

Pl. Syst. Evol. [Suppl.] 8: 31–49 (1994)

# Angiosperm floral structures from the Early Cretaceous of Portugal

ELSE MARIE FRIIS, KAJ RAUNSGAARD PEDERSEN, and PETER R. CRANE

Received November 24, 1993

**Key words:** Angiosperms, *Chloranthaceae*, *Clavatipollenites*-complex, *Laurales*, *Magnoliidae*. – Early Cretaceous, epigyny, fossil flowers, Portugal.

**Abstract:** Rich fossil floras including structurally preserved flowers, fruits, seeds and dispersed stamens have recently been extracted from mid-Early Cretaceous sediments from Portugal. The material illustrates considerable morphological variability in number and arrangement of floral parts, with the most diversity among magnoliid taxa; relatively few eudicots have been identified. Generally the flowers are small with undifferentiated perianth and massive androecium, and include both bisexual and unisexual forms. The most unexpected result of this study of Early Cretaceous flowers from Portugal is the presence of several epigynous forms at this early stage of angiosperm evolution.

Portugal has long been known as one of the classic areas to have yielded Cretaceous angiosperm leaf floras, the earliest comprehensive study being that of SAPORTA (1894) with subsequent works by TEIXEIRA (1945, 1946, 1947, 1948, 1950, 1952). However, more recently studies of Portuguese Cretaceous floras have mainly concentrated on palynological aspects (e.g., DINIZ & al. 1974; KEDVES & PITTAU 1979; KEDVES & DINIZ 1981, 1983; TRINCÃO 1985, 1990; BATTEN 1986, 1989; TRINCÃO & al. 1989). During the past few years field work in Portugal has yielded numerous new Cretaceous floras consisting of structurally preserved angiosperm flowers, fruits, seeds, dispersed stamens and pollen as well as remains of non-angiospermous seed plants and pteridophytes. In a previous study (FRIIS & al. 1992) we described well-preserved Late Cretaceous flowers from the northern part of the Western Portuguese Basin. In this paper we provide the first characterization of angiosperm and angiosperm-like reproductive structures from the Early Cretaceous of Portugal. Some of these specimens provide almost complete details of floral construction, while others are incompletely preserved and floral structure is not fully understood. However, together these fossils illustrate the morphological diversity of Early Cretaceous flowers and document the presence of specific floral features at an early stage in angiosperm evolution.

## Material

The Early Cretaceous fossils described here were extracted from sediment samples collected in the Western Portuguese Basin and comprise material from three localities in the Beira Litoral Region (East of Vale de Agua, Juncal, and Famalicão) and two localities from the Estremadura Region (Catefica and Torres Vedras). The fossil flora from Juncal consists primarily of leaf impressions but it also includes a number of small flowers preserved as imprints. The other four floras consist of structurally-preserved fossils that were extracted from the sediments by sieving. Specimens in these floras are charcoalified or lignitized and are often extremely well-preserved. The fossil floras of Catefica, Vale de Agua and Famalicão include many of the same taxa and are thought to be of similar age. A few structurally preserved reproductive organs are also shared by these three floras and that from Juncal, indicating that the Juncal flora may also be of the same age. Based on the regional geological descriptions (e.g., França & Zbyszewski 1963; Rey 1972, 1982; Teixeira & al. 1968; Zbyszewski & al. 1974) and preliminary palynological data the four floras are probably of mid-Early Cretaceous age, possibly as old as the Early Barremian and not younger than Aptian while the fossil flora from Torres Vedras is probably slightly older (Valanginian or Hauterivian). A more detailed description of the localities and discussion of the ages will be published elsewhere (Friis & al., unpubl.).

## Results

Angiosperm and angiosperm-like reproductive organs from the Early Cretaceous of Portugal include numerous taxa some of which are represented only by fragmentary material. In the present work we describe a selection of these fossil to document the diversity of floral form at this early stage in angiosperm evolution.

**Multistaminate structure** (Fig. 1). This floral type is represented by a single specimen from the Torres Vedras locality (Fig. 1a, b). It is conical in outline, about 0.75 mm long and about 1 mm in diameter, consisting of numerous stamens in a whorled arrangement (Fig. 1a). Five whorls are preserved, but there are no subtending bracts or tepals at the base of the structure and it is unknown whether the specimen is complete. The number of stamens per whorl increases from the apex towards the base (Fig. 1b). There are four stamens in the apical whorl, eight

Fig. 1. SEM-micrographs of unisexual, staminate reproductive structure with many stamens in whorled arrangement from Torres Vedras (S101220) *a* Lateral view; *b* apical view, ×60

in the second, 11 in the third whorl (one apparently missing) and 16 in the fourth. The fifth whorl is incompletely preserved and the number of stamens cannot be established. Each stamen is minute, about 0.3 mm long and 0.15 mm broad, basifixed with an extremely short filament, and with a dithecate, tetrasporangiate anther dehiscing by longitudinal slits. Pollen grains have not been seen in the pollen sacs, but several grains were observed adhering to the surface of the stamens. These are monocolpate and reticulate. The reticulum of most pollen grains is contracted and distorted, probably due to fossilization.

The most appropriate morphological interpretation of the multistaminate structure is uncertain. No bracts or tepals were observed at the base of the individual stamens or at the base of the conical structure. Thus it is possible that the fossil is an inflorescence axis with many unistaminate and naked flowers in a whorled arrangement, but alternatively it could be a single unisexual flower with a multi-

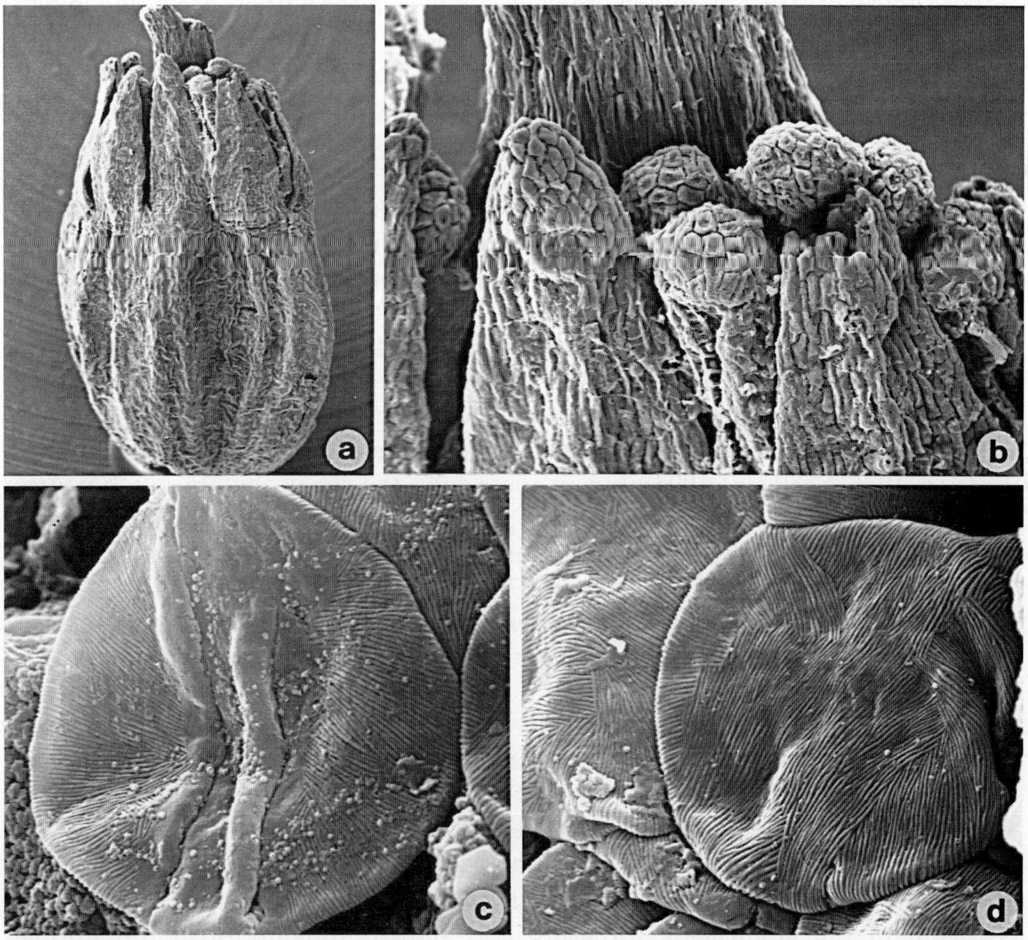


Fig. 2. SEM-micrographs of epigynous flower from Vale de Agua. *a* Lateral view of flower (S101304) showing inferior gynoecium, tepal-lobes, parts of androecium, and central, stylar area, ×35; *b* detail of Fig. 2a showing apical extensions of connectives, ×175; *c–d* finely striate pollen from flower (S101288), ×4500; *c* distal view of pollen grain showing colpus area; *d* proximal view of pollen grain showing fingerprint-like striate sculpture

staminate androecium and stamens arranged in several whorls. The regular pattern of increasing stamen number seen in the fossil structure is unusual both for angiosperm flowers and for inflorescences. The systematic affinities of this fossil are uncertain, but the occurrence of monocolpate and reticulate pollen indicates that the relationships of the fossil are probably among the magnoliid angiosperms.

**Epigynous flower with finely striate pollen** (Fig. 2). This flower type is represented by numerous specimens in the Early Cretaceous floras from the Vale de Agua, Catefica, and Famalicão localities (Friis & al., unpubl.). The flowers are small, 1.3–3.1 mm long, bisexual and actinomorphic. The perianth is undifferentiated with one whorl of seven, or more rarely five, coriaceous tepal lobes. Tepal lobes are imbricate and sometimes of variable size within the same flower. There are more stamens than sepals, possible twice as many. The anthers are dithecate, tetrasporangiate and almost sessile with lateral to slightly extrorse valvate dehiscence. The connective tissue between the two pairs of pollen sacs is massive and has a distinct, globose apical protrusion. Pollen grains in situ within the anthers are monocolpate, circular to elliptical in equatorial outline, about 11–14 µm in diameter, and with a finely striate sculpture that forms a fingerprint-like pattern. The organization of ovary and ovules is not yet clearly established as most of the specimens are brittle and difficult to section, but the gynoecium is apparently unilocular formed from two carpels, and with a single seed.

Morphologically, the fossil flowers show similarities to those of certain *Laurales*, particularly to members of the *Hernandiaceae*, which also includes forms with small epigynous flowers. In the subfamily *Gyrocarpoideae* of the *Hernandiaceae* the perianth is undifferentiated with 4, 5 or 7 lobes that may be of unequal size (Kubitzki 1969b). Pollen of the *Hernandiaceae* is inaperturate with finely spinose sculpture (Shutts 1959, Kubitzki 1969a, b). Striate pollen is uncommon in extant *Magnoliidae*. Among the *Laurales* it is recorded in the *Lauraceae* (*Dahlgrenodendron*, Van der Merwe & al. 1990) and the *Monimiaceae* (*Ephippiandra* and *Hortonia*, Walker 1976, Sampson 1993), but in these taxa the striations are much coarser than in the fossil material. Monocolpate pollen with fine fingerprint-like striations very similar to that observed in the fossil material occurs in extant *Cabomba* (Walker 1976, Schneider & Jeter 1982), but flowers in *Cabomba* are distinctly different in being hypogynous and trimerous with an apocarpous gynoecium.

**Epigynous (?) flower with reticulate pollen** (Fig. 3a, b). This floral type is represented by a single specimen from the Torres Vedras locality. It is incompletely preserved, one side of the fossil being completely abraded, and details of its organization are uncertain. The flower is about 1.3 mm long, bisexual and apparently actinomorphic. Perianth parts are basally fused to each other, and apparently also to the gynoecium. Six stamens are preserved and while 3 or 4 are missing from the abraded side we estimate that the androecium consisted of a single whorl of nine or ten stamens. Anthers are elongate with a small apical extension of the connective, dithecate and tetrasporangiate with longitudinal dehiscence (Fig. 3a). No filament has been observed. Pollen grains found in situ in the anthers are spherical, about 12 µm in diameter, monocolpate and tectate (Fig. 3b). The colpus is long and extends almost to the equator of the grain. The tectum is foveolate to reticulate with lumina up to 1 µm in diameter and muri with rounded profiles. Columellae are densely spaced, simple and of moderate length.

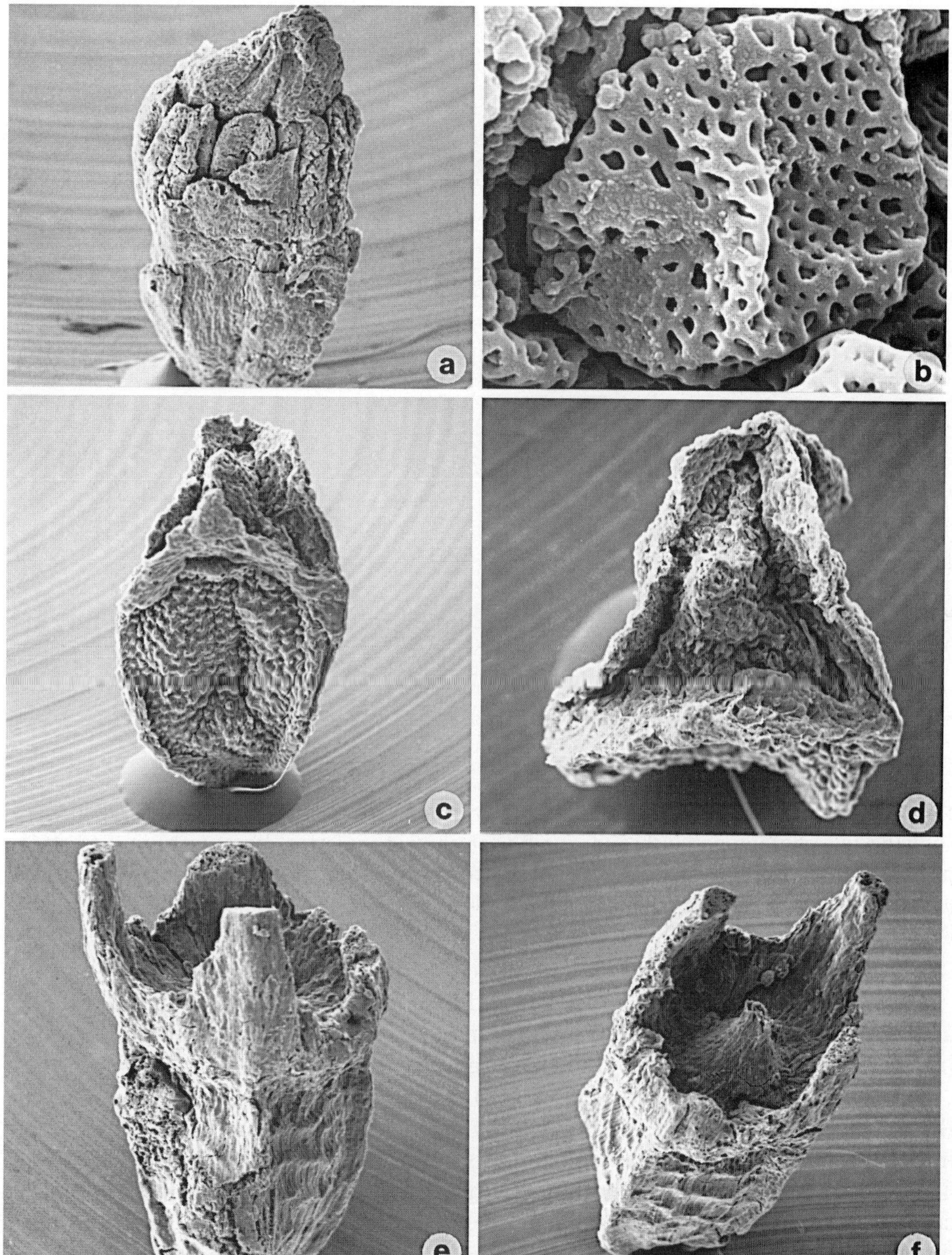


Fig. 3. SEM-micrographs of floral structures from Torres Vedras. *a–d* Flower with reticulate pollen (S101306); *a* lateral view of flower showing ovary, remnants of perianth, and stamens, ×60; *b* proximal view of pollen grain, ×5000; *c–d* trimerous reproductive structure (S101307); *c* lateral view showing position of supposed perianth, ×60; *d* apical view, ×90; *e–f* tetramerous reproductive structure (S100754), ×50; *e* lateral view showing position of supposed perianth; *f* apical view of structure show in *e*.

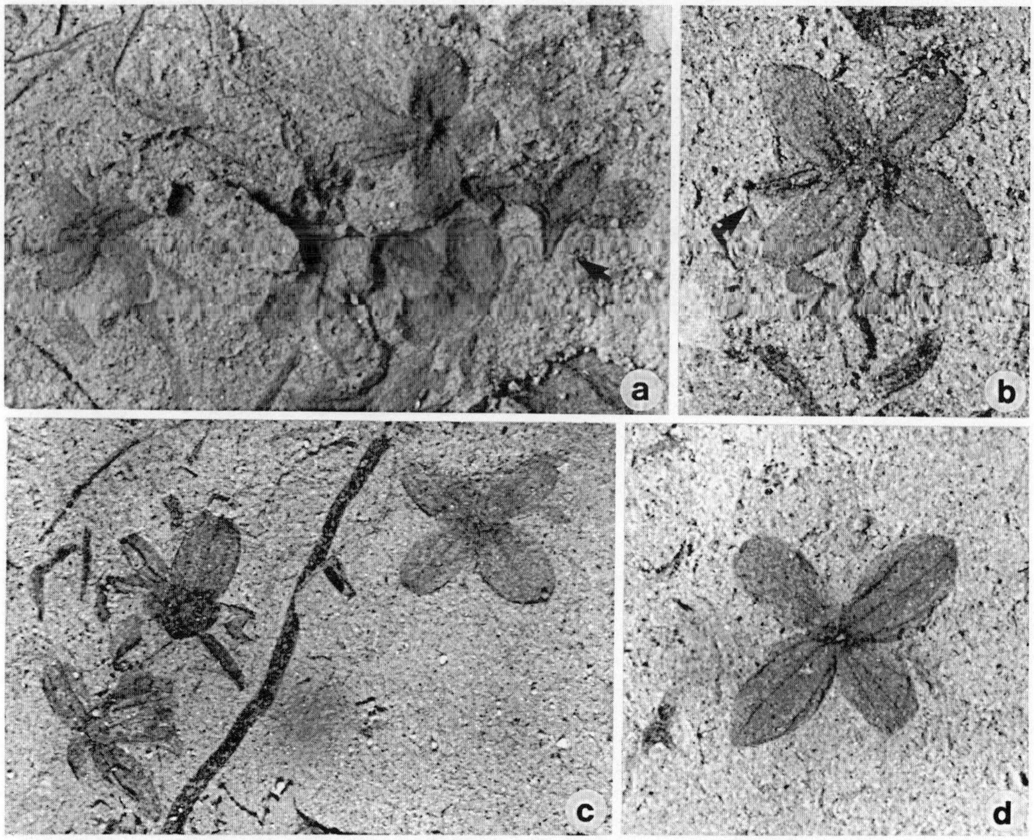

Fig. 4. Reflected light micrograph of epigynous flowers from Juncal. *a* Several flowers showing tetramerous perianth and one flower showing inferior ovary (arrow) (S101451), ×40; *b* flower with one stamen preserved (arrow) (S101449a) ×60; *c* flowers with part of androecium preserved (S101450), ×40; *d* flower showing venation pattern of tepals (S101453), ×60

The monocolpate, reticulate pollen indicate relationship within the magnoliid angiosperms, and the simple gynoecium may suggest a lauralean affinity, but the information presently available does not allow more detailed comparison.

**Epigynous, tetramerous flowers** (Fig. 4). This flower type occurs abundantly at the Juncal locality, but it is known only from impressions and the amount of information that may be obtained from this material is limited. The flower is about 3.5 mm long and about 5 mm in diameter, tetramerous with an epigynous perianth consisting of one whorl of basally fused tepals. Imprints of tepals are typically distinct and without abraded margins, indicating that they were probably tough and leathery. Each tepal lobe is supplied by 3 unbranched veins. Anthers are sessile or with very short filaments. They are elongate, about 1.8 mm long, with a short, pointed apical extension of the connective. Minute resin bodies have been observed over the surface of some of the anthers. The gynoecium is small, about 1.7 mm long, narrow and conical in outline. It appears to consist of a single unit (not apocarpous), but none of the fossils are sufficiently well-preserved to clearly document the organization of gynoecium.

Because of the limited information available detailed discussion of the systematic affinities of this material is not possible, but the presence of resin bodies (perhaps remains of ethereal oil cells) indicates a magnoliid affinity for this taxon, perhaps close to the *Laurales*.

**Triangular and quadrangular epigynous structures** (Fig. 3c–f). In addition to epigynous floral structures of definite angiospermous affinity the Early Cretaceous floras of Portugal include several ovulate structures with an epigynous organization for which the relationships are currently uncertain.

The triangular ovulate structures were discovered at the Torres Vedras locality. The fossils are about 1 mm long and 0.6 mm broad, obovate in longitudinal outline and sharply triangular in transection, pointed apically and topped by three broadly triangular tepal-like organs (Fig. 3c, d). There are no remains of stamens and no pollen grains have been observed at the apical part of the structure. The cuticle of the outer wall of the seed/ovary is finely verrucate.

Similar triangular gynoecia with three epigynous perianth lobes are known in extant *Hedyosmum* of the *Chloranthaceae* (ENDRESS 1971, pers. comm. 1993), but no pollen has been observed on the fossil structures and details of gynoecium wall are unknown impeding further comparison with *Hedyosmum*.

The quadrangular ovulate structures occur abundantly in the Early Cretaceous floras of Portugal and have been recorded in all floras investigated so far. Similar quadrangular fossils are also common in the Early Cretaceous floras of the Potomac Group sediments of eastern North America and several species appear to be present. The material illustrated in this work is from the Torres Vedras locality (Fig. 3e, f). The fossils are about 0.6–1 mm long and 0.4–0.6 mm wide, elliptical to ovate in longitudinal outline, apically pointed, and distinctly quadrangular in transection. Vascular bundles extend from the corners of the structures into four stiff, tepal-like projections near the apex of the structure. The fossils are unilocular with an outer wall of sclerenchyma cells and an inner thin, membraneous layer. The inner layer has a long slender micropylar tube. The sclerenchyma of the outer wall are organized into an outer layer of longitudinally aligned cells and an inner layer of transversely aligned cells. The wall is also ornamented by irregular transverse ridges and may have been covered by an additional fleshy layer.

Sectioning and maceration of many specimens have failed to show a pollen chamber or pollen grains inside the apical projection. Based on this negative evidence we interpret the structure as being a unilocular and one-seeded angiosperm fruit, and the apical sclerenchymatic projection as a style rather than a micropyle. However, we were unable to demonstrate stigmatic tissue in these fossils, no association with stamens or pollen has been established, and the systematic affinities remain uncertain.

An epigynous organization with tepal-like organs at the apex of the ovulate structure is unknown in non-angiospermous seed plants, and indicates that relationships of the triangular and quadrangular ovulate structures were probably within the angiosperms. However, the fossil record of pollen and new studies of Cretaceous seeds and pollen organs indicate the presence of a much wider variety of seed plants during the Mesozoic than previously anticipated, particularly among those groups of seed plants that are probably closely related to the angiosperms (e.g., KRASSILOV 1986, PEDERSEN & al. 1989, VAN KONIJNENBURG-VAN

CITTERT 1992). In extant *Gnetales* and in the extinct *Bennettitales* and *Eucommiidites* groups the nucellus is surrounded by more than one envelope. All of these ovules have an elongated micropylar tube formed from the inner integument. In *Vardekloeftia* the inner integument projects well beyond the outer envelope which may form an indistinct collar near the base of the micropyle. The *Eucommiidites* group includes a variety of small triangular seeds with a two-layered seed wall, while in extant **Gnetales** the nucellus is surrounded by two or three envelopes, and gnetalean pollen has been observed in situ in a variety of four-angled seeds (KRASSILOV 1986; PEDERSEN & al. 1989, FRIIS & al., unpubl.).

**Unilocular, one-seeded fruits with *Clavatipollenites*-type pollen** (Fig. 5a, b). Small, unilocular and one-seeded fruits (or fruitlets) with thin fruit wall and hard seed coat, occur abundantly in all of the Early Cretaceous floras from Portugal as well as in the Early and mid-Cretaceous floras of eastern North America. A number of distinct taxa have been recognized, several of which have pollen grains preserved at, or near, the stigmatic portion of the fruit. The Portuguese fossils include one fruit type with *Clavatipollenites*-type pollen adhering to the fruit surface. This taxon is represented by several fruits from the Vale de Agua locality. The fruits are oval to elliptical in outline and laterally flattened, about 1 mm long and 0.6 mm broad, with a thick outer cuticle and scattered trichomes, about 0.1 mm long (Fig. 5a). The pollen grains are monocolpate, circular to elliptical in equatorial outline, about 12–15 µm in diameter, and coarsely reticulate. The colpus extends almost to the equator. The tectum is loosely attached to the foot layer. Lumina are angular, up to about 2 µm in diameter. Muri are broad, rounded, and ornamented by densely spaced transverse ridges (Fig. 5b). Columellae are high and widely spaced.

*Clavatipollenites*-type pollen has previously been found together with small unilocular and one-seeded fruits from the mid-Cretaceous of eastern North America assigned to the extinct genus *Couperites* (PEDERSEN & al. 1991, CRANE & al., 1994). Among extant plants *Couperites* is most closely related to the

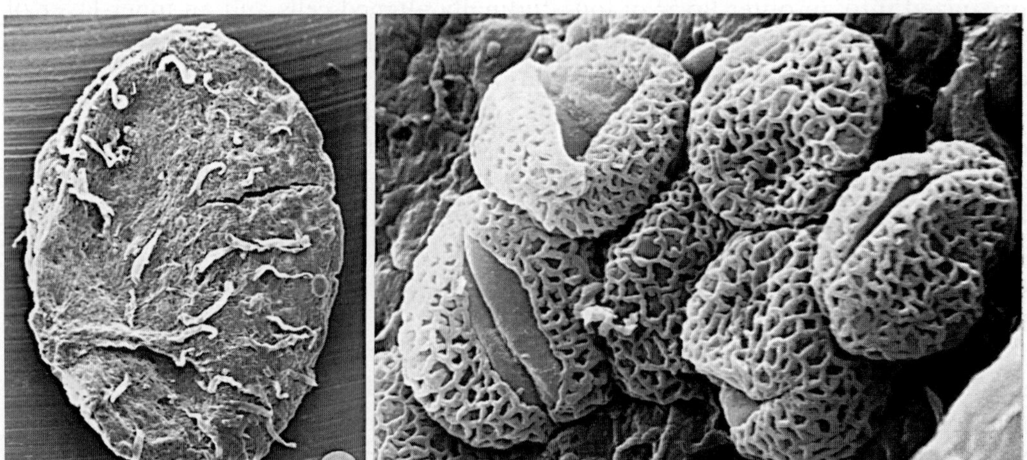


Fig. 5. SEM-micrographs of one-seeded fruit/fruitlets with *Clavatipollenites*-type pollen from Vale de Agua. *a–b* Fruit with trichomes (S101317); *a* lateral view of fruit, × 65; *b* pollen grains from apical part of fruit shown in Fig. 6a, × 2500

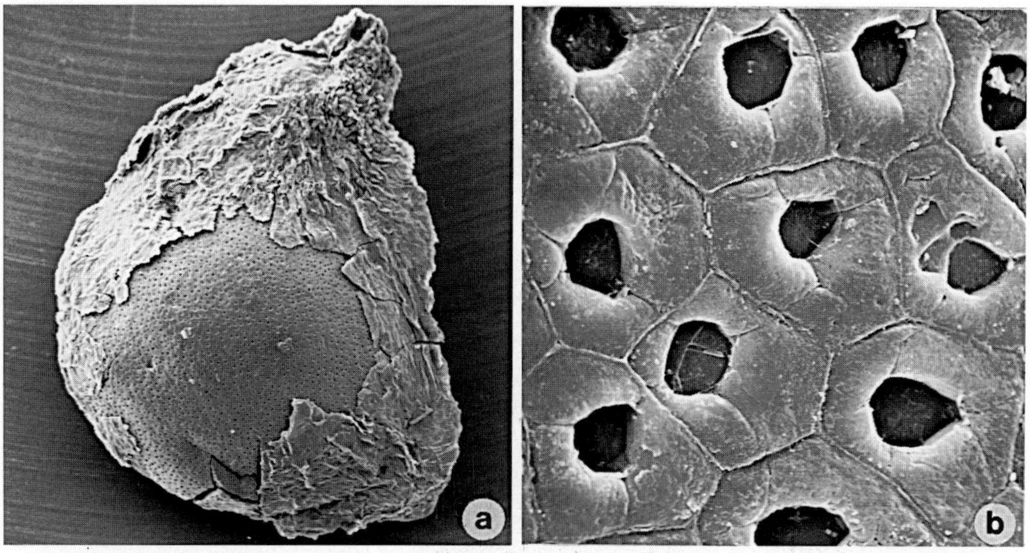


Fig. 6. SEM-micrographs of one-seeded fruit/fruitlet from Vale de Agua (S101297). *a* Lateral view showing abraded fruit wall and seed ×35; *b* details of seed wall showing crystal cells, ×1000

*Chloranthaceae* although details of seed characters indicate that the fossils cannot be included in the *Chloranthaceae* as presently circumscribed. The Portuguese fossils are similar to *Couperites* in general morphology of fruit and seed, but they lack the resin bodies in the fruit wall described for *Couperites*. The pollen grains observed on the Portuguese fossils also differ from those attached to *Couperites* in being smaller and in having a more open reticulum. These differences are, however, minor and the Portuguese fossils are probably also related to the extant *Chloranthaceae*.

Several other fruits with the same general organization as *Couperites* may also belong to the chloranthoid complex, but the absence of pollen currently prevent inclusion of these taxa in the *Clavatipollenites*-group. They may be separated based on differences in morphology and sizes as well as details of fruit and seed wall. One of the most common taxa includes small fruits with resin bodies in the fruit wall similar to those of *Couperites*. The fruits are, however, smaller than *Couperites* and the resin bodies more densely packed. Another common taxon (Fig. 6) is distinguished by the presence of distinct crystal cells in the outer tissue of the seed coat. A variety of *Clavatipollenites*-type pollen has been described from the Early Cretaceous floras of Portugal from dispersed anthers (PEDERSEN & al. 1994a), and also from dispersed palynofloras (e.g., CHAPMAN 1987, TRINCÃO 1990), indicating that this complex was both widespread and diverse.

**Unilocular ribbed fruits** (Fig. 7a). Numerous small, ribbed fruits or fruitlets were discovered from the Famalicão and Vale de Agua localities. The fossils are elongate elliptic in longitudinal outline and laterally flattened, about 3 mm long and about 1.2 mm broad, with slightly concave dorsal and ventral margins. The fruit wall is formed by small, slightly elongate sclerenchyma cells and ornamented by about 5–6 longitudinal ribs on each side. Occasionally the ribs anastomose. The outer

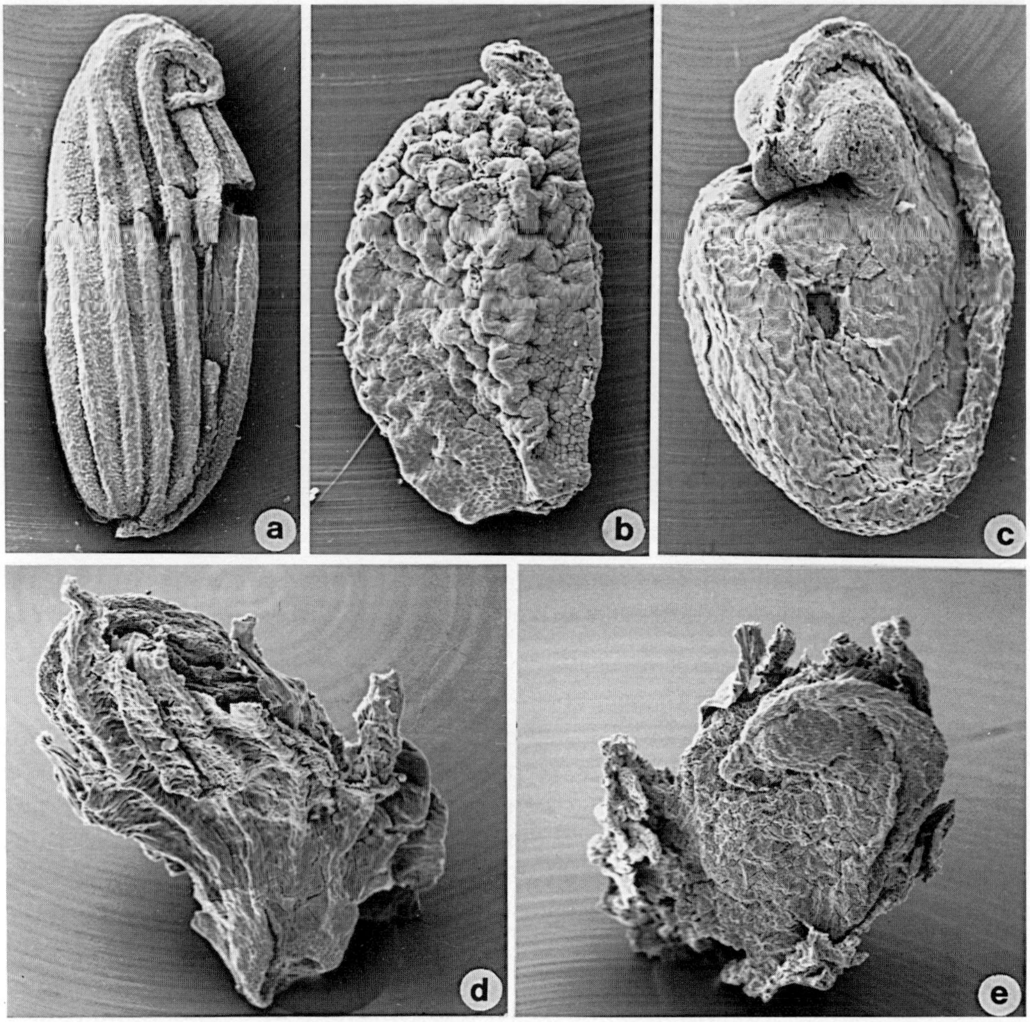

Fig. 7. SEM-micrographs of one-seeded fruits/fruitlets. *a* Fruit from Catefica with ribbed fruit wall and recurved stylar area (S101291), × 50; *b* fruit from Famalicão with rugulate fruit wall and sessile stigmatic area (S101299), × 25; *c* fruit from Vale de Agua with strongly recurved stigmatic area (S101322), × 25; *d–e* fragment of hypogynous flower from Famalicão with ovary/fruit similar to that in *c* (S101443), × 35, *d* lateral view, *e* oblique-apical view

epidermis of the fruit wall is finely verrucate. The stigmatic area is bipartite apically and extended into a long curved style that bends downwards along the ventral margin. There is a single, anatropous seed in each fruit/fruitlet. The seed is apically attached with the micropyle oriented towards the apex and apparently a dorsal raphe. The seed coat is thin and membraneous. No pollen grains have been observed on the stigmatic surface.

The fossils are morphologically similar to fruitlets of the extant genus *Thalictrum* in the *Ranunculaceae*.

**Rugulate fruit** (Fig. 7b). This taxon is represented by a small fruit/fruitlet recovered from the Catefica locality. It is elliptical to ovate in longitudinal outline and laterally flattened, about 1.3 mm long and 0.7 mm broad. The ventral margin is slightly curved and the dorsal margin is strongly convex. The fruit wall is thick and formed by sclerenchyma cells and with irregular, rugulate ornamentation. The outer epidermis has small isodiametric cells. It is possible that an additional parenchymatous tissue was present outside the sclerenchymatous tissue. The stigma is almost sessile, curving backwards towards the dorsal margin. Pollen grains were observed at the stigmatic area embedded in a substance with sticky appearance which may indicate that the stigma was wet or that abundant pollenkitt was produced in the anther. The pollen grains are monocolpate, almost circular in equatorial outline, about 13 μm in diameter, with finely reticulate tectum. There is no information on seed structure.

The presence of monocolpate pollen on the stigmatic surface indicates relationship with magnoliid angiosperms. Similar unilocular, one-seeded fruits with a hard fruit wall are known in extant *Amborellaceae* in the *Laurales*, but a more detailed evaluation of systematic affinities cannot be made on the characters available.

**Unilocular fruit from hypogynous flower** (Fig. 7c–e). This taxon is represented by many specimens from the Famalicão and Vale de Agua localities. The material consists of dispersed fruits (Fig. 7c) as well as rare specimens in which fruits are still supported by the perianth (Fig. 7d–e). The perianth consists of two whorls. The outer whorl has five narrow triangular and membraneous parts that are basally fused to form a shallow tepal cup. Each perianth lobe has three prominent vascular bundles. Perianth parts in the inner whorl are narrow and linear, and alternate with those of the outer whorl. One specimen has a single stamen preserved indicating that the flower was bisexual, but number of stamens cannot be established based on the material available. The anther has a distinct apical extension of the connective. No pollen was observed in situ within the anther, or on the stigmatic surface of the fruits. The gynoecium is monomerous and the fruit unilocular with a single, thin-walled seed. The fruits are broadly elliptical to ovate in longitudinal outline, slightly compressed laterally and about 1.5–3.5 mm long and 1–1.5 mm broad. The dorsal margin is strongly convex. The stigmatic area is recurved into a distinct sinus near the apex of the ventral margin. The fruit wall is thick, and is composed of sclerenchymatous cells and has an indistinct rugulate ornamentation.

Evaluation of the systematic affinity is impeded by incomplete information on floral structure and the lack of pollen characters.

**Dispersed stamen with monocolpate, reticulate pollen** (Fig. 8a–c). This stamen type is represented in the Portuguese floras by a single specimen from the Vale de Agua locality. The stamen is massive, perhaps fleshy in life, and about 3 mm long. It expands gradually distally, is dorsiventrally flattened and there is no differentiation into filament and anther. The anther is dithecate and tetrasporangiate with two lateral pairs of pollen sacs each dehiscing by lateral valves (Fig. 8a, b). Pollen sacs are minute, about 0.6 mm long, and constitute a minor portion of the stamen tissue. In contrast the sterile tissue is abundant forming a broad, massive basal part, a broad connective separating the two pairs of pollen sacs, and a prominent, cuboidal apical extension of the connective. Glands are present in the sterile tissue at the base of the pollen sacs. In situ pollen grains are almost spherical, about 25–30 μm in diameter, monocolpate and coarsely reticulate. The colpus is long, extending to

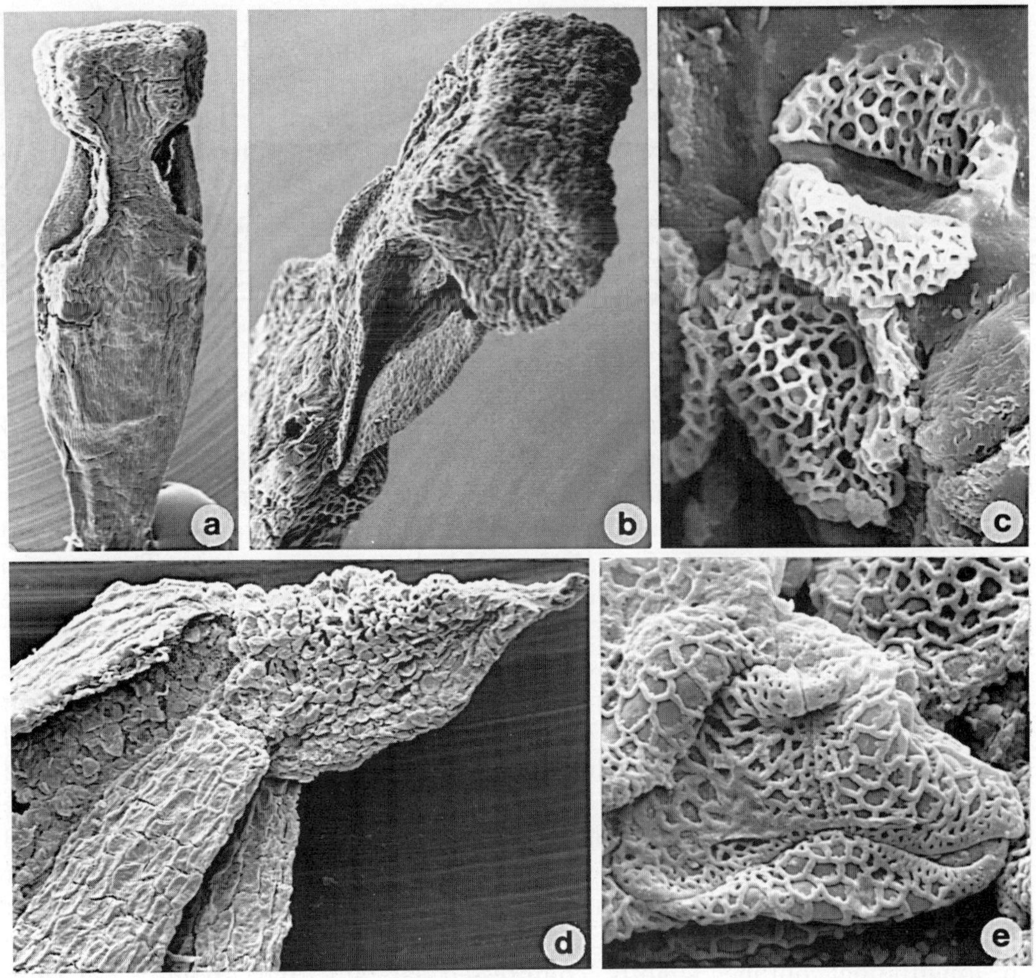

Fig. 8. SEM-micrographs of stamens from Vale de Agua. *a–c* Stamen with monocolpate, reticulate pollen (S101303); *a* dorsal or ventral view of stamen showing small pollen sacs and massive sterile tissue at base and apex, ×26; *b* lateral view of stamen showing apical extension of connective and valvate dehiscence of theca, ×55; *c* pollen grains enlarged showing a single distal colpus and reticulate exine, ×1200; *d–e* stamen with trichotomocolpate pollen (S101425); *d* apical part of stamen showing triangular extension of connective, x 150; *e* distal view of trichotomocolpate pollen grain from stamen in *d*, ×3000

the equator of the grain (Fig. 8c). The tectum is loosely attached to the foot layer. Lumina are typically 2–4 µm in diameter. Muri are narrow with sharply triangular profiles, and are supported by widely spaced, simple columellae.

The valvate dehiscence and the monocolpate pollen grains indicate a magnoliid affinity for this stamen, but a more detailed evaluation of its systematic affinity is not possible based on the material available.

**Dispersed stamen with trichotomocolpate pollen** (Fig. 8d, e). This stamen type is represented by a single specimen from the Vale de Agua locality. The stamen is about 1 mm long, dithecate and tetrasporangiate. Pollen sacs are about 0.7 mm long and constitute the bulk of the stamen. The sterile tissue forms a short, broad, filament and an elongated, triangular, apical extension of the connective (Fig. 8d), but there is poorly developed connective between the two pairs of pollen sacs. In situ pollen grains are triangular in equatorial outline, about 15 µm in diameter, trichotomocolpate, and reticulate. The three arms of the colpus extend almost to the equator. The tectum is loosely attached to the foot layer, and is heterogenous reticulate. Lumina are about 1 µm in diameter over most of the proximal and distal surface. Towards the colpus margin, and around the corners of the grain, the tectum is finely reticulate to foveolate (Fig. 8e). Muri are low with rounded profiles, and are supported by short, densely-spaced, simple columellae.

The valvate dehiscence and the monocolpate pollen grains also indicate a magnoliid affinity for this stamen, but a more detailed evaluation of its systematic affinity is not possible based on the material available.

**Dispersed stamen with monocolpate, coarsely reticulate pollen** (Fig. 9a, b). This stamen type is represented by several specimens from the Torres Vedras locality. Stamens are massive, and were perhaps fleshy in life, about 0.6 mm long, dithecate and tetrasporangiate with valvate dehiscence. The filament is massive and passes gradually into the swollen connective tissue. In situ pollen grains are embedded in a pollenkitt-like substance. They are circular to elliptical in equatorial outline, about 10 µm in diameter, monocolpate and coarsely reticulate. The colpus extends almost to the equator. The tectum is very loosely attached to the foot layer with a homogenous reticulum. The lumina are about 1 µm in diameter and muri are low with rounded profiles, and supported by short and widely spaced columellae.

The valvate dehiscence and pollen morphology indicate a magnoliid affinity of the stamens. In stamen morphology and aspects of pollen characters they are very similar to stamens from the late Albian West Brothers locality, in the Potomac Group of eastern North America, of probable chloranthoid affinity (Crane & al. 1989) and they may belong to the same complex. Pollen grains in the North American material differs, however, in being multiaperturate.

**Dispersed stamen with monocolpate, reticulate pollen** (Fig. 9c, d). This stamen type is represented by a single specimen from the Torres Vedras locality. The stamen is about 1.3 mm long, dithecate and tetrasporangiate with valvate dehiscence. The anther is basifixed on a short, thin filament, about 0.2 mm long. The connective tissue is massive and extends into a cuboidal apical cap. In situ pollen grains are circular to ellipsoidal in equatorial outline, about 12 µm in diameter, and reticulate. The tectum is reticulate over most of the surface with lumina of two size classes, the larger being about 0.7 µm wide and the smaller about 0.1 µm. Muri are broad and low with rounded profiles and are supported by short, densely spaced, simple columellae.

**Dispersed stamen with tricolpate, foveolate pollen** (Fig. 9e, f). This stamen type is represented by several specimens from the Torres Vedras locality. Stamens are about 0.6 mm long, dithecate and tetrasporangiate with valvate dehiscence. The anther is basifixed and the two pairs of pollen sacs are laterally positioned. The short, thin filament passes gradually into the massive connective which has a small dome-shaped apical extension. In situ pollen grains are spherical, about 16 µm in diameter, tricolpate and finely foveolate. The colpi are broad with rounded ends and a finely verrucate colpus membrane. The foveae of the tectum are minute and evenly distributed.

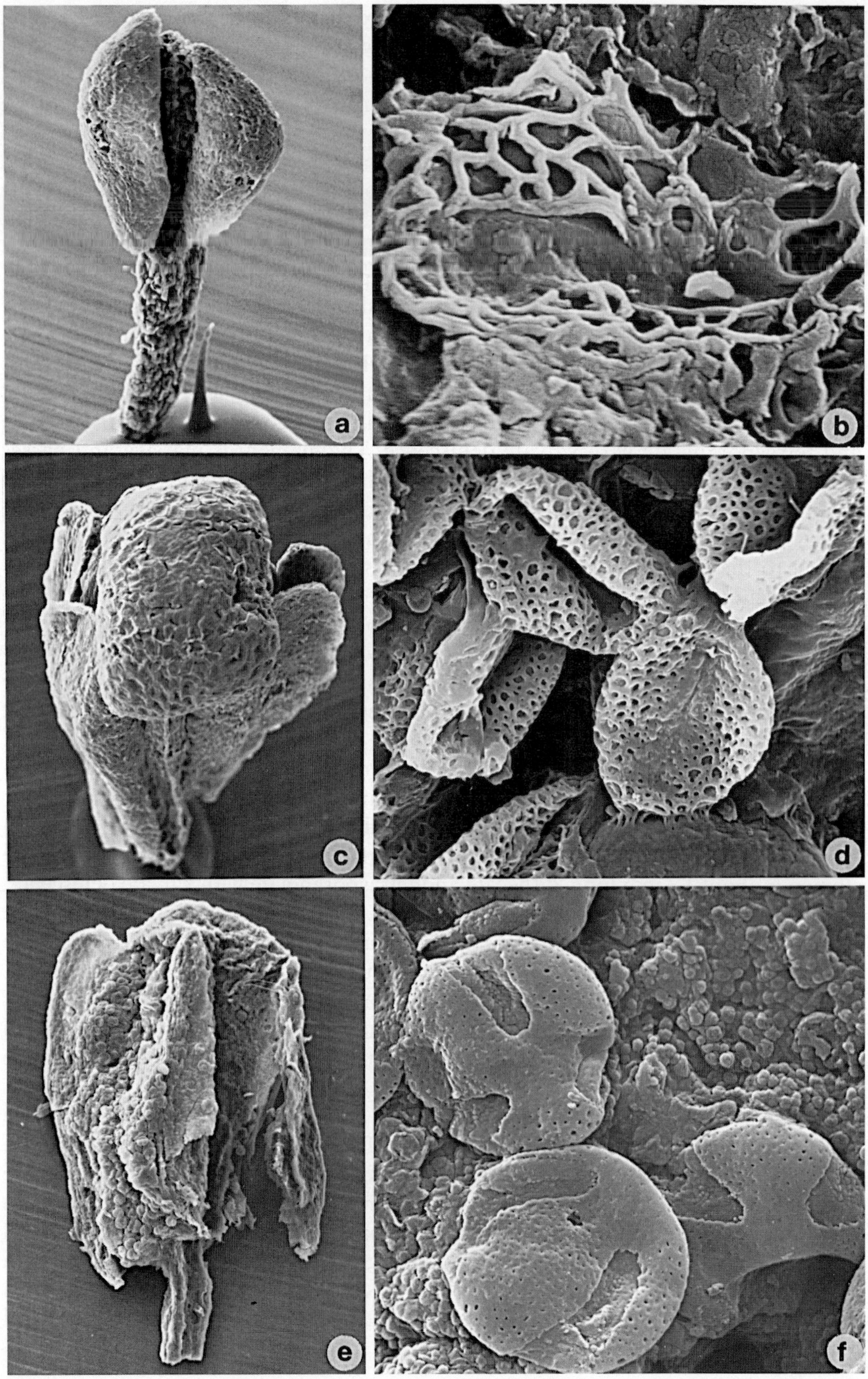
a
b
c
d
e
f

The valvate dehiscence of the anthers, coupled with the presence of tricolpate pollen, indicates a hamamelididean affinity for these fossils (e.g., HUFFORD & ENDRESS 1989), which closely resemble several platanoid stamens described from the Potomac Group sediments of eastern North America (e.g., FRIIS & al. 1988, CRANE & al. 1993, PEDERSEN & al. 1994b). Most of these platanoids have finely to coarsely reticulate pollen, but there is considerable variation in exine sculpture within the group.

## Discussion

**Floral organs and organization.** Most of the floral structures recovered so far from the Early Cretaceous floras of Portugal are fragmentary and consist predominantly of dispersed stamens, fruits and seeds. For most taxa information on other aspects of flower morphology, such as phyllotaxis, number of parts, floral symmetry, and bisexuality/unisexuality, is therefore incomplete. However, a few more complete flowers have also been found, and together with the dispersed organs, they reveal some of the diversity of floral form among these Early Cretaceous angiosperms.

In general, the reproductive organs are minute, 0.5–4 mm, and this is consistent with generalizations made on the basis of previous observations of Early Cretaceous floras from other areas (e.g., FRIIS & CREPET 1987, FRIIS & ENDRESS 1990). It is interesting that none of the Portuguese floral structures shows a distinct spiral arrangement of parts while several taxa show an apparently whorled organization. The most unexpected result, however, is the presence of epigynous flowers in the early history of angiosperms. Although the angiospermous affinity has been established with certainty only for three of these forms (small epigynous flower with finely striate pollen from Famalicão, Catefica and Vale de Agua; tetramerous flower from Juncal; small epigynous (?) flower from Torres Vedras) it is highly probable that the other epigynous structures described here are also angiospermous. In addition to epigynous forms, hypogynous flowers are also present. In one of these the outer perianth parts are fused to form a shallow floral cup. The perianth in the fossil flowers described to date is typically undifferentiated.

Both unisexual (e.g., multistaminate structure) and bisexual (e.g., epigynous flower with finely striate pollen) flowers are present. Most stamens from the Early Cretaceous floras of Portugal have massive sterile tissue, typically with little or no differentiation into filament and anther, and with a prominent apical extension of the connective tissue. Pollen sacs are generally small and in most specimens open by lateral valves. Pollen grains in situ include a wide variety of monoaperturate pollen which may be either monocolpate or trichotomocolpate. Tricolpate pollen is also present but less common. All grains are small, typically 10–20 μm in diameter,

Fig. 9. SEM-micrographs of stamens from Torres Vedras. *a–b* Stamen with monocolpate and coarsely reticulate pollen (S101410); *a* lateral view of stamen showing valvate dehiscence and fleshy basal part, × 120; *b* distal view of pollen grain, × 4000; *c–d* stamen with reticulate, monocolpate pollen (S101408); *c* apical view of stamen showing massive extension of connective and valvate dehiscence, × 75; *d* pollen grains enlarged, × 3000; *e–f* stamen with foveolate, tricolpate pollen (S101407); *e* showing valvate dehiscence and small apical extension of the connective, × 120; *f* tricolpate pollen grains, × 2500 and with reticulate to foveolate tecta. The valvate dehiscence, the presence of abundant sterile tissue with apical connective extensions, and pollen morphology indicate that all dispersed stamens are of magnoliid (monocolpate and trichotomocolpate pollen) or hamamelidid (tricolpate) affinity.

Small unilocular and one-seeded fruits or fruitlets are extremely common and diverse in all of the Early Cretaceous floras. Some are known only from dispersed material and for these we have not been able to establish whether they were derived from monomerous gynoecia or from multicarpellate apocarpous gynoecia. One group of fruits includes taxa in which the supporting tissue is in the seed wall and in which typically only the outer cuticle of the fruit wall is preserved. One of these has pollen grains of the *Clavatipollenites*-type and is probably related to the chloranthoid *Couperites* described from the mid-Cretaceous of eastern North America. Another group of fruits includes taxa in which the supporting tissue is in the fruit wall and in which seed wall is typically delicate. Some of these fruits may be magnoliid, others are probably those of early ranunculids.

## Conclusions

The general conclusion to emerge from this and previous studies of angiosperm reproductive structures is one of steadily decreasing diversity back in time. Campanian floras include a great diversity of angiosperms, among which those of "higher" hamamelidids, rosids and dilleniids are especially common (e.g., Friis & Skarby 1982; Friis 1983, 1984, 1985a, b, 1990) while magnoliid and "lower" hamamelidids are less diverse. In mid-Albian to early Cenomanian floras from North America magnoliids are proportionately more diverse, while the diversity of eudicots (as recognized by triaperturate pollen) is restricted to "lower" hamamelidids and some apparent rosids. The results presented here on Early Cretaceous floras in Portugal from the early history of angiosperms, continue this trend and indicate a diversity of magnoliid taxa with relatively few eudicots. As among extant taxa, the flowers of these Early Cretaceous magnoliids appear to have exhibited considerable variability in the number and arrangement of floral parts, although no large, multiparted floral structures have been encounted so far. Several of these fossils appear to have no counterparts among extant taxa.

We wish to thank P. K. Endress and F. B. Sampson for helpful comments on the fossils; B. Larsen for assistance in the field and for help in preparing the material, and Y. Arremo for help in preparing the plates. Funding from the Swedish Natural Science Research Council (EMF), the Carlsberg Foundation (KRP, EMF) and the National Science Foundation (BSR-9020237, PRC) is gratefully acknowledged.

## References

Batten, D. J., 1986: The Cretaceous *Normapolles* pollen genus *Vancampopollenites*: occurrence, form, and function. – Palaeontol., Spec. Papers **35**: 21–39.

– 1989: Systematic relationships between *Normapolles* pollen and the *Hamamelidae*. – In Crane, P. R., Blackmore, S., (Eds): Evolution, systematics, and fossil history of the *Hamamelidae*, 2 "Higher" *Hamamelidae*, pp. 9–21. Syst. Assoc. Spec. Vol. **40B**. – Oxford: Clarendon Press.

CHAPMAN, J. L., 1987: Comparison of *Chloranthaceae* pollen with the Cretaceous "*Clavatipollenites* complex". Taxonomic implications for palaeopalynology. – Pollen Spores **29**: 249–272.

CRANE, P. R., FRIIS, E. M., PEDERSEN, K. P., 1989: Reproductive structure and function in Cretaceous *Chloranthaceae*.–Pl. Syst. Evol. **165**: 211–226.

– – – 1994: Palaeobotanical evidence on the early radiation of magnoliid angiosperms. – Pl. Syst. Evol. [Suppl.] **8**: 51–72.

– PEDERSEN, K. R., FRIIS, E. M., DRINNAN, A. D., 1993: Early Cretaceous (Early to Middle Albian) platanoid inflorescences associated with *Sapindopsis* leaves from the Potomac Group of Eastern North America. – Syst. Bot. **18**: 328–344.

DINIZ, F., KEDVES, M., SIMONCSICS, P., 1974: Les sporomorphes principaux de sédiments crétacées de Vila Flor et de Carrajao, Portugal. – Comum. Serv. Geol. Portugal **48**: 161–178.

ENDRESS, P. K., 1971: Bau der weiblichen Blüten von *Hedyosmum mexicanum* Cordemoy (*Chloranthaceae*). – Bot. Jahrb. **91**: 39–60.

FRANÇA, J. C., ZBYSZEWSKI, G., 1963: Carta Geológica de Portugal na escala de 1/50 000. Notícia explicativa da Folha 26-B Alcobaça–Lisbon: Serviços Geológicos de Portugal.

FRIIS, E M., 1983: Upper Cretaceous (Senonian) floral structures of juglandalean affinity containing *Normapolles* pollen. – Rev. Palaeobot. Palynol. **39**: 161–188.

– 1984: Preliminary report of Upper Cretaceous angiosperm reproductive organs from Sweden and their level of organization. – Ann. Missouri Bot. Garden **71**: 403–418.

– 1985a: *Actinocalyx* gen.nov., sympetalous angiosperm flowers from the Upper Cretaceous of southern Sweden. – Rev. Palaeobot. Palynol. **45**: 171–183.

– 1985b: Structure and function in Late Cretaceous angiosperm flowers. – Biol. Skr. Dansk Vid. Selsk. **25**: 1–37.

– 1990: *Silvianthemum suecicum* gen. et sp. nov., a new saxifragalean flower from the Late Cretaceous of Sweden. – Biol. Skr. Dansk Vid. Selsk. **36**: 1–35.

– Crepet, W. L., 1987: Time of appearance of floral features. – In FRIIS, E. M., CHALONER, W. G., CRANE, P. R. (Eds): The origins of angiosperms and their biological consequences, pp. 145–179. – Cambridge: Cambridge University Press.

– ENDRESS, P. K., 1990: Origin and evolution of angiosperm flowers. – Adv. Bot. Res. **17**: 99–162.

– SKARBY, A., 1982: *Scandianthus* gen. nov., angiosperm flowers of saxifragalean affinity from the Upper Cretaceous of southern Sweden. – Ann. Bot. **50**: 569–583.

– CRANE, P. R., PEDERSEN, K. R., 1988: Reproductive structure of Cretaceous *Platanaceae*. – Biol. Skr. Dansk Vid. Selsk. **31**: 1–56.

– PEDERSEN, K. R., CRANE, P. R., 1992: *Esgueiria*, new fossil flowers from the Late Cretaceous of Portugal with combretaceous features. – Biol. Skr. Dansk Vid. Selsk. **41**: 1–45.

HUFFORD, L. D., ENDRESS, P. K., 1989: The diversity of stamen structures and dehiscence patterns among *Hamamelididae*. – Bot. Linn. Soc. **99**: 301–346.

KEDVES, M., DINIZ, F., 1981: *Probrevaxones*, a new pollen group for the first *Brevaxones* form-genra from the Upper Cenomanian of Portugal. – Acta Bot. Acad. Sci. Hung. **27**: 383–402.

– – 1983: Les *Normapolles* du Crétacé surpérieur en Europe: Implications paléobiogéographiques. – Geobios **16**: 329–345.

– PITTAU, P., 1979: Contribution a la connaissance des pollens des *Normapolles* du « group papilloide » du Crétacé Supérieur du Portugal. – Pollen Spores **21**: 169–209.

KRASSILOV, V. A., 1986: New floral structures from the Lower Cretaceous of Lake Baikal area. – Rev. Palaeobot. Palynol. **47**: 9–16.

Kubitzki, K., 1969a: Monographie der *Hernandiaceae*. Teil I. – Bot. Jahrb. Syst. **89**: 78–148.
– 1969b: Monographie der *Hernandiaceae*. Teil II. – Bot. Jahrb. Syst. **89**: 149–209.

Pedersen, K. R., Crane, P. R., Drinnan, A. N., Friis, E. M., 1991: Fruits from the mid-Cretaceous of North America with pollen grains of the Clavatipollenites type. – Grana **30**: 577–590.

– – Friis, E. M., 1989: Pollen organs and seeds with *Eucommiidites* pollen. – Grana **28**: 279–294.

– Friis, E. M., Crane, P. R., 1994a: Ultrastructure of pollen from Cretaceous reproductive structure. – In Kurmann, M. H., Doyle, J. A., (Eds): Ultrastructure of fossile spores and pollen and its bearing on relationships among fossil and living groups, pp. 139–159. – Kew: Royal Botanical Gardens.

– – – Drinnan, A. N., 1994b: Reproductive structures of an extinct platanoid from the Early Cretaceous (latest Albian) of eastern North America. – Rev. Palaeobot. Palynol. **80**: 191–303.

Rey, J., 1972: Recherches géologiques sur le Crétacé inférieur de l'Estremadura (Portugal). – Serviços Geol. Portugal, Mem. n.s. **21**: 1–477.

– 1982: Dynamique et Paléoenvironnements du Basin Mésozoique d'Estremadura (Portugal), au Crétacé Inférieur. – Cretac. Res. **3**: 103–111.

Sampson, F. B., 1993: Pollen morphology of the *Amborellaceae* and *Hortoniaceae* (*Hortonioideae*: *Monimiaceae*). – Grana **32**: 154–162.

Saporta, G. de, 1894: Flore fossile du Portugal. Nouvelles contributions à la flore Mésozoique. Accompagnées d'une notice stratigraphique par Paul Choffat. – Lisbon: Imprimerie de l'Académie Royale des Sciences.

Schneider, E. D., Jeter, J. M., 1982: Morphological studies of the *Nymphaeaceae*. XII. The floral biology of *Cabomba caroliniana*. – Amer. J. Bot. **69**: 1410–1419.

Shutts, C. F., 1959: The comparative morphology of the *Hernandiaceae*. – Ph.D. Thesis, Indiana University.

Teixeira, C., 1945: Nymphéacées fossiles du Portugal. – Lisbon: Serviços Geológicos de Portugal.

– 1946: Flora Cretácica de Esgueira (Aveiro). – Portugalieae Acta Biol. **1**: 235–242.

– 1947: Nouvelles recherches et révision de la flore de Cercal. – Brontéia, Sér. de Ciênc. Naturais, Fasc. I, **15**: 5–15.

– 1948: Flora Mesozóica Portuguesa. Part I. – Lisbon: Serviços Geológicos de Portugal.

– 1950: Flora Mesozóica Portuguesa. Part II. – Lisbon: Serviços Geológicos de Portugal.

– 1952: Notes sur quelques gisements de végétaux fossiles du Crétacé des environnements de Leiria. – Revista da Faculdade de Ciências de Lisboa, 2. Ser., C, **2**: 133–154.

– Zbyszewski, G., Torre de Assunção, C., Manupella, G., 1968: Carta Geológica de Portugal na escala de 1/50 000. Notícia explicativa da Folha 23-C Leiria. – Lisbon: Serviços Geológicos de Portugal.

Trincão, P. R. P., 1985: Estudo Palinostratigráfico do Cretácico inferior Português Ante-Albiano. – Coimbra: Centro de Geociências da Universidade de Coimbra (INIC).

– 1990: Esporos e pólenes do Cretácico inferior (Berriasiana-Aptiana) de Portugal: Paleontologia e biostratigrafia. – Unpubl. dissertation, Universidade de Aveiro, Lisbon.

– Pena Dos Reis, R., Pais, J., Proença Cunha, P, 1989: Palinomorfos anti-cenomanianos do "Gres do Buçaco" (Lousã, Portugal). – Ciências da Terra (UNL) **10**: 51–64.

Van der Merwe, J. J. M., Van Wyk, A. E., Kok, P. D. F., 1990: Pollen types in *Lauraceae*. – Grana **29**: 185–196.

van Konijnenburg-van cittert, J. H. A., 1992: An enigmatic Liassic microsporophyll, yielding *Ephedrites* pollen. – Rev. Palaeobot. Palynol. **71**: 239–254.

Walker, J. W., 1976: Comparative pollen morphology and phylogeny of the ranalean complex. – In Beck, C. B., (Ed.): Origin and Early Evolution of Angiosperms, pp. 241–299. – New York: Columbia University Press.

ZBYSZEWSKI, G., MANUPELLA, G., DA VEIGA FERREIRA, O., 1974: Carta Geológica de Portugal na escala de 1/50 000. Notícia explicativa da Folha 27-A Vila Nova de Ourem. – Lisbon: Serviços Geológicos de Portugal.

Addresses of the authors: ELSE MARIE FRIIS, Department of Palaeobotany, Swedish Museum of Natural History, Box 50007, S-104 05 Stockholm, Sweden. – KAJ RAUNSGAARD PEDERSEN, Department of Geology, Universitetsparken, University of Aarhus, DK-8000 Århus C. Denmark. – PETER R. CRANE, Department of Geology, The Field Museum, Roosevelt Road at Lake Shore Drive, Chicago, IL 60605, USA.

Accepted December 20, 1993 by P. K. ENDRESS

Pl. Syst. Evol. [Suppl.] 8: 51–72 (1994)

# Palaeobotanical evidence on the early radiation of magnoliid angiosperms

PETER R. CRANE, ELSE MARIE FRIIS, and KAJ RAUNSGAARD PEDERSEN

Received January 21, 1994

**Key words:** *Calycanthaceae*, *Chloranthaceae*, *Circaeaster*, *Clavatipollenites*-complex, *Laurales*, *Magnoliales*. – Early Cretaceous, epigyny, fossil flowers, Potomac Group, Puddledock locality.

**Abstract:** Early to mid-Cretaceous Potomac Group sediments have yielded a number of well-preserved angiosperm flowers, fruits, seeds, and dispersed stamens. The most diverse assemblage from the Early Cretaceous part of the sequence is from the Puddledock locality, Virginia, eastern USA. This material is of Early to Middle Albian age and is particularly rich in magnoliid and hamamelidid angiosperms. Flowers and fruits from the Puddledock flora are mostly simple and few-parted and so far only two reproductive structures with numerous parts have been recovered. In this paper we focus on floral organs showing magnoliid features, some of which may be compared to extant taxa, at least at the level of order and perhaps family. Other floral organs are more difficult to classify. Fossils from the Puddledock locality document the earliest appearance of calycanthaceous, and possible lauraceous, floral characters. As recorded also from the Early Cretaceous floras of Portugal, several different kinds of epigynous angiosperm and angiosperm-like reproductive structures are present. The discovery of small *Circaeaster*-like spiny fruits with monocolpate pollen is especially significant.

Over the last decade a variety of angiosperm reproductive structures have been described from mid-Cretaceous rocks in North America and Europe. Based on these reports, as well as dispersed leaves and pollen, it is clear that magnoliid dicots, nonmagnoliid dicots (eudicots) and monocots had all begun to diversify by the mid-Cretaceous. In our own studies we have focused particularly on angiosperm reproductive structures from the Potomac Group of eastern North America (e.g., FRIIS & al. 1988, DRINNAN & al. 1990, PEDERSEN & al. 1991, CRANE & al. 1993), as well as from the extensive Cretaceous sequence in Portugal (FRIIS & al., this volume). In many of these floras, flowers, fruits, and stamens of magnoliid angiosperms are particularly diverse, and together with evidence from dispersed pollen, they indicate that several magnoliid groups had already differentiated by around the Early-Late Cretaceous boundary (c. 97 myr BP).

In this paper we provide the first description of fossil reproductive organs from an important new locality in the Potomac Group (Puddledock), focusing particularly on the diversity of probable and possible magnoliid remains. This survey indicates that magnoliids were a diverse and significant component of mid-Cretaceous floras, and also that the diversity of flower types seen in extant magnoliids was manifested at an early stage in angiosperm evolution. We then place these discoveries in a broader context through a brief overview of fossil remains of *Magnoliidae* from the Potomac Group and elsewhere.

### Material and methods

The material described in this paper is from several clay and silt horizons in the Tarmac Lone Star Industries sand and gravel pit, located south of Richmond and east of the Appomattox River in Prince George County, Virginia, which we term the Puddledock locality. Based on palynological analyses (Christopher in Dischinger 1987) the sediments exposed in the Lone Star Gravel Pit were assigned to the basal part of subzone IIB in the palynological zonation established for the Potomac Group by Brenner (1963) and others (Doyle 1969, Doyle & Hickey 1976, Hickey & Doyle 1977, Doyle & Robbins 1977). Subzone IIB is of Middle Albian age but may extend down into the Early Albian (Doyle 1992). More than 38 samples have been collected during two field seasons. Samples 1–7, 72–84 were collected by the authors and A. N. Drinnan in 1988. Samples 142–159 were collected by the authors in 1991. The content of angiosperm reproductive structures in individual samples varies considerably. Samples 4 and 73, for example, contain numerous angiosperm stamens.

The Puddledock flora is extremely rich, and contains a very diverse and abundant assemblage of angiosperm reproductive structures. In addition to a variety of magnoliid angiosperms (only a selection of which are described and illustrated here), the Puddledock flora also contains a possible liverwort, a variety of pteridophytes (e.g., *Onychiopsis*, schizaeaceous ferns, dispersed sporangia, and various smooth and reticulate megaspores), several conifers (e.g., *Frenelopsis*, *Pseudofrenelopsis*, *Glenrosa*, Srinivasan 1992) and various presumed gymnosperm seeds, some of which resemble *Erdtmanispermum* (Pedersen & al. 1989). Eudicots include dispersed probable platanoid fruits and stamens, as well as several other fruits and stamen types.

## Results

**Magnolialean gynoecium** (Fig. 1). This fossil taxon is one of the two multipartite angiosperm floral structures from the Puddledock flora and is also the largest of the angiosperm fossils recovered. It is known from two charcoalified specimens (Fig. 1a–c) and three strongly compressed specimens (Fig. 1d). The larger specimens are up to about 3.5 mm long and 2 mm in diameter. The smallest specimen is about 1.5 mm long and 0.6 mm in diameter. The fossils consist of a broad, stout peduncle and an elongate, conical receptacle bearing numerous, spirally arranged fruitlets. There is no information on the number or organization of seeds. None of the fossils show attached stamens or perianth parts, but the presence of an androecium and perianth is indicated by scars at the base of the gynoecium. Supposed perianth scars are slightly transversely elongate, while supposed staminate scars are circular and prominent, indicating that stamens had a stout filament. Pollen grains were observed on several specimens in the androecial region and also on the gynoecium. Grains are circular to triangular in equatorial outline, about 12 μm in diameter, monocolpate and reticulate (Fig. 1e–g). The colpus is wide and extends almost to

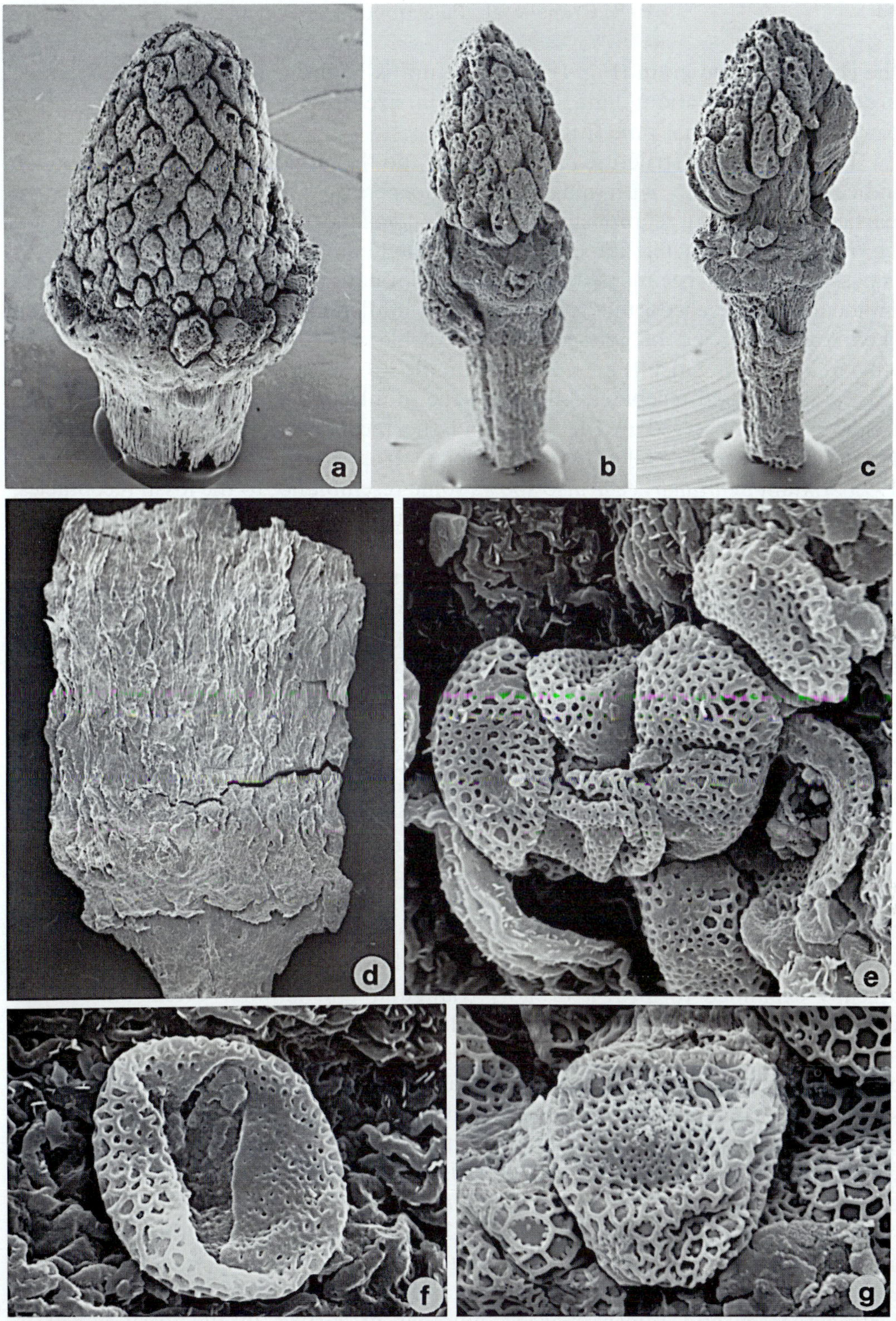


Fig. 1. SEM-micrographs of magnolialean floral structure from the Puddledock locality, Virginia, USA. *a* Larger, charcoalified specimen with spirally arranged fruitlets and scars from probable androecium and perianth (PP43827, sample 82), ×25; *b–c* smaller, charcoalified specimen viewed from two sides (PP43780, sample 156), ×45; *d* strongly compressed lignitized specimen (PP437701, sample 01), ×20; *e* clump of pollen grains (PP43701, sample 01), ×2500; *f* distal view of pollen grain (PP43701, sample 01), ×3500; *g* proximal view of pollen grain (PP43828, sample 04), ×3500

the equator of the grain (Fig. 1f). The tectum is heterogenous reticulate. Lumina are angular up to about 1 μm, decreasing in size towards the colpus margin and towards the proximal pole (Fig. 1e–g).

The multipartite structure of these fossils and the monocolpate, reticulate pollen indicate relationship with magnoliid angiosperms. The conical gynoecium in particular suggests a relationship with the *Magnoliales*, but the lack of information on ovule/seed characters impedes more detailed comparison with extant forms. It is possible that all specimens studied here are immature. There are a variety of small unilocular, one-seeded fruits/fruitlets in the Puddledock flora, that may have been shed from this kind of gynoecium. A possible analogue in the *Laurales* is the *Amborellaceae* in which there are several unilocular and one-seeded fruitlets borne on a small conical receptacle, although the number of parts in extant *Amborella* is fewer than observed in the fossil and the receptacle is only slightly convex. However, *Amborella* pollen is quite different from that described here (Sampson 1993). Extant *Schisandraceae* also shows some resemblance to the fossil receptacles, but pollen of *Schisandraceae* differs in the aperture configuration by being tricolpate or hexacolpate (Walker 1974).

***Calycanthus*-like flower** (Fig. 2). The flower is known from a single specimen that shows a flower with a cup-shaped hypanthium borne on a stout peduncle. The peduncle was poorly preserved and was therefore removed before mounting the specimen for SEM to avoid charging (Fig. 2a, b). The hypanthium of the flower contains numerous carpels in the center and numerous stamens around its upper margin (Fig. 2). External to the stamens are remnants of a perianth. The floral cup has at least two narrow "bracts" at the base and about 6–8 broad ribs (Fig. 2a, b). The surface of the cup has numerous elongated, simple hairs and also shows bulging protuberances that may indicate the positions of oil glands. Stamens are extrorse and arranged in dense series around the margin of the floral cup (Fig. 2c). They are laminar and more or less square in outline, tetrasporangiate with the four parallel pollen sacs embedded in the adaxial surface in a single plane (Fig. 2d–f). The two pairs of sporangia are close together, separated only by a narrow strip of connective (Fig. 2f). Ventrally the connective is well-developed and apically expanded to form a broadly triangular and laminar extension. Anther dehiscence is valvate with laterally attached valves (Fig. 2e). Pollen grains occur abundantly in the stamens. They are more or less circular in equatorial outline, about 15 μm in diameter, monocolpate and reticulate. The colpus is broad and short. The tectum has a more or less homogeneous reticulum with lumina up to about 0.8 μm in diameter and narrow, smooth muri. Staminodes are apparently present between stamens and gynoecium. Gynoecium consists of numerous free carpels enclosed in the floral cup and spirally arranged.

Affinity with magnoliid angiosperms is indicated both by floral morphology and pollen structure. The deep urceolate hypanthium with densely arranged extrorse stamens at the hypanthium margin indicates close relationship with *Calycanthaceae* and *Idiospermaceae*. Number of carpels in the fossil flower is apparently high and comparable to that in *Calycanthaceae*, whereas the gynoecium in the *Idiospermaceae* consists of one to three carpels only. In androecium features the fossil flower is very similar to extant *Idiospermum* while extant *Calycanthaceae* differs in details of stamen morphology, and in particular have stamens with more

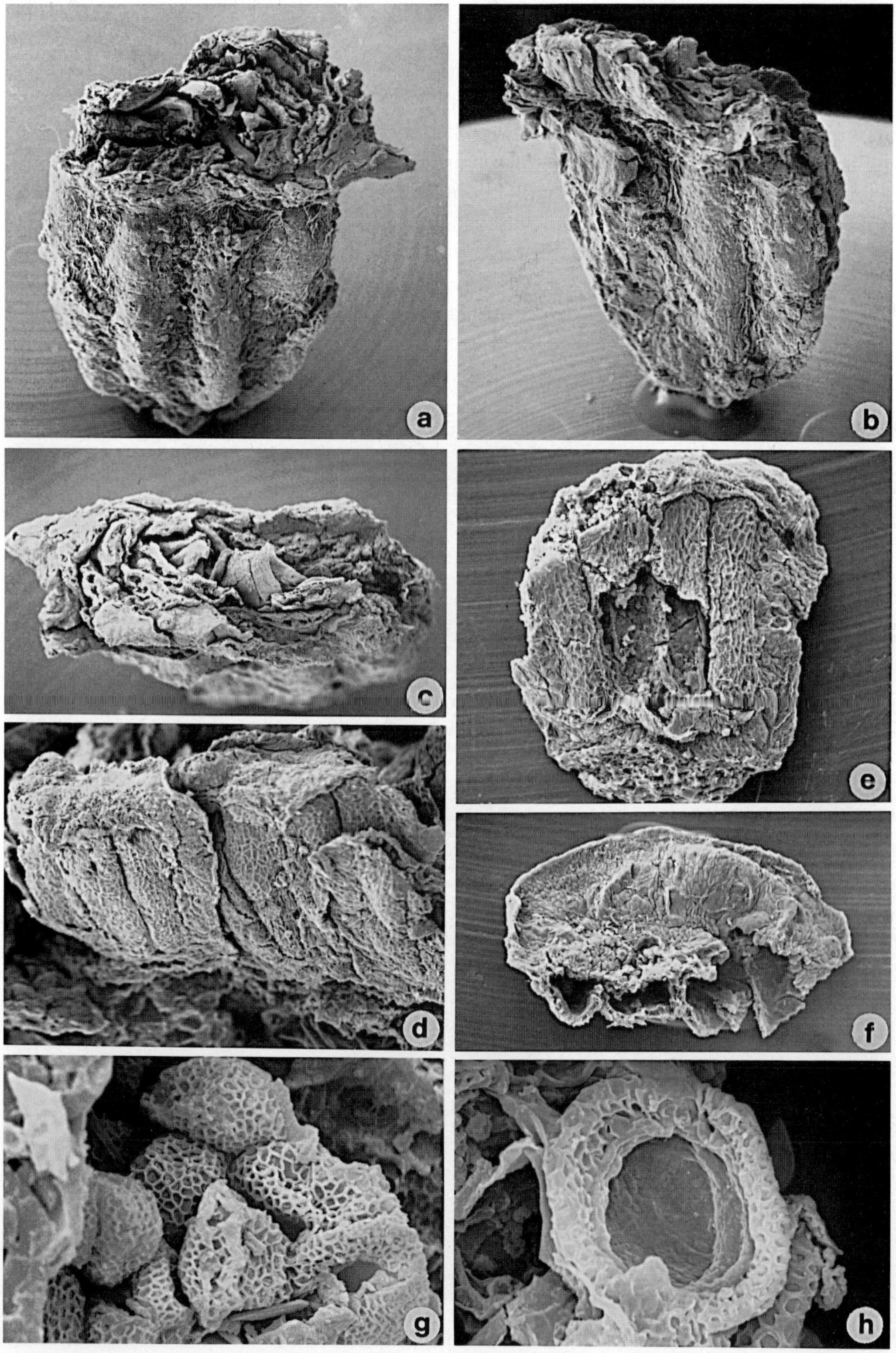

Fig. 2. SEM-micrographs of *Calycanthus*-like flower from the Puddledock locality, Virginia, USA (PP43703, sample 83). *a–b* Flower cup viewed from two sides, ×20; *c* apical view of flower showing spiral arrangement of stamens and central carpels, ×25; *d* dorsal view of stamens at the margin of flower cup, ×75; *e* dorsal view of stamen removed from the flower, ×100; *f* section of a stamen showing the four sporangia in a linear arrangement, ×100; *g* clump of pollen grains in situ in stamen, ×1500; *h* distal view of single grain showing broad colpus, ×2500

protruding pollen sacs and simple longitudinal dehiscence. Both *Calycanthaceae* and *Idiospermaceae* are small families with only few species and the extant species probably do not represent the full diversity of these groups. The material provides the earliest record of calycanthoid fossil flowers. A detailed description and discussion of the fossil flower will be published separately (Friis & al. 1994b). Flowers of calycanthaceous affinity have also been described from sediments about 10–15 million years younger in the Turonian of New Jersey (Crepet & Nixon 1994).

**Lauralean flower** (Figs. 3, 4). This floral type is represented in the Puddledock flora by two fragments of a flower, one is preserved as a compressed lignitic fossil and the other is charcoalified. The charcoalified specimen has remnants of three

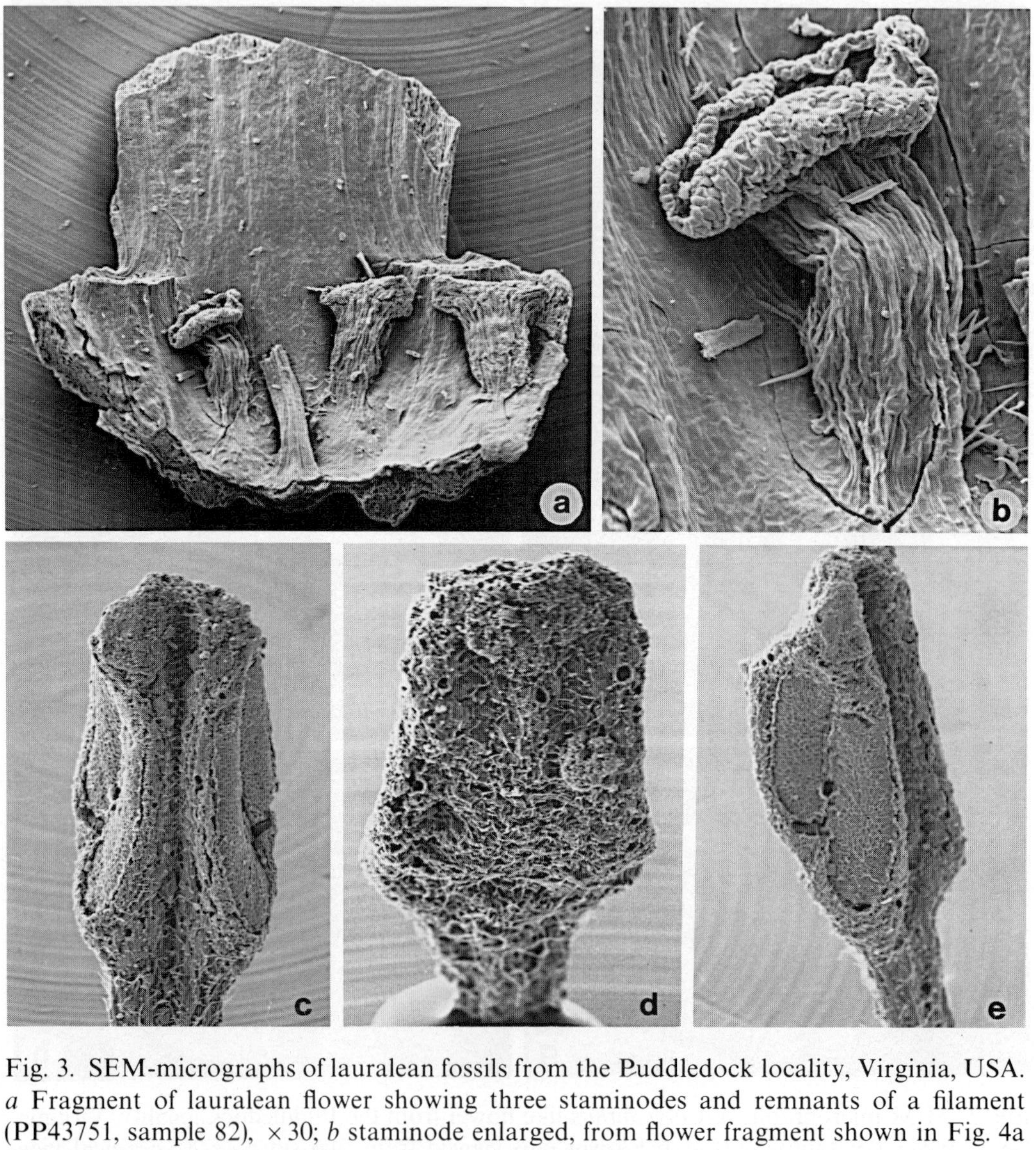

Fig. 3. SEM-micrographs of lauralean fossils from the Puddledock locality, Virginia, USA. *a* Fragment of lauralean flower showing three staminodes and remnants of a filament (PP43751, sample 82), × 30; *b* staminode enlarged, from flower fragment shown in Fig. 4a (PP43751, sample 82), × 120; *c–e* dispersed lauralean stamen in dorsal/ventral view (*c, d*) and lateral view (*e*) (PP43768, sample 73). – c–d × 50, e × 70

tepals with parts of the androecium attached (Fig. 3a). The tepals include one broad, five-veined tepal from an outer whorl and two narrower, three-veined tepals from an inner whorl. Fused to the perianth are one filament of a stamen standing in front of the tepal of the outer whorl and three staminal appendages. The compression fossil is more complete and has remnants of six tepals, three filaments and five lateral staminal appendages (Fig. 4a). The central part of the flower is not preserved in any of the fragments and it is unknown whether a gynoecium was present. It is also uncertain whether all whorls of the androecium are preserved. However, based on the available fragments a tentative floral diagram has been constructed (Fig. 4b). We reconstruct the flower as trimerous with six tepals fused at the base into a well-developed cup. The three outer tepals are very broad, while the three inner tepals are narrow. The androecium has an outer (?) whorl of six staminal appendages, apparently in three pairs, each of which seems to be associated with a single stamen. Both stamens and staminal appendages are fused to the floral cup. Anthers are not preserved attached to the stamen filaments. The staminal appendages consist of a short filament-like stalk that expands apically into a kidney-shaped head (Fig. 3a, b). Simple trichomes are densely spaced on outer and inner surfaces of the flower, but preserved only in patches in the charcoalified specimen (Fig. 3b).

Although the material is fragmentary, the affinity with the Laurales is well-supported by the trimerous organization of the flower and the distinct appendages. These features are basic in several lauralean families and a more detailed evaluation of the systematic position of the fossils will require additional information. The fossils are important in showing that trimerous floral construction

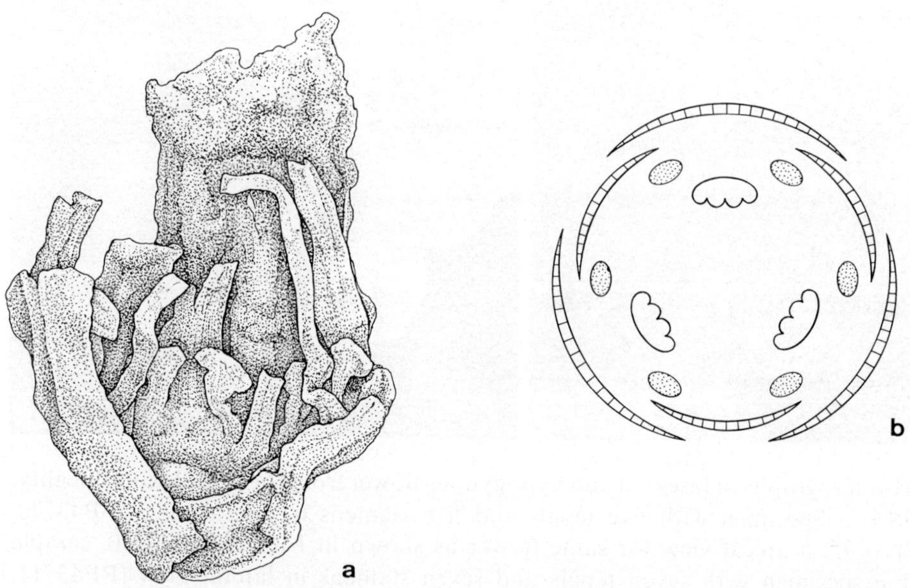


Fig. 4. Line drawing and floral diagram of lauralean flower from the Puddledock locality, Virginia, USA. *a* Compressed flower (PP43735, sample 82), × 5; *b* floral diagram based on two fragments of lauraceous flower

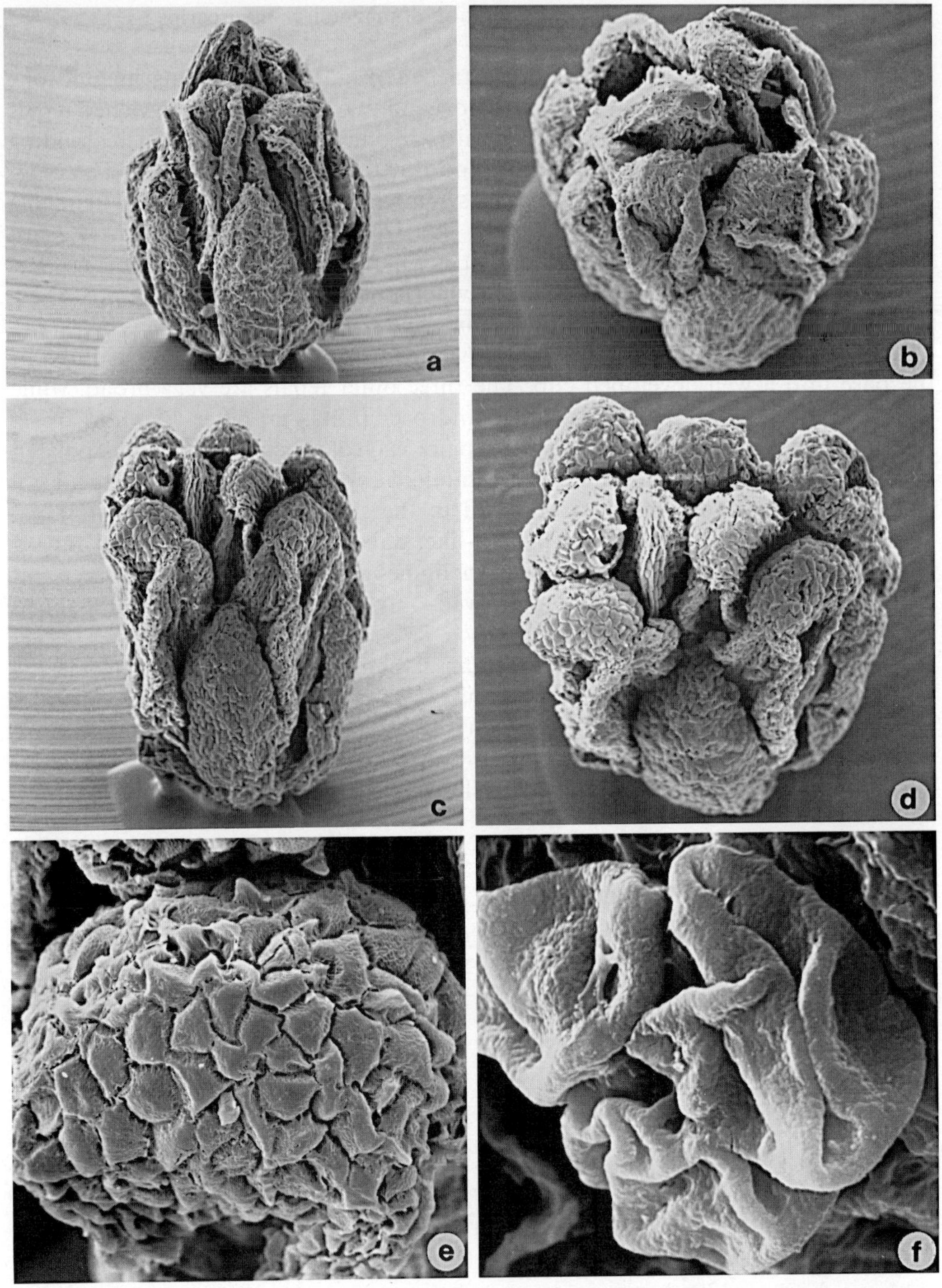


Fig. 5. SEM-micrographs of bisexual and hypogynous flower from the Puddledock locality, Virginia, USA. *a* Specimen with five tepals and five stamens in lateral view (PP43770, sample 152), ×75; *b* apical view for same flower as shown in Fig. 6a (PP43770, sample 152), ×100; *c* specimen with seven tepals and seven stamens in lateral view (PP43711, sample 156), ×75; *d* apical view of same flower as shown in Fig. 6c (PP43771, sample 156), ×100; *e* apical extension of connective enlarged, same specimen as Fig. 6c–d (PP43711, sample 156), ×450; *f* clump of in situ pollen grains from anther (PP43710, sample 156), ×3000

and staminal appendages were already developed at a very early stage in the history of the *Laurales*.

**Dispersed lauralean stamen** (Fig. 3c–e). This stamen type is represented by a single dispersed specimen. The filament is compact and grades into the basifixed anther without a joint. The anther is tetrasporangiate and dithecate with the laterally placed pairs of pollen sacs separated by abundant sterile tissue. The sterile tissue is more abundantly developed on one side (Fig. 3d) than on the other (Fig. 3c) and has an irregular surface perhaps indicating the presence of oil glands. The two pollen sacs of each lateral pair are elongate, but slightly different in size and arranged at different levels with the smaller pollen sacs in the more distal position. Dehiscence is apparently valvate and with each valve attached by a distal hinge. (Fig. 3e). Trichomes on the stamens are densely spaced particularly on the filaments.

The valvate dehiscence with four apically hinged valves clearly indicates the lauralean affinity of this dispersed stamen, and this is also supported by the position of the two sporangia of each pair at different levels. However, in four-loculed valvate anthers of extant lauralean taxa, the sporangia are distinctly separated with one distal and one proximal pollen sac in each pair (e.g., Endress & Hufford 1989), and thus the fossil is more or less intermediate between the normal situation in most other angiosperms and the typical lauralean type.

The similar size of the filaments and the presence of trichomes may indicate that this stamen was produced by the lauralean flower described previously (Fig. 3a, b), but so far there is no conclusive evidence for such a link.

**Bisexual and hypogynous flowers** (Fig. 5). This floral type is represented in the Puddledock flora by numerous specimens all of which are dispersed flowers or flower fragments and there is no information on inflorescence structure. The flowers are hypogynous and bisexual, about 0.8 mm long and 0.6–0.7 mm in diameter, with pentamerous to heptamerous perianth and androecium, and monomerous or dimerous gynoecium. Tepals are narrow triangular and appear leathery. They are free from each other and are arranged in a single whorl. Stamens are apparently in a single whorl alternating with the tepals and have elongated basifixed anthers on short stout filaments. Anthers are tetrasporangiate, dithecate and apparently open by laterally attached valves. The connective between the pollen sacs is massive and extends apically into a spherical to conical protrusion. The gynoecium consists of one or two free carpels. Each carpel is unilocular and contains a single seed. Pollen grains in situ in the anthers are monocolpate, more or less elliptical in equatorial outline, finely granular, and about 19 µm in diameter. They are typically strongly folded indicating that the pollen wall is thin.

The monocolpate pollen and floral organization indicates a relationship with extant magnoliids, and the unilocular and one-seeded carpels suggest relationships with the *Laurales*, but we have been unable to identify the combination of characters seen in these fossils in any modern magnoliid family.

**Pistillate flower with five carpels** (Fig. 6). This taxon is represented by two charcoalified specimens. Both have a stout, broad peduncle bearing a flat receptacle with five free carpels alternating with five small tepals (Fig. 6a–c). There are no remains of an androecium and we interpret the specimens as unisexual, pistillate flowers. The full length of the specimens including the stalk is about 1.5 mm. The carpels are about 1 mm long, elongate ovoid with a very indistinct stigmatic area

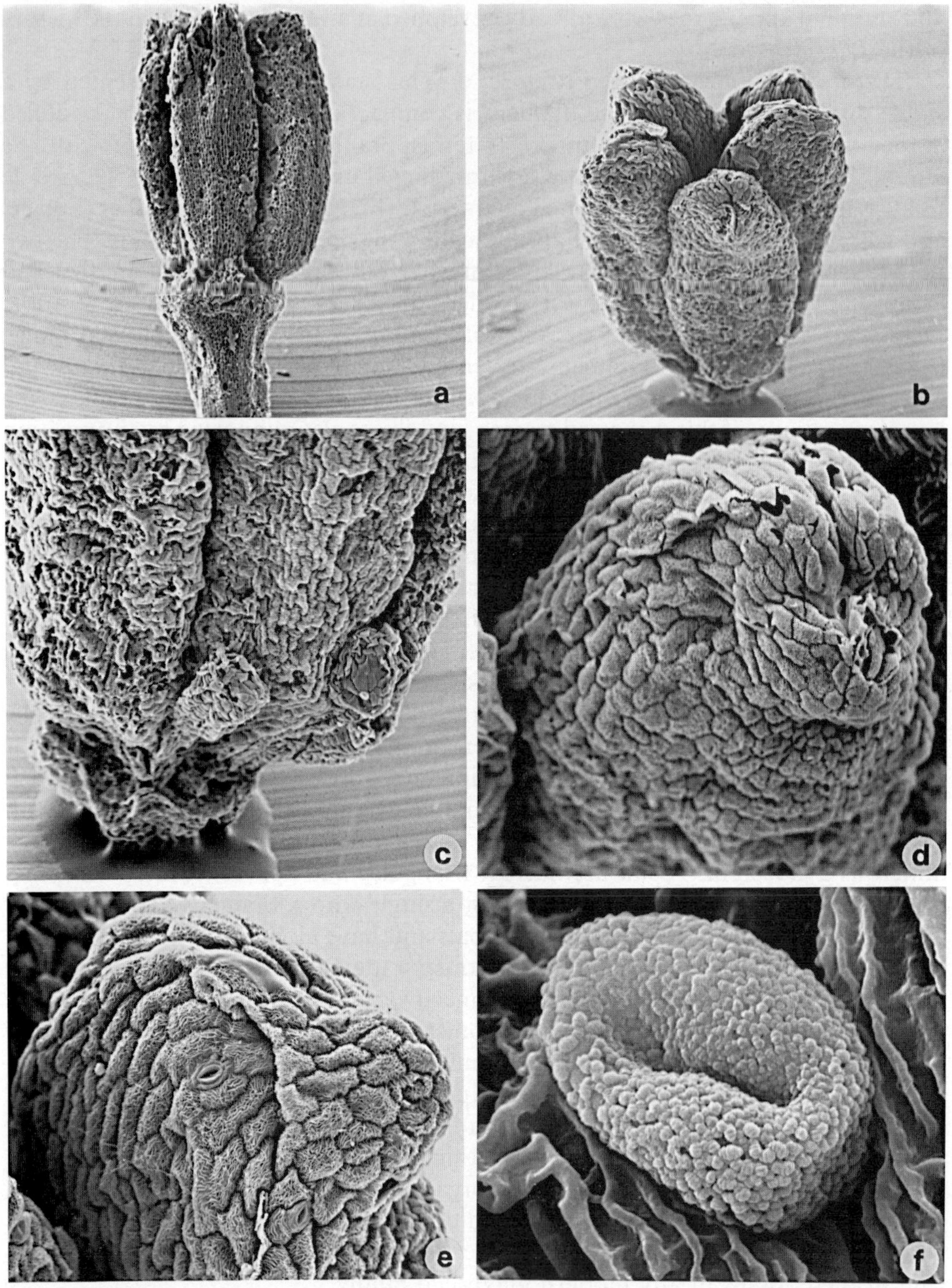


Fig. 6. SEM-micrographs of pistillate, pentamerous flower from the Puddledock locality, Virginia, USA. *a* Lateral view of flower showing stout pedicel and carpels (PP43741, sample 82), × 40; *b* apical view of flower showing the five free carpels (PP43740, sample 82), × 50; *c* basal part of flower showing small tepal lopes (PP43740, sample 82), × 100; *d* apical part of carpel showing indistinct stigmatic area and ventral suture (PP43740, sample 82), × 150; *e* apical part of carpel showing stomata (PP43740, sample 82), × 250; *f* pollen grain from apical part of fruit (PP43740, sample 82), × 4000

(Fig. 6b, d) and with scattered stomata in the apical part (Fig. 6e). The ventral carpel margin is distinct and may not have been completely sealed (Fig. 6d). There is no information on number or organization of seeds. Two pollen grains, both of the same type have been observed on the carpels (Fig. 6f). They are monocolpate, circular to elliptical in equatorial outline and about 12 μm in diameter. The colpus is short with a coarsely granular colpus membrane. The tectum has a dense and irregular reticulum with beaded muri that vary considerably in shape.

Apocarpous unisexual flowers of the same general structure are known in several angiosperm subclasses. The presence of monocolpate pollen on the carpel may indicate a magnoliid affinity. However, only two grains were observed and additional evidence is needed to establish the link between the pollen and the pistillate flower.

**Staminate flower with five stamens** (Fig. 7). This taxon is represented by a single staminate flower consisting of a small receptacle bearing five stamens (Fig. 7a). No remains of perianth have been observed. Stamens are of different sizes and are apparently borne in a spiral. The complete structure is about 0.75 mm long and the individual anthers up to about 0.6 mm long. Anthers are tetrasporangiate and dithecate, slightly extrorse and apparently with valvate dehiscence. The connective is massive and there is a distinct apical extension. Epidermis of the apical connective has distinct polygonal and isodiametric cells and scattered stomata (Fig. 7b). None of the anthers had dehisced and pollen grains are unknown.

Although this taxon is incompletely known it is included in the present study to document the presence of unisexual, staminate flowers in the Puddledock flora. There is also some resemblance to the pentamerous, pistillate flowers described above. Both are very small, have five reproductive parts, polygonal and isodiametric

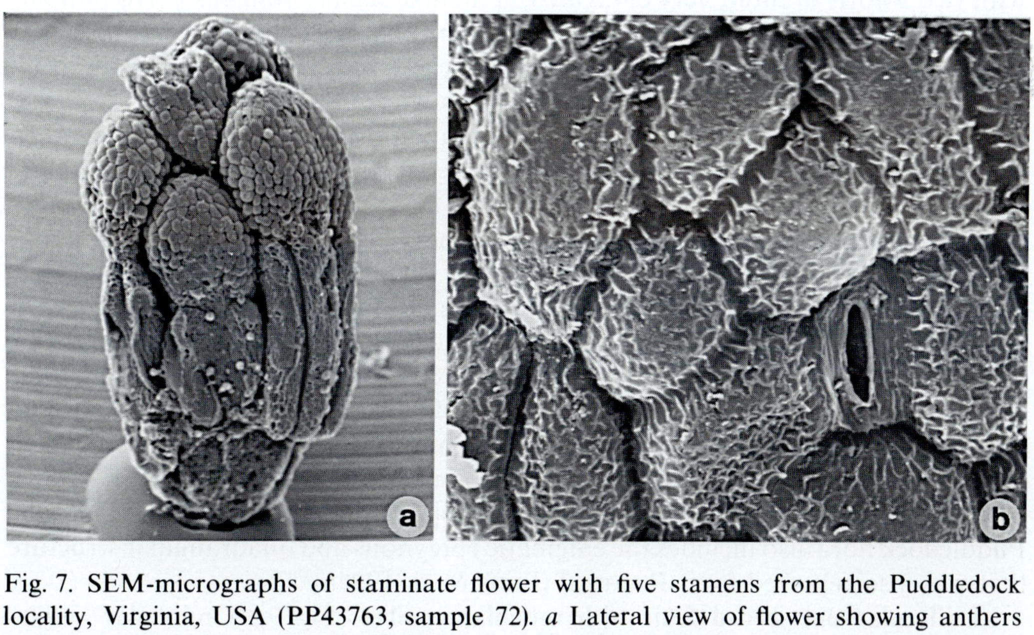

Fig. 7. SEM-micrographs of staminate flower with five stamens from the Puddledock locality, Virginia, USA (PP43763, sample 72). *a* Lateral view of flower showing anthers and prominent apical connective extensions, × 85; *b* detail of connective extension showing epidermal cells and stomata, × 1200

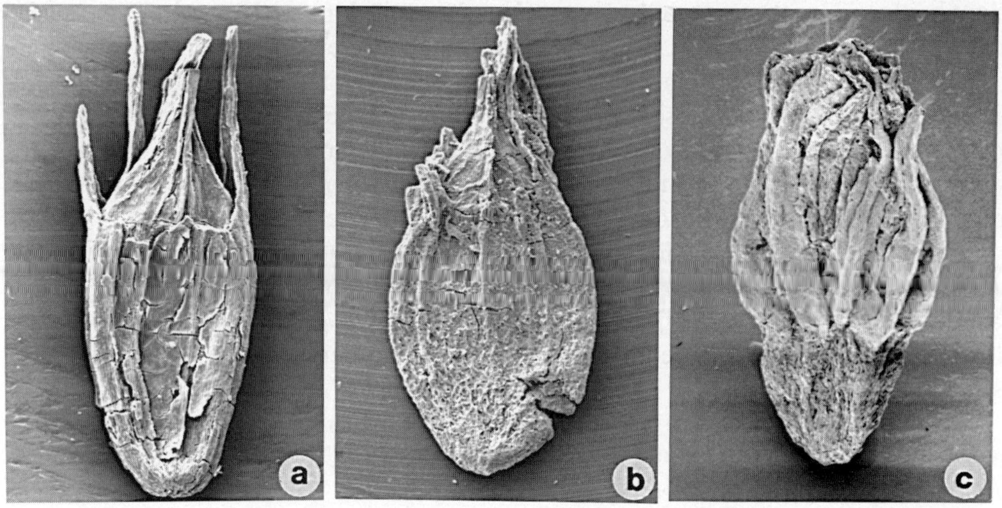


Fig. 8. SEM-micrographs of epigynous flowers from the Puddledock locality, Virginia, USA. *a* (PP43755, sample 82), ×27; *b* (PP43762, sample 73), ×30; *c* (PP43829, sample 73), ×42

epidermal cells and similar scattered stomata at the apical part of the reproductive organs. However, the material so far available is insufficient to establish a link between the two organ types.

**Epigynous flowers** (Fig. 8). Several kinds of epigynous flowers are common in the Puddledock samples. **Epigynous flower type 1** (Fig. 8a, b) is represented by numerous specimens typically 2.5 mm long. The flowers are apparently trimerous with two whorls of three, very elongated, spine-like tepals. Stamens corresponding to these flowers are unknown. The ovary is probably unilocular and distinctly ribbed both above and below the insertion of the tepals. The surface of the ovary is also covered with sharply pointed trichomes with inflated bases.

**Epigynous flower type 2** (Fig. 8c) is also common in the Puddledock samples. It is typically about 1.4 mm long, with eight tepals (? in two whorls of four) and apparently eight stamens. Stamens have short filaments, elongated anthers and a distinct apical protrusion of the connective. The ovary is unilocular and contains a single reticulate seed.

The occurrence of several kinds of epigynous flowers in the Puddledock samples is consistent with their diversity and abundance in other mid-Cretaceous floras. The two epigynous flowers described here show some resemblance to possible lauralean flowers from the Early Cretaceous of Portugal (Friis & al. 1994a) and the Puddledock flowers may also be of lauralean affinity. However, no pollen has been observed and the affinities of the fossils cannot be established based on available information. In addition to the two epigynous forms described here the Puddledock flora also includes the enigmatic epigynous and quadrangular structure that was also reported from Portugal (Friis & al. 1994a).

**Unilocular one-seeded fruit with reticulate pollen** (Fig. 9). Unilocular single-seeded fruits with thin fruit wall and hard seed coat are very characteristic elements of Early Cretaceous angiosperm assemblages (see Friis & al. 1994a) and are

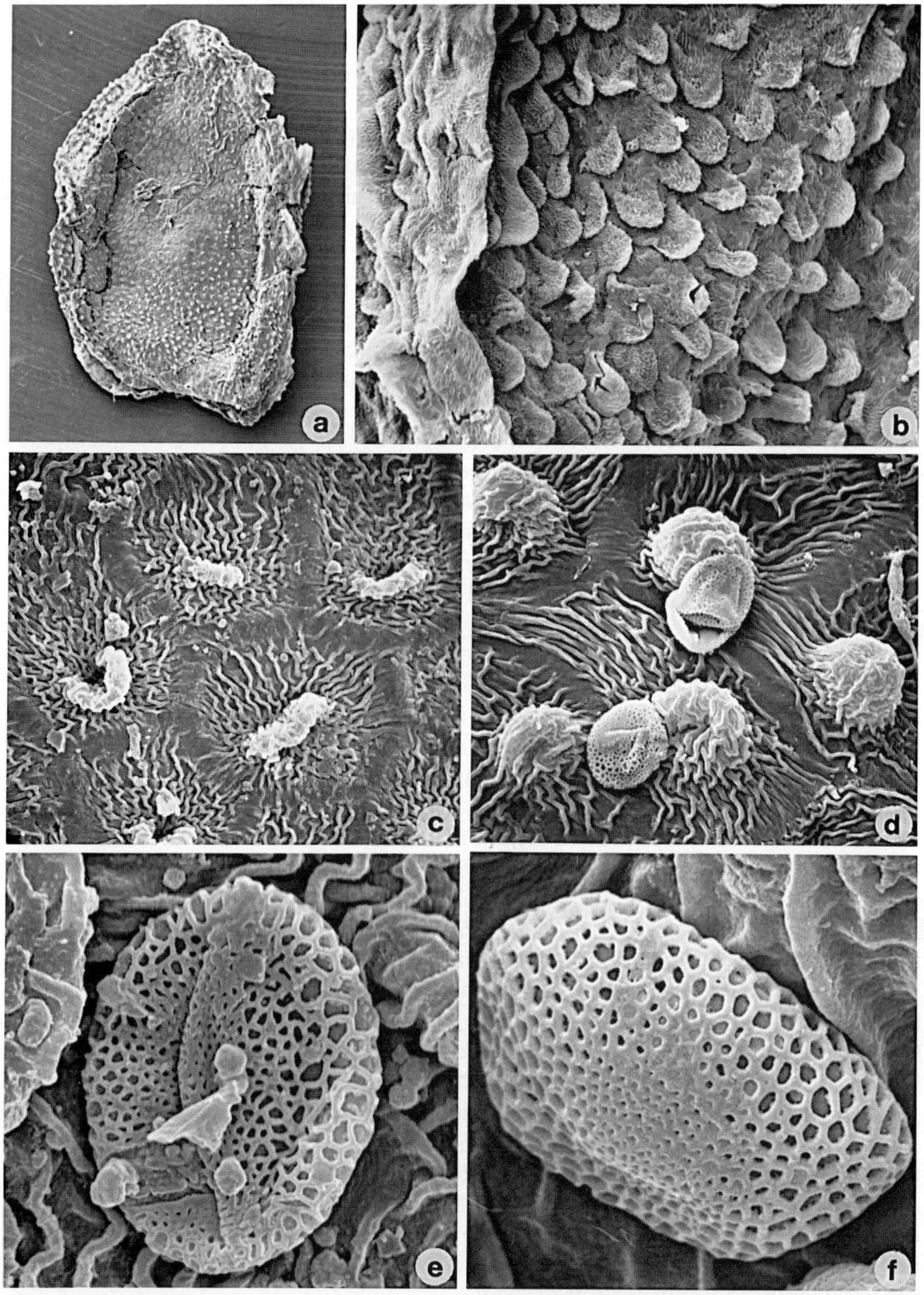

Fig. 9. SEM-micrographs of unilocular and one-seeded fruit/fruitlet from the Puddledock locality, Virginia, USA. *a* Lateral view of fruit (PP43706, sample 04), ×35; *b* details of fruit surface (PP43705, sample 04), ×300; *c* detail of fruit surface showing polygonal and isodiametric cells, central trichomes twisted and collapsed (PP43708, sample 73), ×800; *d* pollen grains adhering to fruit surface (PP43706, sample 04), ×1000; *e* distal view of pollen grain (PP43708, sample 73), ×4000; *f* proximal view of pollen grains showing decrease in lumina size towards the pole (PP43706, sample 04), ×5000

also present in the Puddledock flora. Several distinct taxa occur that may be distinguished by details of fruit and seed anatomy as well as the morphology of associated pollen. One taxon, described and illustrated here (Fig. 9), includes small, laterally compressed fruits that lack a clearly differentiated stigmatic area. Fruits are about 1.4 mm long and 1.1 mm broad. One margin (probably ventral) is straight or slightly convex and the other (probably dorsal) is strongly convex, and semi-circular in outline. The thick, wrinkled cuticle is the only part of the fruit wall preserved indicating that most of the original tissue was soft. The epidermal cells of the fruit wall, as seen on the cuticle, are polygonal and almost isodiametric, about 0.04 mm in diameter, each with a distinct, central papillose trichome (Fig. 9b, c). The seeds are apparently anatropous and basally attached, and the seed coat hard with a dark shiny surface. Pollen grains of one type have been observed on the surface of several specimens (Fig. 9d–f). They are monocolpate, circular to elliptical in equatorial outline, about 15 μm in diameter, and heterogeneous reticulate. The colpus extends almost to the equator. The reticulum is coarse over most of the surface with angular lumina up to about 1 μm in diameter. Muri are broad and rounded without ornamentation (Fig. 9e, f). Lumina decrease in size towards the colpus margin and at the proximal pole grade into an almost solid tectum (Fig. 9f).

Similar fruits, probably related to the Puddledock material at the generic level, have been identified from another Potomac group assemblage (Kenilworth) of approximately the same age (Friis & al., unpubl.). Pollen grains in this species are trichotomocolpate, but otherwise identical in tectum structure to those observed on the Puddledock fossils. Pollen and fruit structure indicate a magnoliid affinity, probably close to the *Laurales* or *Piperales*, and the *Couperites-Clavatipollenites* complex. *Couperites mauldinensis* with *Clavatipollenites*-pollen on the stigma has been described from younger (lower Cenomanian) strata in the Potomac Group of Maryland and has been interpreted as closely related, but not identical, to extant *Chloranthaceae* (Pedersen & al. 1991). *Couperites* is similar to the Puddledock fossils in general organization, but has an apically, rather than basally, attached seed, and the pollen grains have ornamented muri. *Couperites* is also distinguished from the Puddledock fossil described here by the presence of abundant resin bodies (?oil glands) in the fruit wall, a feature observed in at least two other kinds of unilocular, single-seeded fruits from Puddledock that are not described here. A single fruit of *Couperties* has also been identified from the Puddledock flora. It differs from *C. mauldinensis* in details of the fruit epidermis and should be assigned to a new species. The *Couperites* fruit from Puddledock is abraded at the stigmatic area and no pollen were observed on the fruit. However, several anthers with *Clavatipollenites*-type pollen have been found associated with the fruit and may represent the same taxon. The anthers are elongate, tetrasporangiate with small spherical resin bodies in the anther wall.

***Circaeaster*-like fruit with monocolpate pollen** (Fig. 10). This taxon occurs in most of the Puddledock samples and is represented by numerous specimens. All specimens are dispersed fruits and there is no information on floral or inflorescence structure. The material comprises smaller immature ovaries as well as larger mature fruits. Most specimens are compressed, but a few charcoalified fruits show the original three-dimensional form. The fruits are ellipsoidal and densely spiny except

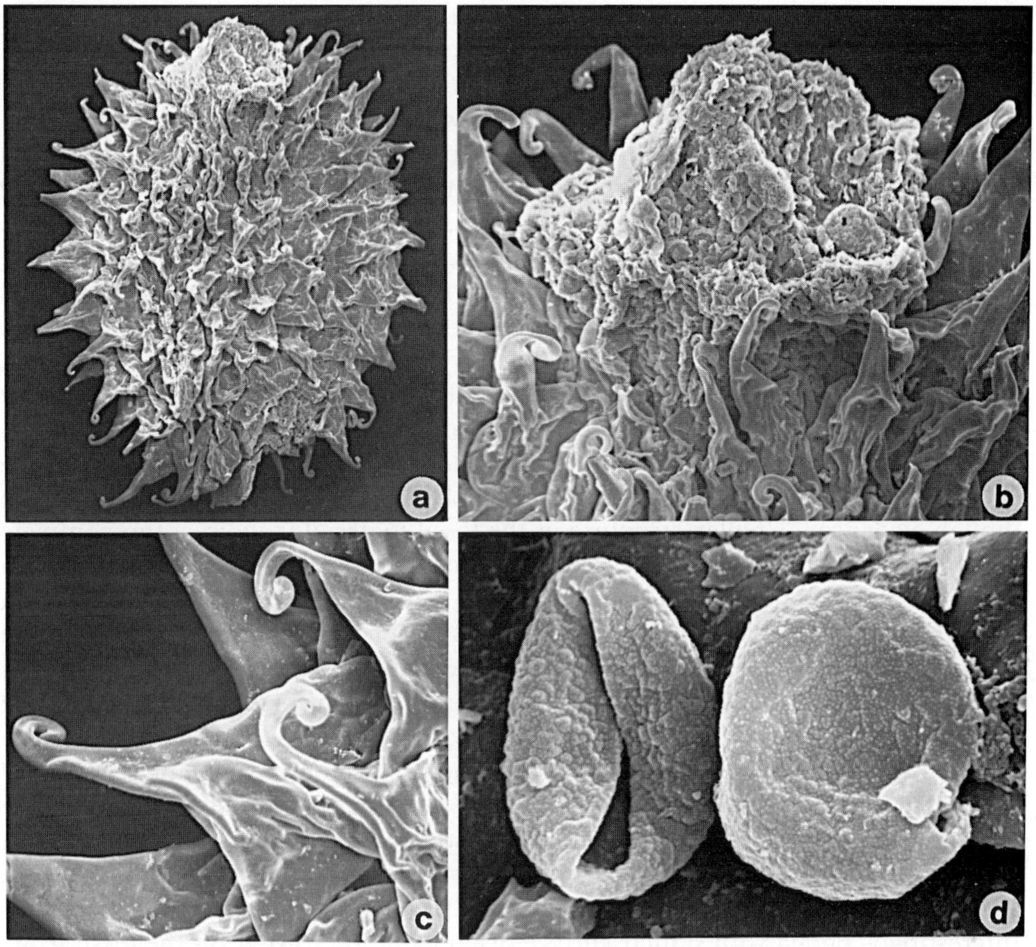

Fig. 10. SEM-micrographs of *Circaeaster*-like fruit from the Puddledock locality, Virginia, USA (PP43778, sample 152). *a* Complete fruit, ×45; *b* stigmatic area free of spiny trichomes, ×150; *c* details of spiny trichomes, ×225; *d* pollen grains from stigmatic area, distal (left) and proximal (right) view. ×2500

at the apical stigmatic area. Mature fruits are up to 2.0 mm long and 1.5 mm broad. Immature ovaries are narrow ellipsoidal in outline, about 0.6 mm long and 0.5 mm broad. The ovary is borne on a narrow stalk attached near the base of the (supposed) ventral margin. A distinct rib passes along the ventral margin to the stigma. The stigmatic area is distinct and swollen (Fig. 10b), and extended apically into a short narrow tip which is rarely preserved. The fruit wall and seed coat are thin, membraneous with little or no mechanical tissue. The fruits are unilocular with a single seed. The seeds are apically attached, orthotropous with the chalaza apical and the micropyle pointing downwards. The point of attachment of the seed is close to the stigma. The fruits are indehiscent and rupture irregularly at the base at germination. The outer epidermal cells of the fruit wall are large, polygonal, almost isodiametric, and about 0.2 mm in diameter. Each cell has a central, unicellular trichome, about 0.2 mm long, with a large swollen base and usually distinctly recurved to coiled tips (Fig. 10b, c). Pollen grains of one type occur abundantly on the stigmatic area of the fruits and between the spines. They are monocolpate, circular to elliptical in equatorial outline, about 16–19 μm in diameter. The tectum has low, irregular verrucae covered by minute and densely spaced echinae. The colpus is short and broad with a granular colpus membrane.

In fruit and seed characters this fossil material is somewhat similar to extant *Circaeaster*, although comparisons with extent taxa reveal a mosaic of magnoliid and possible ranunculid characters (FRIIS & al., unpubl.). *Circaeaster* is a monotypic genus with simple, few-parted flowers; fruits are one-seeded, conspicuously spiny, and seeds orthotropous, pendulous (e.g., FOSTER 1963). There are small differences in fruit morphology between the fossil material and extant *Circaeaster*, thus the modern fruits are slightly larger and narrower in outline, trichomes are longer, without the swollen base, and epidermal cells of the fruit wall are smaller. The most significant difference between the fossils and extant *Circaeaster*, however, is pollen morphology. Pollen in extant *Circaeaster* is tricolpate and striate (NOWICKE & SKVARLA 1982), while pollen observed on the fossil fruits is monocolpate. Even though evidence for pollen structure in the fossil material is indirect (no complete flowers or stamens have been recovered) we consider the link between the pollen and the fruits as rather strong. Pollen grains were observed on many fruits from several different samples. In several specimens pollen grains were abundant on the surface, and in all specimens pollen grains were of the same kind. The discovery of monocolpate pollen on *Circaeaster*-like fruits may be significant in clarifying the evolution of tricolpate pollen and elucidating the phylogenetic relationships between the magnoliids and the eudicots. The systematic position of *Circaeaster* has been much debated. In most modern classifications it is placed in the *Ranunculales* (e.g., TAKHTAJAN 1980, CRONQUIST 1981), but a relationship with *Chloranthaceae* and *Piperales* has previously been suggested (e.g., MAXIMOWICZ 1881, BENTHAM & HOOKER 1883) based on the orthotropous, pendulous seeds. Interestingly, pollen with similar surface ornamentation to that on the fossil fruits has been reported for extant *Piperales* (WALKER 1974) and spiny fruits with orthotropous seeds also occur in the *Piperales* (*Circaeocarpus*), but the spines differ in being multicellular and glochidiate.

**Dispersed stamens with pollen** (Fig. 11). A variety of dispersed stamens with in situ pollen have been recovered from the Puddledock locality but only four different types are illustrated here. **Dispersed stamen type 1** (Fig. 11a, b) is approximately 2.7 mm long, distinctly laminar with the two pairs of pollen sacs apparently embedded in the extensive connective. There is no indication of a filament but the stamen is perhaps incompletely preserved. The connective is well-developed laterally as well as apically to the pollen sacs. The anther is folded up laterally over the pollen sacs and the exact arrangement of the theca and patterns of dehiscence is difficult to study. In situ pollen is elliptical in equatorial outline, approximately 11 μm in diameter, monocolpate and has a smooth surface. Stamen and pollen morphology indicate a magnolialean affinity. Similar laminar stamens with pollen sacs embedded in or protruding from an extensive connective were illustrated by ENDRESS & HUFFORD (1989) for several magnolialean families (e.g., *Magnoliaceae*, *Himantandraceae*). A magnolialean affinity of this stamen is also supported by pollen morphology with similar smooth, thin-walled pollen occurring in *Magnolia* (e.g., WALKER 1974).

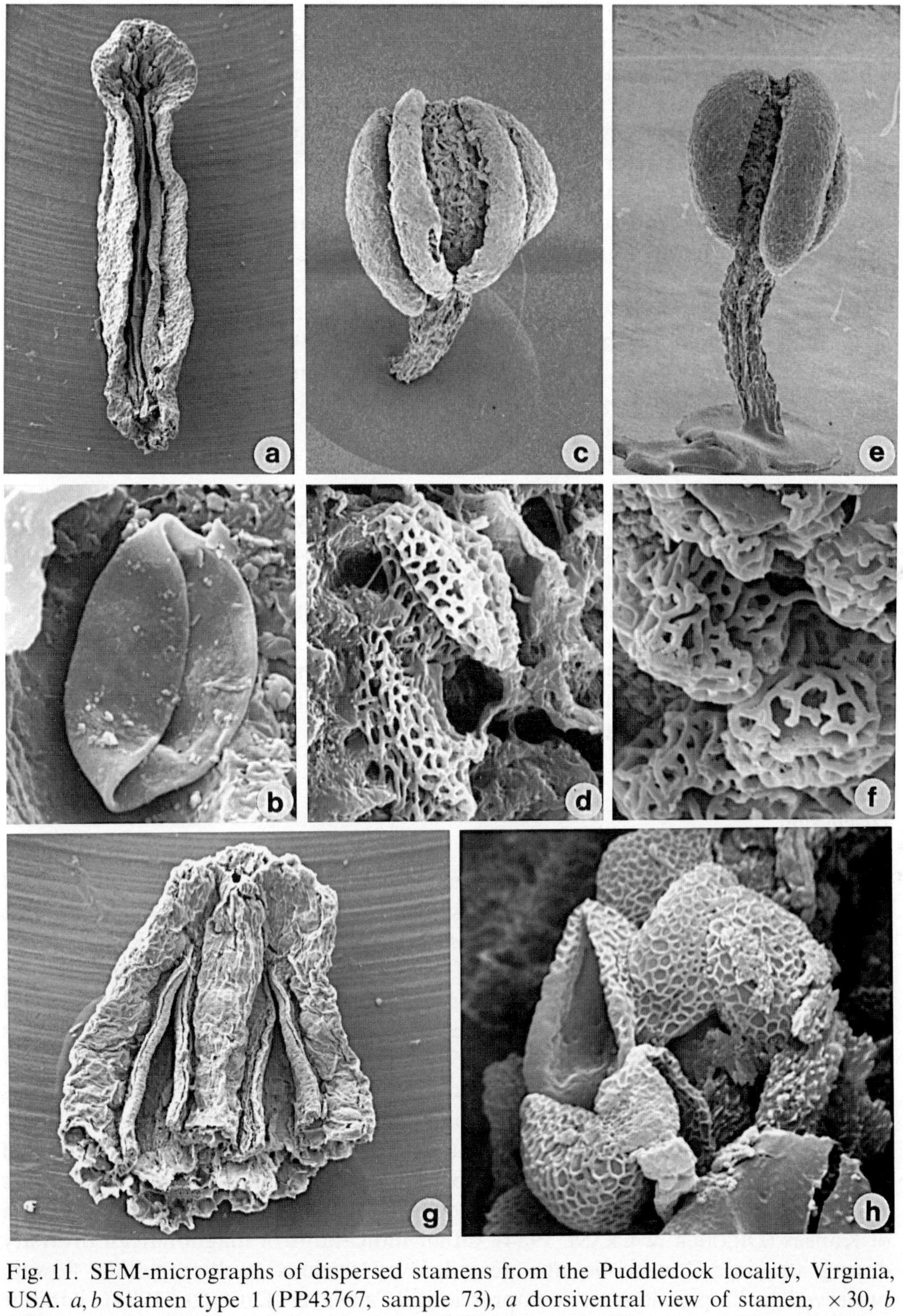

Fig. 11. SEM-micrographs of dispersed stamens from the Puddledock locality, Virginia, USA. *a*, *b* Stamen type 1 (PP43767, sample 73), *a* dorsiventral view of stamen, ×30, *b* pollen grain, ×4000; *c*, *d* stamen type 2 (PP43830, sample 73), *c* lateral view of stamen, ×110, d pollen grains, ×2200; *e*,*f* stamen type 3 (PP35234, sample 73), *e* lateral view of stamen, ×60, *f* pollen grains, ×2000; *g*, *h* stamen type 4 (PP43765, sample 73), *g* dorsiventral view of stamens, ×70, *h* pollen grains, ×4000

**Dispersed stamen type 2** (Fig. 11c, d) is about 0.5 mm long and shows clear differentiation into a short (?incompletely preserved) filament and a distinct tetrasporangiate anther with a short, rounded, apical extension of the connective. The two pairs of pollen sacs are separated by a massive swollen connective. Dehiscence of the two thecae is valvate into two lateral valves that become strongly reflexed revealing the inner surface of the pollen sacs (Fig. 11c). Pollen is embedded in a pollenkitt-like substance and thus difficult to study (Fig. 11d). The pollen grains are about 15 μm long, apparently monocolpate. The tectum is coarsely reticulate with narrow muri and large angular lumina. An androecium formed of three stamens very similar to that described here has been described from another Potomac Group locality (West Brothers) of late Albian age and has been tentatively interpreted as of chloranthoid affinity (CRANE & al. 1989).

**Dispersed stamen type 3** (Fig. 11e, f) is about 0.9 mm long and very similar to stamen type 2 in morphology, particularly in the strongly valvate anther dehiscence, but it differs slightly in the form of the in situ pollen. Pollen in stamen type 3 is also monocolpate, but apparently more spherical in form and the reticulum is somewhat coarser.

**Dispersed stamen type 4** (Fig. 11g, h) is known only from an incomplete distal fragment of a laminar stamen. The anther is tetrasporangiate with the two thecae embedded in and separated by the extensive connective. Each dehisces by a single longitudinal slit. Above the point at which the two thecae terminate there is a rounded, triangular extension of the connective. In situ pollen is elliptical in equatorial outline, approximately 7 μm long, monocolpate and coarsely reticulate with narrow muri and lumina varying considerably in size. Stamen and pollen morphology indicate a magnolialean affinity.

## Discussion

**Systematic affinities of mid-Cretaceous magnoliids**. The different kinds of magnoliid remains recovered from the Puddledock locality are indicative of the systematic diversity attained at this level of angiosperm evolution by the end of the Early Cretaceous. Reproductive structures of probable magnolialeans, calycanthoids, lauraleans and possible chloranthoids are all present, along with several floral structures that cannot currently be referred to extant groups. These discoveries are consistent with previous indications of the mid-Cretaceous diversity of magnoliid angiosperms. They also extend the stratigraphic range of certain groups and expand our knowledge of the morphology of others.

The magnolialean flower from Puddledock provides further evidence of mid-Cretaceous *Magnoliales* that had been inferred previously from the flowers and fruits of *Archaeanthus* preserved as compressions from the early Cenomanian of Kansas (DILCHER & CRANE 1984). Other indications of magnolialean diversity during the mid-Cretaceous are provided by various multicarpellate flowers preserved as impressions or compressions (CRANE & DILCHER 1984), laminar stamens and rows of dispersed seeds from the early Cenomanian of Maryland (FRIIS & al. 1991, CRANE & al., unpubl.), and dispersed pollen grains of *Lethomasites fossulatus* from the early Aptian of the Potomac Group (WARD & al. 1989).

Among *Laurales* the Puddledock flower and stamen provide the earliest unequivocal evidence for this group and extend previous records of mid-Cretaceous *Lauraceae* based on flowers and inflorescences of *Mauldinia mirabilis* from the early Cenomanian of Maryland (Drinnan & al. 1990) and flowers of *Perseanthus crossmanensis* from the Turonian of New Jersey (Herendeen & al. 1994). Other evidence of lauralean plants during the mid-Cretaceous is provided by inflorescences and flowers of *Prisca reynoldsii* (Retallack & Dilcher 1981, see also Drinnan & al. 1990), abundant dispersed leaves (Upchurch & Dilcher 1990), wood (Herendeen 1991), and perhaps flowers with affinities to *Hernandiaceae* (Friis & al. 1994a). Similarly, the putative calycanthoid from Puddledock complements the recent discovery of a calycanthoid flower from the Turonian of New Jersey (Crepet & Nixon 1994).

The abundance and diversity of *Laurales* sensu stricto during the mid-Cretaceous was apparently matched by that of chloranthoids judging both from the dispersed pollen grains that are thought to be related to this group (e.g., *Asteropollis*, *Brenneripollis*, *Clavatipollenites*, *Stephanocolpites*, *Transitoripollis*, *Tucanipollis*; Walker & Walker 1984, Doyle & Hotton 1991) and the abundance of unilocular single seeded fruits in Early and mid-Cretaceous floras. The link between these pollen grains and fruits was first established based on *Couperites mauldinensis* fruits with attached *Clavatipollenites* pollen from the early Cenomanian of Maryland, and is further supported by the unilocular one-seeded fruits from Puddledock (Fig. 9) and possibly also the spiny *Circaeaster*-like fruit (Fig. 10). Similar fruits with other kinds of pollen attached to their stigmatic surface, including *Liliacidites*/*Similipollis* type grains, are common in many fossil floras from both the Potomac Group and the mid-Cretaceous of Portugal. Under current phylogenetic interpretations these "chloranthoid" plants probably include forms closely related to extant *Chloranthaceae* as well as a variety of other "basal" *Laurales* (Doyle & Hotton 1991). More secure evidence of the early differentiation of the *Chloranthaceae* is provided by characteristic three-lobed androecia known from the Santonian/Campanian of southern Sweden (*Chloranthistemon endressii*; Crane & al. 1989) and from the Turonian of New Jersey (*Chloranthistemon crossmanensis*; Herendeen & al. 1994). The gross form of these androecia, including the arrangement of pollen sacs, is identical to that in extant *Chloranthus*.

Other magnoliid remains recorded from the mid-Cretaceous, but not known from the Puddledock locality, include pollen tetrads assigned to *Walkeripollis* that are very similar to those of extant *Winteraceae* (Doyle & al. 1990a, b), fruits preserved as compressions that resemble those of *Ceratophyllum* (Dilcher 1989) and dispersed pollen grains perhaps with similarities to those of extant *Myristicaceae* (aff. *Clavatipollenites*; Walker & Walker 1984) or *Winteraceae*-*Illiciaceae* (*Afropollis*, *Schrankipollis*; Doyle & al. 1990a, b).

**Structural features of mid-Cretaceous angiosperm flowers.** Evidence from Puddledock and other Potomac Group localities support generalizations based on fossil floras from Portugal and elsewhere (Friis & al. 1994a) concerning the structural features that are common among early angiosperm flowers. Mid-Cretaceous angiosperm floral structures are generally small, there is evidence of bisexual and unisexual forms and there is little evidence for marked differentiation of perianth whorls.

Stamens are frequently massive with substantial development of the connective, particularly between the thecae and the stamen apex (Friis & al. 1991). Valvate dehiscence is also general in almost all the magnoliid stamens so far identified in fossil floras from both the Potomac Group and Portugal (Friis & al. 1991, 1994a). In situ pollen is typically monocolpate and often with reticulate exine sculpture.

Among the gynoecia recognized in the Puddledock flora the relative paucity of forms with numerous helically-arranged parts is consistent with information from other mid-Cretaceous floras (Friis & al. 1994a). Much more common are unilocular, single-seeded gynoecia which may be either similar to *Couperites* in having a thick testa, thin fruit wall, no style and relatively poor differentiation of the stigmatic surface, or associated with epigyny in which case the fruit wall is thicker and an elongated style is present. Both of these kinds of fruits are also widespread in other fossil floras from the Potomac Group and Portugal. While some conclusions are possible as to the systematic affinities of the *Couperites*-type fruits the relationships of the epigynous forms remain to be determined.

We thank Dr G. R. Upchurch, Jr and Dr A. N. Drinnan for invaluable assistance in the field; Dr J. A. Doyle, Dr P. K. Endress, and Dr F. B. Sampson for helpful comments and discussion; Dr V. Srinivasan for help in sorting the material, Y. Arremo and P. Lidmark for help in preparing the illustrations. This work was supported in part by NSF research grant BSR-90-20237 (PRC), Swedish National Science Council grant G-BU/GU9381-300-312 (EMF), and the Carlsberg Foundation (KRP and EMF).

**References**

Bentham, G., Hooker, J. D., 1883: *Circaeaster* – Genera plantarum **3(2)**: 1220–1221.

Brenner, G. J., 1963: The spores and pollen of the Potomac Group of Maryland. – Maryland Dept. Geol., Mines Water Res. Bull. **27**: 1–215.

Crane, P. R., Dilcher, D. L., 1984: *Lesqueria*: an early angiosperm fruiting axis from the Mid-Cretaceous. – Ann. Missouri Bot. Gard. **71**: 384–402.

– Friis, E. M., Pedersen, K. R., 1989: Reproductive structure and function in Cretaceous *Chloranthaceae*. – Pl. Syst. Evol. **165**: 211–226.

– Pedersen, K. R., Friis, E. M., Drinnan, A. N., 1993: Early Cretaceous (middle Albian) platanoid inflorescences and leaves from the Potomac Group of eastern North America. – Syst. Bot. **18**: 328–344.

Crepet, W. L., Nixon, K. C., 1994: Flowers of Turonian *Magnoliidae* and their implications. Pl. Syst. Evol. [Suppl.] **8**: 73–91.

Cronquist, A., 1981: An integrated system of classification of flowering plants. – New York: Columbia University Press.

Dilcher, D. L., 1989: The occurrence of fruits with affinities to *Ceratophyllaceae* in Lower and mid-Cretaceous sediments. – Amer. J. Bot. **76**: 162.

– Crane, P. R., 1984: *Archaeanthus*: An early angiosperm from the Cenomanian of the western interior of North America. – Ann. Missouri Bot. Gard. **71**: 351–383.

Dischinger, J. B., 1987: Late Mesozoic and Cenozoic stratigraphic and structural framework near Hopewell, Virginia. – US Geol. Surv. Bull. **1567**: 1–48.

Doyle, J. A., 1969: Cretaceous angiosperm pollen of the Atlantic Coastal Plain and its evolutionary significance. – J. Arnold Arbor. **50**: 1–35.

– 1992. Revised palynological correlations of the lower Potomac Group (USA) and the Cocobeach sequence of Gabon (Barremian-Aptian). Cretaceous Res. **13**: 337–349.

– Hickey, L. J., 1976: Pollen and leaves from the mid-Cretaceous Potomac Group and their bearing on early angiosperm evolution. – In BECK, C. B., (Ed.): Origin and early evolution of angiosperms, pp. 139–206. – New York: Columbia University Press.

– ROBBINS, E. I., 1977: Angiosperm pollen zonation of the continental Cretaceous of the Atlantic Coastal Plain and its application to deep wells in the Salisbury Embayment. – Palynology **1**: 43–78.

– HOTTON, C. L., 1991: Diversification of angiosperm pollen in a cladistic context. – In BLACKMORE, S., BARNES, S. H., (Eds): Pollen and spores – patterns of diversification, pp. 197–224. – Oxford: Clarendon Press.

– – WARD, J. V., 1990a: Early Cretaceous tetrads, zonosulcate pollen, and *Winteraceae*. I. Taxonomy, Morphology, and ultrastructure. – Amer. J. Bot. **77**: 1544–1557.

– – – 1990b: Early Cretaceous tetrads, zonosulcate pollen, and *Winteraceae*. II. Cladistic analysis and implications. – Amer. J. Bot. **77**: 1558–1568.

DRINNAN, A. N., CRANE, P. R., FRIIS, E. M., PEDERSEN, K. R., 1990: Lauraceous flowers from the Potomac Group (mid-Cretaceous) of eastern North America. – Bot. Gaz. **151**: 370–384.

FOSTER, A. S., 1963: The morphology and relationships of *Circaeaster*. – J. Arnold Arbor. **44**: 299–327.

ENDRESS, P. K., HUFFORD, L. D., 1989: The diversity of stamen structures and dehiscence patterns among *Magnoliidae*. – Bot. J. Linn. Soc. **100**: 45–85.

FRIIS, E. M., CRANE, P. R., PEDERSEN, K. R., 1988: The reproductive structures of Cretaceous *Platanaceae*. – Biol. Skr. Dansk Vid. Selsk. **31**: 1–55.

– – – 1991: Stamen diversity and in situ pollen of Cretaceous angiosperms. – In BLACKMORE, S., BARNES, S. H., (Eds): Pollen and spores – patterns of diversity, pp. 197–224. – Oxford: Clarendon Press.

– PEDERSEN, K. R., CRANE, P. R., 1994a: Angiosperm floral structures from the Early Cretaceous of Portugal. – Pl. Syst. Evol. [Suppl.] **8**: 31–49.

– EKLUND, H., PEDERSEN, K. R., CRANE, P. K., 1994b. *Virginianthus calycanthoides* gen. et sp. nov. – A calycanthaceous flower from the Potomac Group (Early Cretaceous) of eastern North America. – Intern. J. Pl. Sci. (in press).

HERENDEEN, P. S., 1991: Lauraceous wood from the mid-Cretaceous Potomac Group of eastern North America: *Paraphyllanthoxylon marylandense* sp. nov. – Rev. Palaeobot. Palynol. **69**: 277–290.

– CREPET, W. L., NIXON, K. C., 1994: Fossil flowers and pollen of *Lauraceae* from the Upper Cretaceous of New Jersey. – Pl. Syst. Evol. **189**: 29–40.

HICKEY, L. J., DOYLE, J. A., 1977: Early Cretaceous fossil evidence for angiosperm evolution. – Bot. Rev. **43**: 3–104.

MAXIMOWICZ, C. J., 1881: Diagnoses plantarum novarum asiaticarum. IV. – Bull. Acad. Imp. Sci. St. Petersburg III, **27**: 556–558.

NOWICKE, J. A., SKVARLA, J. J., 1982: Pollen morphology and the relationships of *Circaeaster*, of *Kingdonia*, and of *Sargentodoxa* to the *Ranunculales*. – Amer. J. Bot. **69**: 990–998.

PEDERSEN, K. R., CRANE, P. R., FRIIS, E. M., 1989: Pollen organs and seeds with *Eucommiidites* pollen. – Grana **28**: 279–294.

– – DRINNAN, A. N., FRIIS, E. M., 1991: Fruits from the mid-Cretaceous of North America with pollen grains of the *Clavatipollenites* type. – Grana **30**: 577–590.

– FRIIS, E. M., CRANE, P. R., Drinnan, A. N., 1994: Reproductive structures of an extinct platanoid from the Early Cretaceous (latest Albian) of eastern North America. – Rev. Palaeobot. Palynol. **80**: 291–303.

RETALLACK, G., DILCHER, D. L., 1981: Early angiosperm reproduction: *Prisca reynoldsii*, gen. et sp. nov. from mid-Cretaceous coastal deposits in Kansas, U.S.A. – Palaeontographica Abt. B **179**: 103–137.

SAMPSON, F. B., 1993: Pollen morphology of the *Amborellaceae* and *Hortoniaceae* (*Hortonioideae*: *Monimiaceae*). – Grana **32**: 154–162.

SRINIVASAN, V., 1992: Two new species of the conifer *Glenrosa* from the Lower Cretaceous of North America. – Rev. Palaeobot. Palynol. **72**: 245–255.

TAKHTAJAN, A. L., 1980: Outline of the classification of flowering plants (*Magnoliophyta*). – Bot Rev. **46**: 225–359.

UPCHURCH, G. R., DILCHER, D. L., 1990: Cenomanian angiosperm leaf megafossils, Dakota Formation, Rose Creek locality, Jefferson County; southeastern Nebraska. – US Geol. Surv. Bull. **1915**: 1–85.

WALKER, J. W., 1974: Comparative pollen morphology and phylogeny of the Ranalean Complex. In BECK, C. B. (Ed.): Origin and early evolution of angiosperms, pp. 241–299. – New York: Columbia University Press.

– WALKER, A. G., 1984: Ultrastructure of Lower Cretaceous angiosperm pollen and the origin and early evolution of flowering plants. – Ann. Missouri Bot. Gard. **71**: 464–521.

WARD, J. V., DOYLE, J. A., HOTTON, C. L., 1989: Probable granular magnoliid angiosperm pollen from the Early Cretaceous. – Pollen Spores **33**: 101–120.

WEAVER, K. N., CLEAVES, E. T., EDWARDS, J., GLASER, J. D., 1968: Geologic map of Maryland. – Maryland Geological Survey, Baltimore.

WOLFE, J. A., PAKISER, H. M., 1971: Stratigraphic interpretations of some Cretaceous microfossil floras of the middle Atlantic states. – US Geol. Surv. Prof. Pap. **750** B: 35–47.

Addresses of the authors: PETER R. CRANE, Department of Geology, The Field Museum, Roosevelt Road at Lake Shore Drive, Chicago, IL 60605, USA. – ELSE MARIE FRIIS, Department of Palaeobotany, Swedish Museum of Natural History, Box 50007, S-104 05, Stockholm, Sweden. – KAJ RAUNSGAARD PEDERSEN, Department of Geology, University of Aarhus, DK-8000 Århus C, Denmark.

Accepted February 11, 1994 by P. K. ENDRESS

Pl. Syst. Evol. [Suppl.] 8: 73–91 (1994)

# Flowers of Turonian *Magnoliidae* and their implications

WILLIAM L. CREPET and KEVIN C. NIXON

Received March 31, 1994

**Key words:** *Magnoliidae*. Turonian, Cretaceous, angiosperm evolution, pollination, fossil flowers.

**Abstract:** Recently discovered fossil flowers of *Magnoliidae* from Atlantic Coastal Plain deposits of Turonian age are considered in the context of the overall fossil record of magnoliid reproductive remains. The assemblage of fossils from these Turonian localities is diverse and affects our understanding of the history of reproductive structures in the *Magnoliidae* (and in the angiosperms in general). Within *Piperales* (sensu CRONQUIST 1981), they include the earliest stamens similar to those of modern *Chloranthus*. Turonian fossil *Magnoliales* document early diversity of complicated flowers with numerous spirally arranged organs and laminar stamens and also reveal diversity in *Magnoliales* with cupulate floral receptacles. In *Laurales*, there are two taxa of fossil flowers of *Lauraceae*, and a fossil flower sharing some, but not all, characters with *Calycanthaceae*. In the context of the overall Cretaceous record of angiosperms, these fossils have important implications with respect to the sequence of appearance of disparate types of floral morphology in *Magnoliidae*. They also reveal diversity in character complexes now associated with several different modes of insect pollination and suggest that during the Turonian, passive (e.g., wind) dispersal of seeds was important in magnoliid taxa.

## Historical context

Intact anther contents from the Aptian of Argentina and dispersed flowers, fruits, seeds and stamens from Aptian or older deposits in Portugal are the earliest megafossil evidence of *Magnoliidae* and of angiospermous reproductive structures (ARCHANGELSKY & TAYLOR 1993, FRIIS & al. 1994). The fossil anthers from Argentina consist of dense masses of *Clavatipollenites*-type pollen grains that retain the shape of the pollen sacs, but without anther wall preservation. Dispersed *Clavatipollenites* pollen grains are known from the Aptian and earlier. *Clavatipollenites hughesii*, similar to modern *Ascarina* (*Chloranthaceae*) pollen grains, is one of the earliest dispersed pollen grains assignable to a modern family (WALKER & WALKER 1984) and is similar to the Aptian dispersed anther contents from Argentina. *Asteropollis asteroides* HEDLUND & NORRIS, similar to pollen grains of modern *Hedyosmum* (*Chloranthaceae*), and *Stephanocolpites fredericksburgensis* HEDLUND & NORRIS, similar to extant *Chloranthus*, are known from Albian deposits (see Couper 1958, Kemp 1968, Hughes & al. 1979, Walker & Walker 1984, Chapman 1987, Chlonova & Surova 1988, Crane & al. 1989). Extant *Chloranthaceae* are usually divided into two groups of genera, the *Ascarina-Hedyosmum* group and the *Chloranthus-Sarcandra* group, and both are known as pollen grains from the early Cretaceous. Magnoliid stamens reported by Friis & al. (1994) are diverse including dorsiventrally flattened ones with small lateral anther sacs and reticulate monosulcate pollen. Other *Magnoliidae* have been reported from the early fossil record of dispersed angiosperm pollen.

Pollen grains assignable to *Winteraceae* are known from Aptian deposits (Walker & al. 1983) and atectate pollen grains, *Lethomasites*, having a character suite compatible with magnolialean affinities, are also found in the dispersed pollen record of the Aptian (Ward & al. 1989, Doyle & Hotton 1991). Similarities between the exine of *Lethomasites* and the exine of *Williamsoniella* (Crepet unpubl.) and dangers inherent in inferring floral morphology from dispersed pollen grains (based on observed mosaicism in Cretaceous fossil flowers; Crepet & Nixon, unpubl., and discussion by Doyle & Hotton 1991) add uncertainty to any inferences about the floral morphology that might have characterized the taxon producing *Lethomasites* (see Nixon & Wheeler 1992a for a discussion of extinction, mosaics, and sampling). Such considerations also apply to interpretations of dispersed pollen grains similar to that of modern *Winteraceae* (Walker & al. 1983). Also ambiguous with respect to the history of angiosperm reproductive structures are megafossil axes with leaves and fruits similar to modern *Piperales* reported by Taylor & Hickey (1990). These fossils do not have clearly preserved reproductive structures and inferences on floral details would be unwarranted at this time.

*Magnoliidae* are more diverse in the Albian megafossil structures than in older deposits. Branched or compound stamens (see discussion in Herendeen & al. 1993, Crepet & Nixon 1995) with both similarities and differences in relation to those of modern *Chloranthaceae* have been described by Friis & al. (1986) and Crane & al. (1989). They are tripartite like those of modern *Chloranthus* (Endress 1990, Endress & Hufford 1989, Kubitzki & al. 1993) and have similar pollen grains (Crane & al. 1989), but they have terete rather than flattened segments and have four, rather than two, pollen sacs on each of the lateral lobes (Crane & al. 1989). Early Cenomanian fruits (*Couperites mauldinensis*; Pedersen & al. 1991) with pollen grains of the *Clavatipollenites*-type still adherent to their apices are unilocular and unicarpellate as in modern *Chloranthaceae*, but, unlike those of modern *Chloranthaceae*, have a single pendulous, anatropous, bitegmic ovule per carpel.

A recently discovered Early or mid-Albian locality in the Potomac Group of Virginia, USA, includes laminar stamens that might represent the earliest megafossil evidence of *Magnoliales* (Crane & al. 1994) and also includes the earliest evidence of *Calycanthus*-like flowers.

Megafossil floral evidence of *Magnoliales* also appears in the fossil record of the early Cenomanian, but only a fragment of one laminar stamen with valvate dehiscence, an extended connective, and monosulcate, reticulate pollen has so far been discovered in these sediments (Friis & al. 1991). Fossil reproductive remains of *Lauraceae* also appear for the first time in the Cenomanian where well-preserved flowers with tepals, four whorls of stamens (one with lateral appendages), and a single pistil have been described by Drinnan & al. (1990).

There is even more Cenomanian magnoliid diversity in the slightly younger deposits of interior North America. Two taxa of magnolialean fruits with associated leaves and bracts, but without associated or attached stamens, have been described by DILCHER & CRANE (*Archaeanthus*, 1984) and CRANE & DILCHER (*Lesqueria*, 1984; see also DILCHER & al. 1976).

## Turonian fossil *Magnoliidae*

*Magnoliidae* are an interesting and conspicuous element of Turonian deposits (e.g., DOYLE & ROBBINS 1977) of the Atlantic Coastal Plain (HERENDEEN & al. 1993, 1994; CREPET & NIXON 1995 and unpubl.). Preservation is three dimensional and the mode of preservation has been interpreted as either charcoalification or lignification (see discussion in MCGINNES & al. 1971, 1974). The quality of preservation varies depending on the mode of preservation and, in charcoalified specimens, the apparent degree to which specimens have been exposed to heat. Charcoalified specimens do not always have as uniformly well-preserved pollen as lignified specimens, but often have perfect cellular preservation and three-dimensionally preserved morphology.

Turonian fossil *Magnoliidae* include representatives of the *Magnoliales*, *Laurales*, and *Piperales* (sensu CRONQUIST 1981). Fossils representing *Piperales* are dorsiventrally flattened, tripartite stamens similar to those of modern stamens of *Chloranthus* spp. (*Chloranthistemon crossmanensis*; HERENDEEN & al. 1993). These stamens are the earliest fossil evidence of stamens having the same arrangement of microsporangia as stamens of *Chloranthus* in extant *Chloranthaceae*, with two sporangia on each lateral lobe and four sporangia on the central lobe (Fig. 1a, b). Dehiscence is longitudinal and pollen is preserved in the thecae (Fig. 1c). Pollen micromorphology is reticulate and aperture configuration varies among pollen grains with the same anther sac (Fig. 1c). As with other Cretaceous chloranthoid stamens, these stamens have not been found in situ in flowers, and pistillate structures are unknown.

Three magnolialean taxa have so far been identified and two have been described. These are represented by well-preserved flowers with character complexes that include characters now found in *Magnoliales* and *Laurales* (sensu CRONQUIST 1981) and they share features with modern *Calycanthaceae* and *Eupomatiaceae* (CREPET & NIXON, unpubl.). One, taxon "A" (CREPET & NIXON, unpubl.), has remarkably complete floral preservation. The flowers are enclosed by distally expanded concave bracts that are attached at the cupule rim (Fig. 1d, e). The expanded bracts overlap, but each covers at least one third of the flower (Fig. 1d) and they might be interpreted as having functioned in protection of the floral bud much as the calyptra in modern *Eupomatiaceae* or *Himantandraceae*. Alternatively, the bracts along the rim may be morphologically equivalent to petals (or tepals) and may have opened up and served as an attractant to pollinators (but no fossils have been seen with "open" bracts). These bracts presumably abscised because most floral specimens are preserved with exposed stamens and what appear to be bract attachment scars at the rim of the cupules, but without attached bracts (Figs. 1f, 2a). Flowers have a well-developed receptacular cupule (Fig. 2b). Inside of the bracts on the adaxial distal portion of the cupule, are spirally arranged stamens (Fig. 2b–e). Stamens are

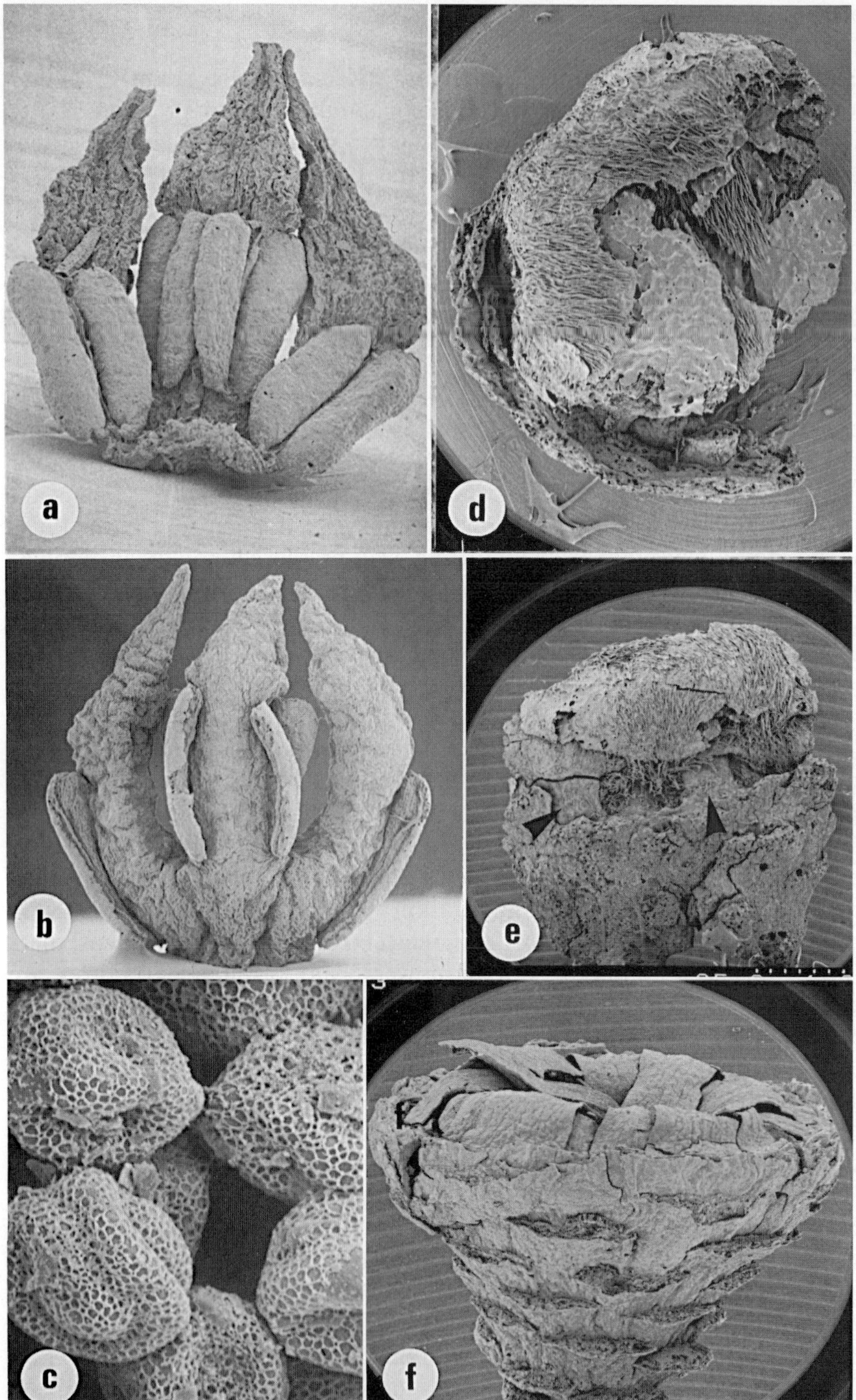


Fig. 1. *a–c Chloranthistemon*. *a* Presumed adaxial view ×70. *b* Presumed abaxial view of *a* ×70. *c* Pollen, ×2000. *d–f* Taxon "A". *d* Top view of the flower showing calyptroid bract complex attached to the flower on the left side, ×25. *e* Lateral view of flower showing the enclosing expanded bracts and their narrow bases inserted at the cupule rim (arrows), ×18. *f* Lateral view of another flower after abscission of calyptrate bracts illustrating the incurved laminar stamens and scale bases on the external cupule, ×17.5

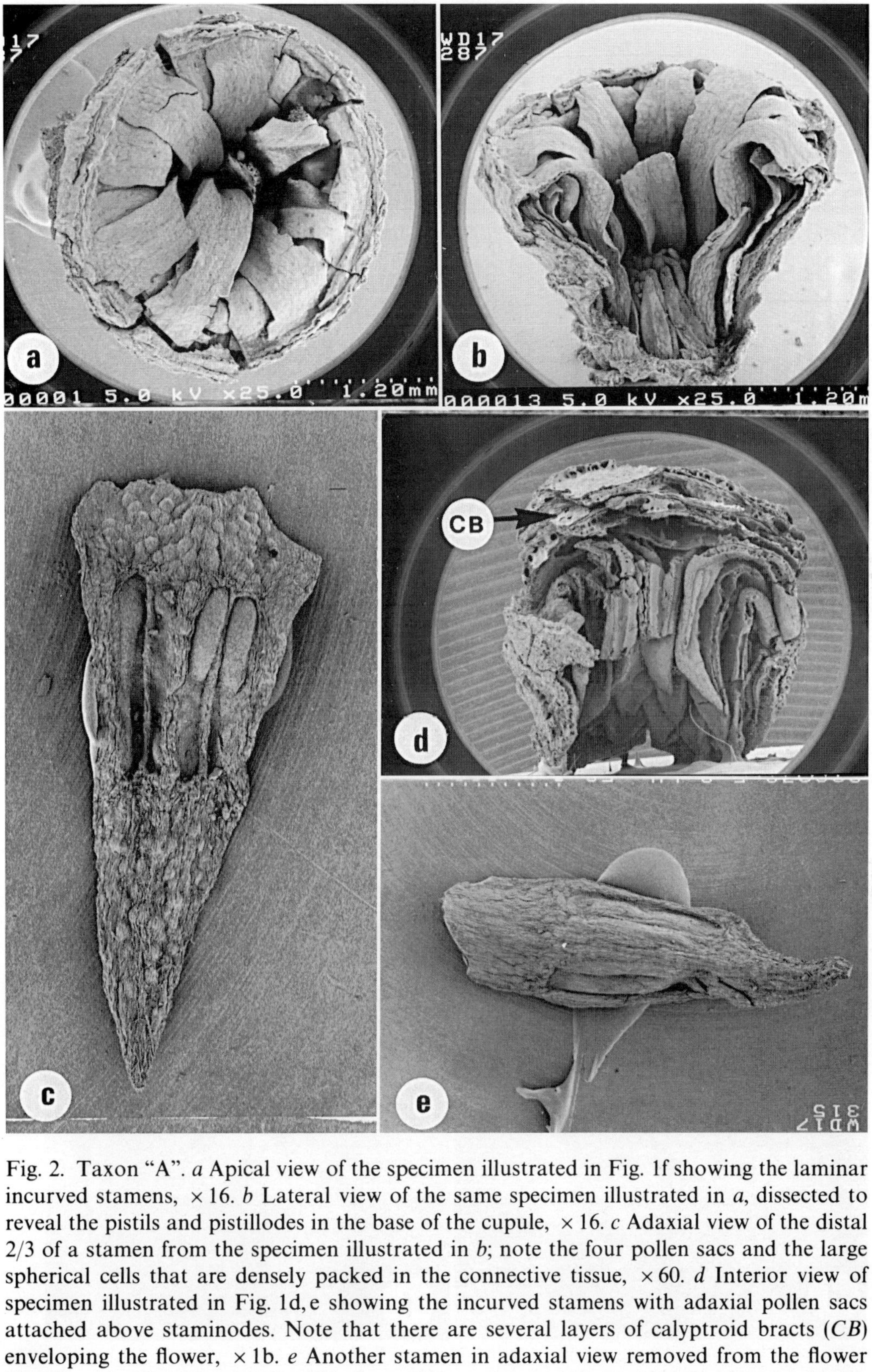


Fig. 2. Taxon "A". *a* Apical view of the specimen illustrated in Fig. 1f showing the laminar incurved stamens, × 16. *b* Lateral view of the same specimen illustrated in *a*, dissected to reveal the pistils and pistillodes in the base of the cupule, × 16. *c* Adaxial view of the distal 2/3 of a stamen from the specimen illustrated in *b*; note the four pollen sacs and the large spherical cells that are densely packed in the connective tissue, × 60. *d* Interior view of specimen illustrated in Fig. 1d, e showing the incurved stamens with adaxial pollen sacs attached above staminodes. Note that there are several layers of calyptroid bracts (*CB*) enveloping the flower, × 1b. *e* Another stamen in adaxial view removed from the flower illustrated in Fig. 1e and in *a* and *b* above, illustrating variation in distal connective length and in the apparent uniformity in pollen maturation within the pollen sacs relative to *c*, X35

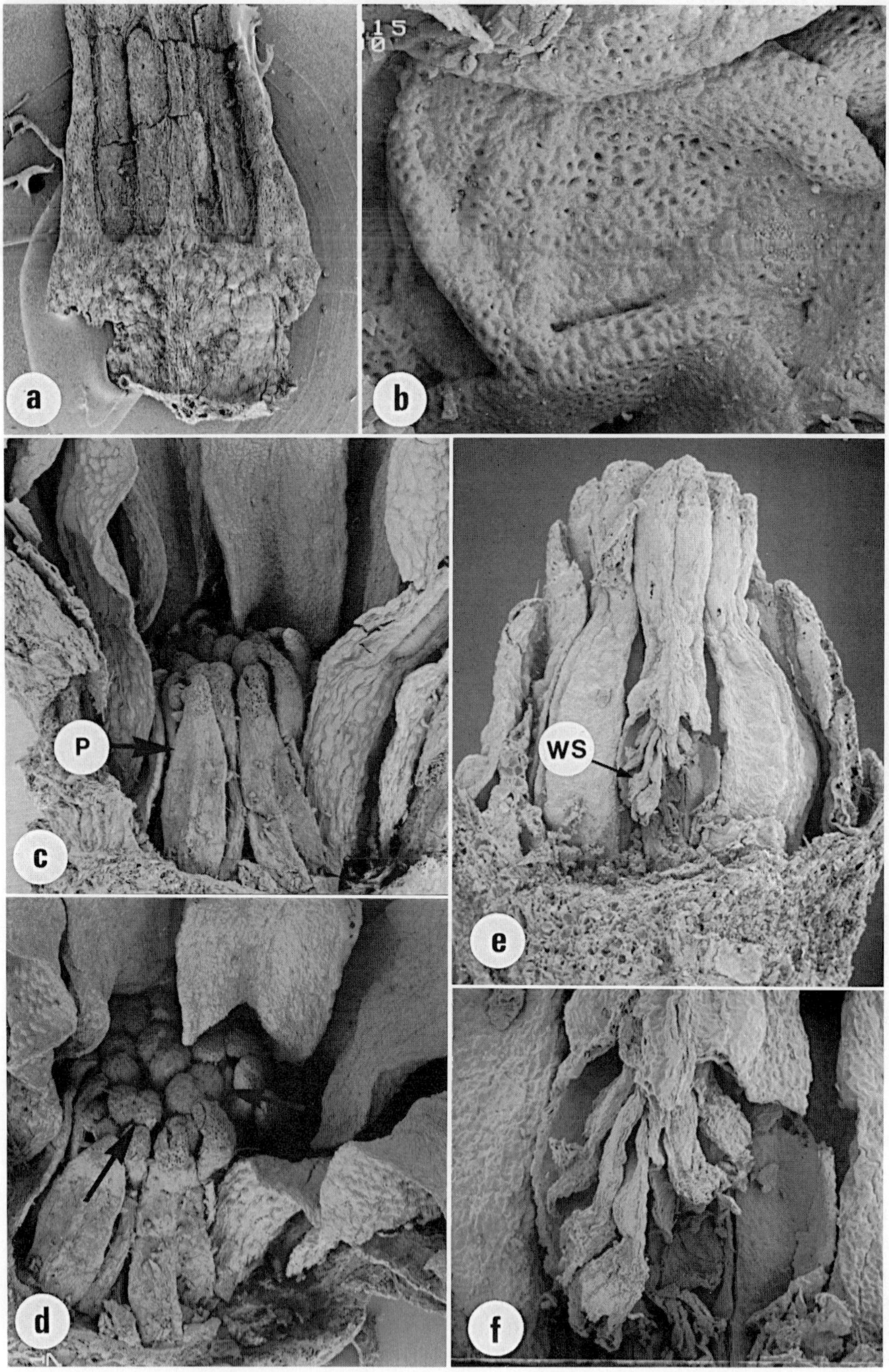
a
b
P
c
WS
e
d
f

laminar, auriculate and linear-lanceolate (Figs. 2c, e, 3a). They incurve acutely between the points of attachment of the pollen sacs and have elongate connective extensions (Fig. 2b–e). There are four adaxial valvate sporangia that sometimes show evidence of uneven pollen development (Fig. 2c). Inner stamens are smaller and tend to have narrower pollen sacs, but even the innermost stamens have pollen sacs and there are no definitive staminodes. Pollen grains are globose with reticulate, apparently tectate-columellate exine structure (Fig. 3b). The long, often distally folded tips of the stamens extend well into the flower where they are adjacent to the pistillodes (Fig. 3c, d). There are whorled carpels at the base of the cupule (Figs. 2b, 3c–e). These are conduplicate and slightly constricted before terminating in a clavate bilobed stigma (Fig. 3e). The stigma surface has scattered large blister-like cells that frequently lose their outer walls to leave holes in the epidermis (Fig. 4a). Seeds are borne marginally in two rows and are winged distally (Fig. 3e, f). The carpels are surrounded by flattened pistillodes that are well separated from the stamens by a zone on the inner surface of the cupule that bears no appendages (Fig. 3c, d). Sometimes masses of stamens are preserved separately, but these seem to be remnants of flowers that have been broken apart by mechanical forces and do not provide evidence that the stamens abscised as a unit as in modern *Eupomatia* (Fig. 4b).

Flowers very similar to this fossil taxon in floral organization, but much smaller and narrower with fewer stamens and carpels have also been discovered in the same deposits. Because of the early stage of the investigation, they will be described elsewhere, but they are additional evidence of the diversity of cupulate *Magnoliales* during the Turonian.

Another taxon (taxon "B") shares eupomatioid and calycanthoid characters (CREPET & NIXON, unpubl.). It has cupulate floral receptacles attached to a robust pedicel (Fig. 4c). The pedicels are eustelic with limited secondary wood or radial metaxylem developed in the individual bundles (Fig. 4d). Vessels have acutely angled ends and scalariform perforation plates with 10–12 bars and scalariform lateral pitting (Fig. 4e, f). The floral receptacular cup is smooth internally and externally covered with imbricate bracts (Fig. 4c). The bracts are ovate and basally auriculate with apiculate apices (Fig. 4c). The adaxial bract surfaces are glabrous and on the abaxial surfaces there are simple trichomes near the apices. While stamens have not been found attached to the cupules, a suite of morphologically similar stamens has been isolated from the same small samples of sediment that contained some of these flowers. Several of the anthers have pollen grains similar

◀

Fig. 3. Taxon "A". *a* Adaxial view of a stamen from near rim of receptacle showing four anther sacs and its auriculate nature, ×35. *b* Reticulate pollen within a stamen, ×3150. *c*. Higher magnification view of specimen illustrated in Fig. 2b illustrating the carpels at the base of the cupule surrounded by flattened pistillodes (*P*) and then by infolded stamen extremities, ×24,5. *d* Three-quarter view of the specimen illustrated in *c* at higher magnification. Note the folded distal tips of the stamens, the flattened pistillodes and the bilobed stigmas (arrows), ×32,6. *e* A flower with cupule, anthers, and pistillodes removed illustrating carpels with bilobed fusiform stigmatic regions and winged seeds (*WS*) within a broken carpel, ×56,7. *f* Closeup of broken carpel illustrated in *e*, showing winged seeds in two rows, ×168

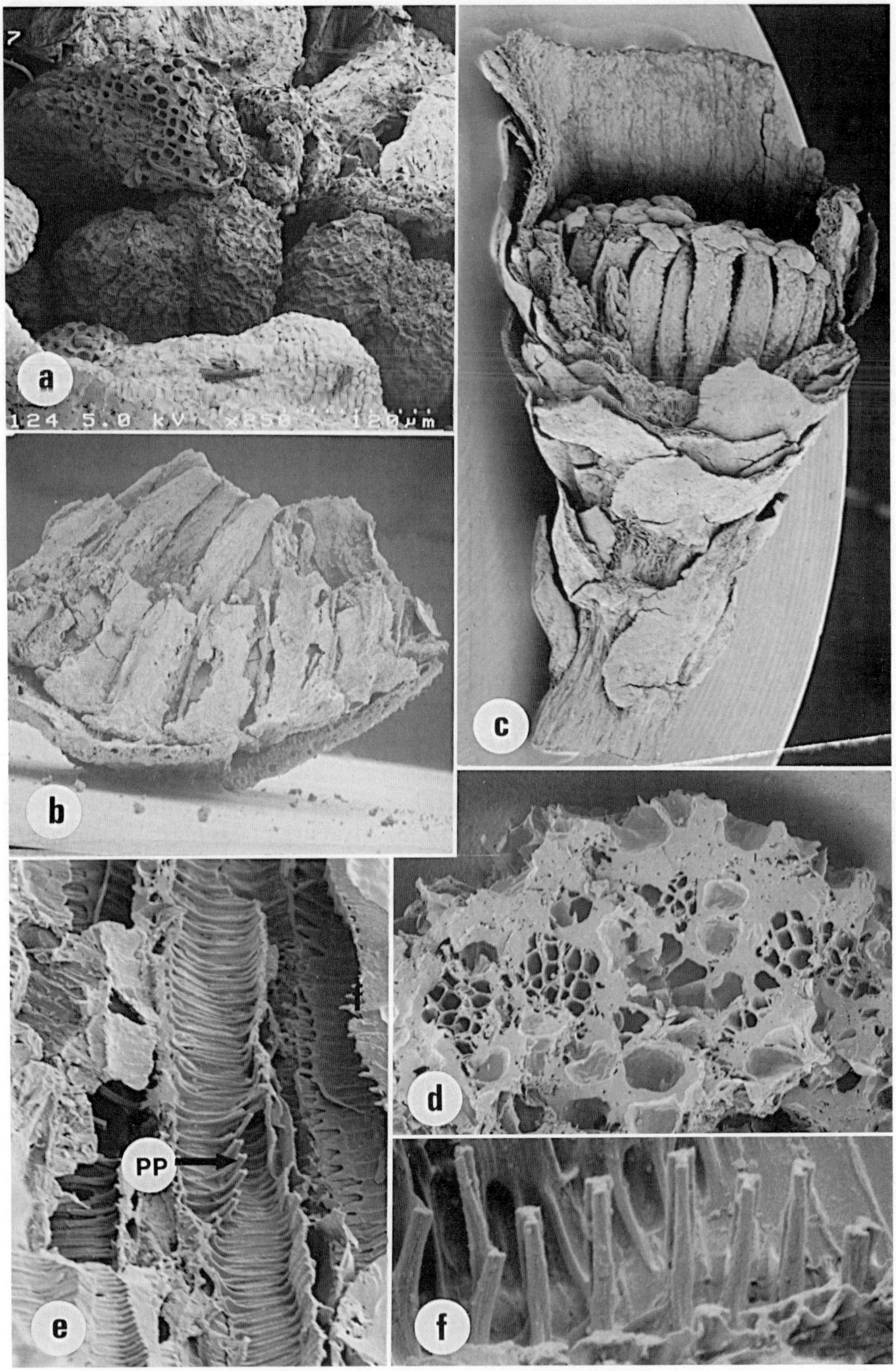


Fig. 4. *a, b* Taxon "A". *a* Closeup of stigmas showing the large epidermal cells that often lose their outer periclinal walls, ×175. *b* Dispersed stamen mass, ×36,4. *c–f* Taxon "B". *c* Floral receptacle, presumably with abscised stamens, showing cupule, carpels at the base of the cupule, robust pedicel, auriculate bracts borne on the pedicel, and external surface of the cupule, ×11,2. *d* Pedicel of the specimen illustrated in *c*, in transverse section illustrating part of a eustelar ring of bundles. Note the radial alignment of the vessels and the large more or less spherical cells in the pith and cortex, ×280. *e* Longitudinal view through one of the bundles illustrated in *d*, showing vessel elements with relatively short, oblique scalariform perforation plates (*PP*), ×1050. *f* High magnification view of vessel elements broken accross the juxtaposed perforation plates, ×4900

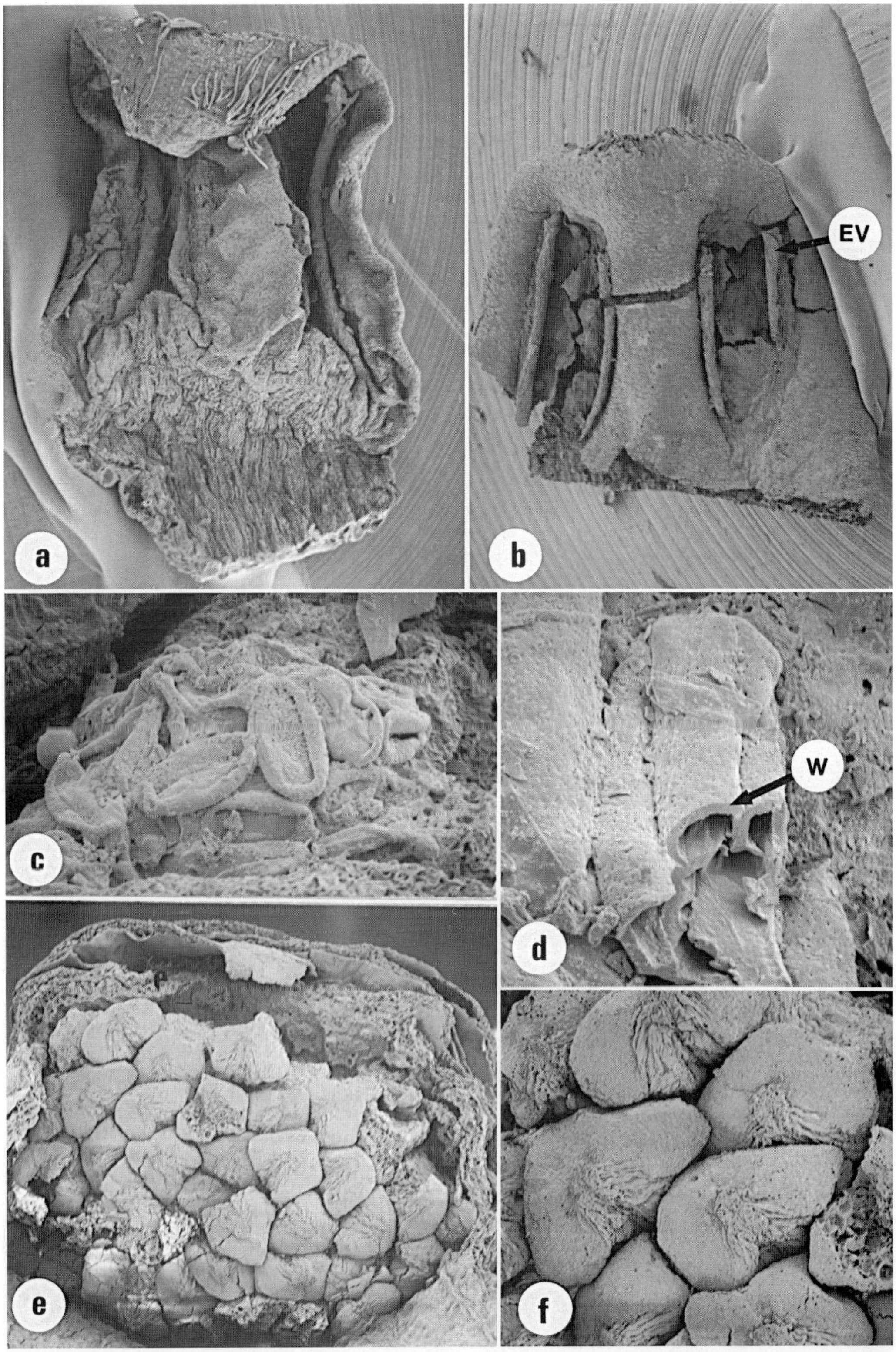


Fig. 5. Taxon "B". *a* Auriculate stamen with four pollen sacs and incurved connective extension with simple elongate hairs, ×77. *b* Stamen broken off above auricles showing four pollen sacs with enrolled valves (*EV*). Note the simple hairs at the tip of the only slightly incurved connective extension, ×38,5. *c* A cluster of pollen within pollen sac, ×1,190. *d* A broken folded pollen grain illustrating the solid (atectate) wall structure (*W*) ×1,533. *e* Top view of the specimen illustrated in Fig. 4c, showing the "plateau" defined by the juxtaposed flattened stigmatic surfaces surrounded by the cupule, ×24,5. *f* Closeup of several stigmatic surfaces of the specimen illustrated in *e* showing the more sculptured epidermis where the sutures meet the stigmatic surfaces, ×84

a
b
c
d
CE
RV
e
f

to those found on the carpels and we assume that these represent the same taxon as the cupulate receptacles (Fig. 5a, b). The stamens are laminar and their auriculate nature can be observed when they are completely preserved (Fig. 5a). They are ovate and sometimes the distal portion of the stamen connective is incurved (Fig. 5a). Stamens are glabrous except for a few simple trichomes near the apex on the abaxial surfaces similar to the pattern found on the bracts on the external surfaces of the cupules (Fig. 5a, b). There are four presumably adaxial thecae in two pairs with valvate dehiscence (Fig. 5a, b). Pollen grains are elongate and monosulcate with occasional folds (Fig. 5c). The exine is smooth and perforate and atectate sensu WALKER & DOYLE (1975; CREPET & NIXON 1995). One isolated stamen similar to the one illustrated in Fig. 5b has pollen grains with the same micromorphological features as those found in the stamens illustrated in Fig. 5a, b, but they are larger and more frequently demonstrate longitudinal folds (Fig. 5d).

Carpels are borne spirally at the slightly concave receptacle base (Figs. 4c, 5e, f). They have an incompletely sealed suture and are constricted slightly distally before expanding to a peltate stigmatic area that overhangs the lateral carpel walls slightly (Fig. 6a), and differs from carpel wall also in epidermal features. Lateral carpel epidermis is pustulate due to numerous large spherical cells similar in appearance to ethereal oil cells in modern *Magnoliidae* (Fig. 6a). On the stigmatic surfaces, the epidermal pattern is complex and lacks the pustulate texture of the epidermis on the lateral walls of the carpels (Fig. 5f). There is a central area of the stigma with more textured epidermis that is continuous with the suture of the carpel (Fig. 5f), while the mostly smooth remaining surface appears to have regularly distributed protruding somewhat angular cells. In aggregate, the juxtaposed stigmatic surfaces of the carpels form a plateau-like surface within the cupule (Figs. 5f, 6a). Within the carpels, there are several seeds that are marginally attached in two rows (Fig. 6b). The seeds are small and elongate with distal wings and small flanges proximally that may be interpreted as arils (Fig. 6b).

Another cupulate magnoliid floral type has many characters in common with modern *Calycanthus* (in *Laurales* sensu CRONQUIST 1981; Fig. 6c–f). Cupulate receptacles have spirally arranged tomentose bracts on the outside with a whorl of stamens with abaxial pollen sacs near the rim (Fig. 6c, arrow). Carpels are spirally arranged with elongate styles extending into the internal hairs that fill the opening of the cupule (Fig. 6d). Carpels bear one mature seed and the seeds are circumscribed by small wings (Fig. 7a). Stamens have expanded connectives and

◄

Fig. 6. *a, b* Taxon "B". *a* Closeup of carpels illustrating the flattened stigmas and the plane that they compose in aggregate, the carpel constriction with slight overhang just beneath the stigmatic areas, and the pustulate lateral walls, ×11,2. *b* Closeup of one of the broken carpels illustrated in *a*, showing the two rows of winged seeds, ×175. *c* External view of a calycanthoid fossil flower showing the abaxially hairy, spirally arranged bracts and one stamen (arrow) near the rim, ×26,6. *d* Internal view of the same specimen showing incompletely sealed conduplicate carpels and the receptacular cup. Note the broken styles at the carpel extremities and the elongate hairs filling the cupule opening, ×33,6. *e* Abaxial view of a slightly eroded stamen. Note the broad, abaxially hairy connective (*CE*) and recurved pollen sac valves (*RV*) ×98. *f* Pollen on recurved anther valve mingled with adjacent hairs, ×1400

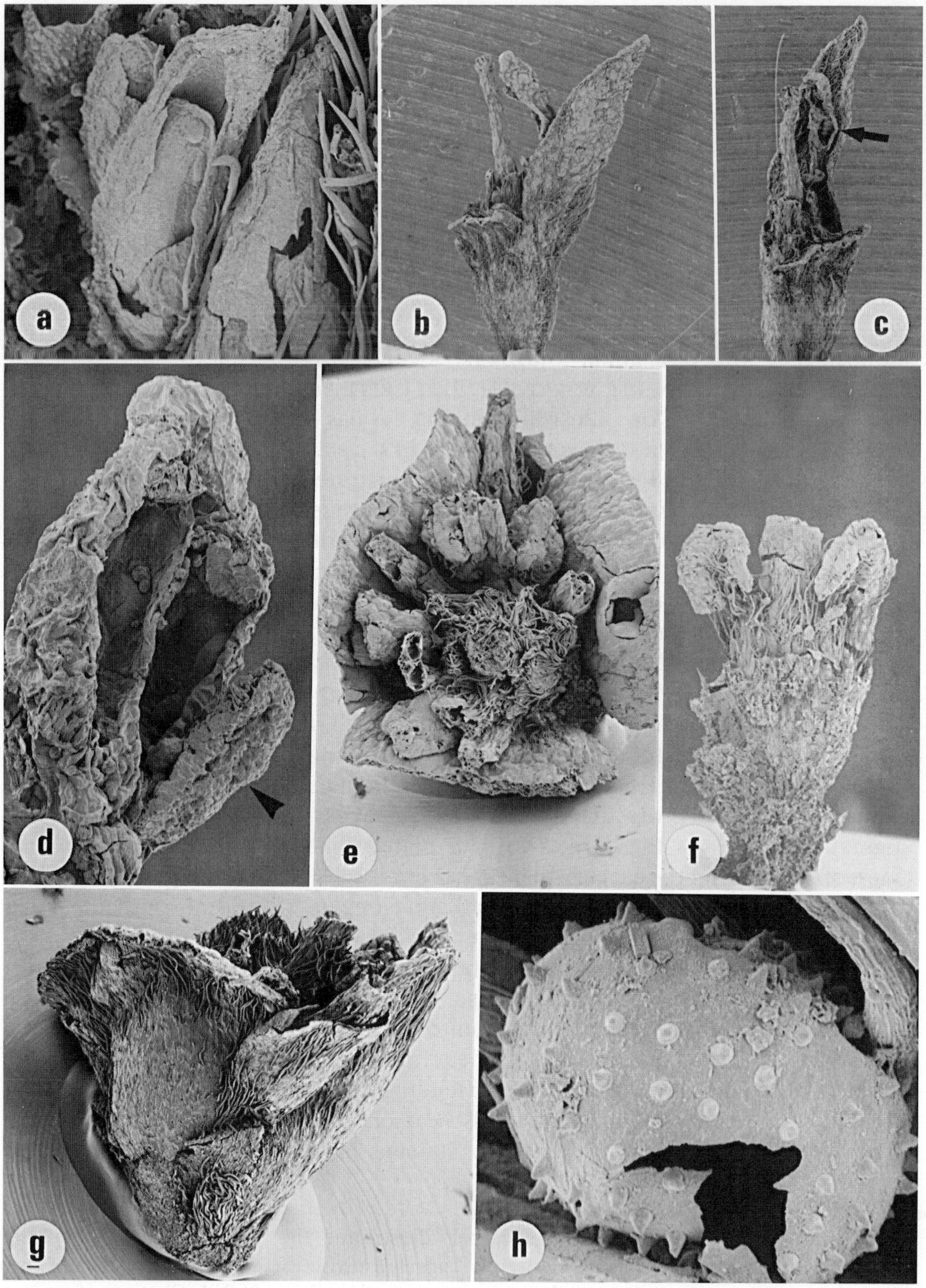


Fig. 7. *a* Broken carpel of calycanthoid flower illustrating the one seed within and its narrow marginal "wing", ×98. *b* A flower of *Lauraceae* with one intact large tepal, one attached complete stamen, and an elongate style, ×21. *c* The other side of the flower illustrated in *b*, note that the anther (arrow), although twisted slightly, shows two tiers of pollen sacs, ×105. *d* Closeup of the top tier of two anther sacs from the flower shown in *b*, *c*; with one valve (arrow) still attached, ×175. *e–h Perseanthus*. *e* Top view of flower with tomentose pistil and several cycles of stamens, ×35. *f* Filament with appendages from the third cycle of stamens in the flower illustrated in *e* ×33. *g* Side view of the floral envelope showing two cycles of tepals, ×35. *h* Inaperturate, thin-walled pollen grain in situ within flower, ×1400

pollen grains are reticulate and globose (Fig. 6e, f; CREPET & NIXON 1995). These flowers are similar to *Calycanthus* and *Chimonanthus* in their apparent functional morphology (KUBITZKI & al. 1993), but there is no evidence of large petals/bracts characteristic of modern calycanthoids nor do the stamens have terminal "food bodies" indicative of extant *Calycanthus*. Although the fossils share several features with modern *Calycanthaceae*, they differ significantly in other characters that will be treated in greater detail in a separate publication.

The Turonian deposits of the Atlantic Coastal Plain also include at least two types of flowers of lauraceous affinity (Fig. 7b, e; HERENDEEN & al. 1994; and unpublished data). One of these lauraceous flowers has stamens that dehisced by four ventral valves in two tiers (Fig. 7b–d). The other taxon (*Perseanthus*) is represented by an incompletely preserved flower (Fig. 7e–h). Anthers are absent from the broken stamen filaments in this fossil, but pollen grains adhere to various parts of the flower. The flowers are trimerous, actinomorphic and cosexual (= bisexual; Fig. 7e). They have two cycles of three tepals basally fused into a short hypanthium (Fig. 7g). Inner tepals are larger than the outer ones and tepals have abaxially borne elongate hairs (Fig. 7g). The androecium includes three cycles of three stamens each and a fourth innermost cycle of staminodes. The third cycle of stamens bears paired lateral substipitate appendages on their filaments (Fig. 7f). The superior unilocular ovary bears a single ovule (HERENDEEN & al. 1994). Pollen grains were found adhering to simple hairs of the ovary and on the bases of the filaments. The grains are spheroidal, inaperturate and have very thin exine with scattered simple spinules (Fig. 7h). Dispersed pollen grains of *Lauraceae* are rare in the fossil record presumably because lauraceous pollen grains typically have a very thin exine with relatively little sporopollenin (e.g., HESSE & KUBITZKI 1983, KUBITZKI 1981, KUBITZKI & al. 1993). The character complex embodied in these fossils suggests affinities with *Lauraceae* and within the family, androecial and pollen characters help limit the possible tribal affinities to *Perseeae* (see discussion in HERENDEEN & al. 1994).

Other fossil flowers of possible magnoliid affinity occur in the Turonian deposits, but lack sufficient details of characters for definite determination of taxonomic affinity at this time. With continued work at these localities, the number of magnoliid taxa known from the Turonian should increase, as well as the overall phylogenetic diversity and number of major clades represented (see NIXON & WHEELER 1992b for a discussion of quantitative measures of phylogenetic diversity).

## Summary of Turonian *Magnoliidae*

1. These deposits contain the earliest dispersed stamens that have the same configuration of microsporangia as those of modern taxa of *Chloranthus*. Earlier chloranthoid stamens are unlike any extant *Chloranthaceae* in the arrangement of microsporangia; thus, the Turonian fossils constitute the earliest evidence of diversification of at least one of the modern lineages within the family, or possibly, the Turonian type of stamen may be a basal form, marking the minimal divergence time of modern *Chloranthaceae* from earlier chloranthoid ancestors.

2. Within *Laurales*, there are two taxa of *Lauraceae* represented by fossil flowers. At least one of these shows strong similarity to the extant genus *Persea*.

3. Three taxa of "magnolialean" affinity are the earliest floral evidence of *Magnoliales* with cupulate (= strongly concave) floral receptacles. These are also the oldest examples of complete specimens of bisexual many-parted flowers with indefinite symmetry that can be presumed to have had complex modes of insect pollination.

4. Taxon "A", with a series of free bracts that form a tightly invested envelope over the flower, provides a model for the derivation of modern calyptrate magnoliids (e.g., *Eupomatia*). Calyptras of some extant magnoliid taxa might be homologous with one of these bracts as suggested by ENDRESS (1986), or with several fused bracts. It is also possible that these bracts were petaloid and served an attractive role. As these structures became modified to form a protective covering over the bud, some of the stamens might have become modified into staminodes taking over the attractive role. In any case, a receptacular or de novo origin of the calyptra is not supported by these fossils.

5. These aggregate characters of the cupulate fossil magnoliids indicate that specialization for various kinds of complex beetle-pollination had been attained in some lineages of magnoliids by no later than the Turonian. However, the presence of laurads with simpler, few-parted, actinomorphic flowers with whorled parts in both earlier and contemporary deposits suggests that diversification in pollination in "magnoliid" lineages was developing simultaneously along very different lines with very different pollinators. The much higher diversity of tricolpate/tricolporate taxa in the Turonian deposits (CREPET & al. 1991, NIXON & CREPET 1993, and unpubl.), with inferred highly specialized hymenopteran, dipteran and/or lepidopteran pollinators suggests that by this time the magnoliid lineages, while perhaps increasing in absolute diversity relative to earlier times, were declining in terms of overall percentage of angiosperm taxa. By the Turonian, angiosperm taxa were mostly of the tricolpate/tricolporate lineages, and thus the Turonian angiosperm flora likely had a very "modern" composition in terms of phylogenetic diversity, with the notable exception that both the monocots and "higher" asterid lineages (i.e., excluding sympetalous dilleniids of CRONQUIST 1981, such as *Ericales*) were rare or absent and subsequently have not yet been recorded in the Turonian flora.

6. The nature of young fruits and examples of winged seeds suggests that seed dispersal was passive in the magnoliids thus far investigated from the Turonian. The broader fossil record of angiosperm flowers suggests that wind or passive dispersal of seeds was the dominant mode during and previous to the Turonian (see relevant discussion in TIFFNEY 1984, CREPET & NIXON 1995). Among tricolpate taxa, a large number of capsular fruits with small unwinged seeds have been found, and a wide variety of small dispersed dicotyledonous seeds, with various types of ornamentation, have been found (NIXON & CREPET, unpubl.). The modes of dispersal for these "higher" dicotyledons in the Turonian is not apparent at this time, and may have been passive or involved various types of animals.

7. Complexity in structure and numbers was not always accompanied by increasing size in flowers of Cretaceous magnoliids. All taxa of Turonian magnoliids are tiny, especially in comparison with their presumed closest modern relatives (see the figures). The relative small size of Turonian magnoliids in consistent with the small size of hamamelid, rosid, and dilleniid taxa from Turonian deposits (CREPET & al. 1991, 1992; CREPET & NIXON 1995; NIXON & CREPET 1993) and

younger Cretaceous localities (FRIIS & ENDRESS 1990, and references therein) as well as with magnoliid and hamamelidid taxa from earlier deposits (e.g., FRIIS & al. 1986, CRANE & al. 1989, DRINNAN & al. 1990, PEDERSEN & al. 1991).

## Discussion

The pattern of Cretaceous diversification in *Magnoliidae* inferred from the fossil record of reproductive structures (DILCHER 1979, DILCHER & CRANE 1984, CRANE & DILCHER 1984, FRIIS & al. 1986, CRANE & al. 1989, FRIIS & al. 1991) and terminating with the Turonian taxa discussed herein is becoming better understood as more fossil taxa are discovered and described (and as correlations with the growing knowledge of the dispersed leaf flora proceed, e.g., UPCHURCH 1984, UPCHURCH & DILCHER 1990). Currently, the observed historical pattern of diversification in magnoliid taxa, while based on relatively few specimens, is consistent with angiosperm stem groups having relatively small, simple flowers, possibly with whorled parts such as those found in the earliest known fossil flowers and floral parts, resembling modern *Chloranthaceae*, *Platanaceae*, *Buxaceae*, and *Lauraceae* (e.g., FRIIS & al. 1991). In the broadest sense, this pattern is consistent with hypotheses of seed plant phylogeny that place modern gnetopsids as the probable sister group of angiosperms (e.g., COULTER & CHAMBERLAIN 1971, NIXON & al. 1994), rather than fossil bennettitoids (CRANE 1985, DONOGHUE & DOYLE 1989 and references cited therein). Under such a hypothesis, angiosperm flowers would have been derived from simpler structures (e.g., gnetopsid strobili or fertile appendages) that were aggregated into "pre-flowers" resembling the "flowers" of modern *Chloranthaceae* and underwent further aggregations to produce actinomorphic, few-parted flowers such as found in *Lauraceae* and various other taxa including monocotyledons, trimerous magnoliids, and many lower *Hamamelididae* (DAHLGREN 1983; see also BURGER 1977, 1981). The more complex, many-parted, often beetle-pollinated flowers of many magnoliids and ranunculids (sensu (TAKHTAJAN 1980, THIEN 1980) would then represent various derived lineages (see GOTTSBERGER 1974 for germane discussion), as opposed to the traditional interpretation (e.g., BESSEY 1915, CRONQUIST 1981) that such flowers represent the archetypal form in angiosperms.

One new insight provided by the Turonian cupulate magnoliids, independent of which phylogenetic hypothesis is favored, is that these types of "classical" beetle-pollinated magnoliid flowers were not always large in relation to higher dicots, as they are in most modern magnoliids. The scale of the flowers and the pollinators in Turonian deposits at least, seems to have been smaller than in modern magnoliids. Average flowers are only a few millimeters in diameter while *Coleoptera* from the same deposits are also tiny (D. GRIMALDI and Q. WHEELER, pers. comm.); see also discussion in CREPET & NIXON, unpubl.). Non-magnoliid flowers from the Turonian deposits are also consistently small relative to modern counterparts (CREPET & al. 1992; NIXON & CREPET 1993, and unpubl.).

Inferences about mode of pollination in Cretaceous angiosperms must be cautiously advanced in view of the relatively sparse fossil record of flowers and insects, limitations on characters that may be discerned from fossils, and the relatively small, but growing, literature on pollination in extant *Magnoliidae*. In fact, there are so many structural characters associated with certain modes of insect pollination that the potential for all of them being observable in any fossil flowers might have once been regarded as minimal. Nonetheless, in an increasing number of cases, it is possible to document germane associations of characters in fossil flowers because of the extraordinary quality of preservation and then to determine the compatibility of the particular flower and its aggregate characters with a given mode of pollination. We consider constraints on determining history of pollination and its significance in greater depth in another paper (CREPET & NIXON, unpubl.).

Because of the overall structural similarity between the cupulate receptacular fossil flowers of the Turonian and extant *Eupomatia*, a discussion of the documented pollination syndrome of *Eupomatia* may be relevant. In extant *Eupomatia*, pollination by the curculionid (*Coleoptera*) genus *Elleschodes* is effected through an interaction of rapid floral development, chemical attractants (ethereal oils), and complex cupulate floral structure (ARMSTRONG & IRVINE 1990, BERGSTRÖM & al. 1991). The fossil flower taxon "A" possesses the cupulate floral structure of modern *Eupomati*a with numerous incurved stamens that have large spherical cells, suggesting that they were laden with ethereal oils (CREPET & NIXON, unpubl.). However, important differences exist between taxon "A" and modern *Eupomatia*. There are no obvious staminodes immediately internal to the stamens in flowers of taxon "A", and there is no evidence of a fused calyptra in taxon "A". With respect to potential pollinators of Turonian magnoliids, *Curculionidae* are diverse by the Turonian (CARPENTER 1992) as well as being represented in fossils from the same sediments as these flowers (work in progress). Thus, the complex of floral characters found in taxon "A" is compatible with, but not unequivocal evidence of, a mode of insect pollination similar to what has been observed within modern *Eupomatia*. Nevertheless, even though it is tempting to interpret taxon "A" as beetle-pollinated, its pollination syndrome may well have differed considerably from that of modern *Eupomatia*.

While not as completely preserved, flowers of taxon "B" also suggest features of beetle-pollinated modern taxa as well. Other Turonian fossil flowers with calycanthoid characters additionally suggest that beetle pollination in *Magnoliales* may have been somewhat diverse by the Turonian (*Calycanthus* is today pollinated by *Nitidulidae*; GRANT 1950). We consider the implications of these flowers and others with respect to the evolution of insect pollination in greater detail in a separate publication (CREPET & NIXON, unpubl.).

We thank M. A. GANDOLFO for assistance with laboratory work, photography and literature searches; JENNIFER SVITKO for electron microscopy. QUENTIN D. WHEELER, and JOE V. MCHUGH, of the department of Entomology at Cornell University and DAVID GRIMALDI, Department of Entomology of the American Museum of Natural History, New York for discussion and tentative identification of fossil *Coleoptera*. Research supported by NSF DEB92-01179 and CALS, Cornell University, Ithaca, New York.

## References

ARCHANGELSKY, S., TAYLOR, T. N., 1993: The ultrastructure of in situ *Clavatipollenites* pollen from the Early Cretaceous of Patagonia. – Amer. J. Bot. **80**: 879–885.

ARMSTRONG, J. E., IRVINE, A. K., 1990: Functions of staminodia in the beetle-pollinated flowers of *Eupomatia laurina*. – Biotropica **22**: 429–431.

BERGSTRÖM, B., GROTH, I., PELLMYR, O., ENDRESS, P. K., THIEN, L. B., HÜBENER, A., FRANCKE, W., 1991: Chemical basis of a highly specific mutualism: chiral esters attract pollinating beetles in *Eupomatiaceae*. – Phytochemistry **30**: 3221–3225.

BESSEY, C. E., 1915: The phylogenetic taxonomy of flowering plants. – Ann. Missouri Bot. Gard. **2**: 109–164.

BURGER, W. C., 1977: The *Piperales* and the monocots. Alternate hypotheses for the origin of monocotyledonous flowers. – Bot. Rev. **43**: 345–393.

– 1981: Heresy revised: the monocot theory of angiosperm origin. – Evol. Theory **5**: 189–225.

CARPENTER, F. M., 1992: Superclass *Hexapoda*. – In KAESLER, R. L., (Ed.): Treatise on invertebrate paleontology, **3,4**, pp. – Geological Society of America. – University of Kansas Press.

CHAPMAN, J. L., 1987: Comparison of *Chloranthaceae* pollen with the Cretaceous "*Clavatipollenites* complex": taxonomic implications for palaeopalynology. – Pollen Spores **29**: 249–272.

CHLONOVA, A. F., SUROVA, T. D., 1988: Pollen wall ultrastructure of *Clavatipollenites incisus Chlonova* and two modern species of *Ascarina* (*Chloranthaceae*). – Pollen Spores **30**: 29–44.

COULTER, J. M., CHAMBERLAIN, C. J., 1917: Morphology of gymnosperms. – Chicago: University of Chicago Press.

COUPER, R. A., 1958: British Mesozoic microspores and pollen grains. – Palaeontogr. **103B**: 75–179.

CRANE, P. R., 1985: Phylogenetic analysis of seed plants and the origin of angiosperms. – Ann. Missouri Bot. Gard. **72**: 716–793.

– DILCHER, D. L., 1984: *Lesqueria*: An early angiosperm fruiting axis from the mid-Cretaceous. – Ann. Missouri Bot. Gard. **71**: 384–402.

– FRIIS, E. M., PEDERSEN, K. R., 1989: Reproductive structure and function in Cretaceous *Chloranthaceae*. – Pl. Syst. Evol. **165**: 211–226.

– – – 1994: Palaeobotanical evidence on the early radiation of magnoliid angiosperms. – Pl. Syst. Evol. [Suppl.] **8**: 51–72.

CREPET, W. L., NIXON, K. C., 1995: The fossil history of stamens. – In D'ARCY, W. C., KEATING, R. G., (Eds): The anther, form function and phylogeny. – New York: Cambridge University Press (in press).

– FRIIS, E. M., NIXON, K. C., 1991: Fossil evidence for the evolution of biotic pollination. – Philos. Trans. Roy. Soc. London **333B**: 187–195.

– NIXON, K. C., FRIIS, E. M., FREUDENSTEIN, J. V., 1992: Oldest fossil flowers of hamamelidaceous affinity, from the Late Cretaceous of New Jersey. – Proc. Natl. Acad. Sci. USA **89**: 8986–8989.

CRONQUIST, A., 1981: An integrated system of classification of flowering plants. – New York: Columbia University Press.

DAHLGREN, R., 1983: General aspects of angiosperm evolution and macrosystematics. – Palaeobot. Palynol. **27**: 291–328.

– CRANE, P. R., 1984: *Archaeanthus*: an early angiosperm from the Cenomanian of the Western Interior of North America. – Ann. Missouri Bot. Gard. **71**: 351–383.

– CREPET, W. L., BEEKER, C. B., REYNOLDS, H., 1976: The reproductive and vegetative morphology of a Cretaceous angiosperm. – Science **191**: 854–856.

DONOGHUE, M. J., DOYLE, J. A., 1989: Phylogenetic analysis of angiosperms and the relationships of the *Hamamelididae*. – In CRANE, P. R., BLACKMORE, S., (Eds): Evolution, systematics, and fossil history of the *Hamamelididae*, **I**: Inroduction and "lower" *Hamamelididae*, pp. 17–45. – Oxford: Clarendon.

Doyle, J. A., Hotton, C. L., 1991: Diversification of early angiosperm pollen in a cladistic context. – In Blackmore, S., Barnes, S. H., (Eds): Pollen and spores: patterns of diversification, pp. 168–195. – Oxford: Clarendon.

– Robinson, E. I., 1977: Angiosperm pollen zonation of the continental Cretaceous of the Atlantic Coastal Plain and its application to deep wells in th Salisbury Embayment. – Palynology **1**: 43–78.

Drinnan, A. N., Crane, P. R., Friis, E. M., Pedersen, K. R., 1990: Lauraceous flowers from the Potomac Group (mid-Cretaceous) of eastern North America. – Bot. Gaz. **151**. 370–380.

Endress, P. K., 1986: Reproductive structures and phylogenetic significance of extant primitive angiosperms. – Pl. Syst. Evol. **152**: 1–28.

– 1990: Evolution of reproductive structures and functions in primitive angiosperms (*Magnoliidae*) – Mem. New York Bot. Gard. **55**: 5–34.

– Hufford, L. D., 1989: The diversity of stamen structures and dehiscence patterns among *Magnoliidae*. – J. Linn. Soc. Bot. **100**: 45–85.

Friis, E. M., Crane, P. R., Pedersen, K. R., 1986: Floral evidence for Cretaceous chloranthoid angiosperms. – Nature **320**: 163–164.

– – – 1991: Stamen diversity and in situ pollen of Cretaceous angiosperms. – In Blackmore, S. Barnes, S. H., (Eds): Pollen and spores: patterns of diversification, pp. 197–224. – Oxford: Clarendon.

– Endress, P. K., 1990: Origin and evolution of angiosperm flowers. – Adv. Bot. Res. **17**: 99–162.

– Pedersen, K. R., Crane, P. R., 1994: Angiosperm floral structures from the Early Cretaceous of Portugal. Pl. Syst. Evol. [Suppl.] **8**: 31–49.

Gottsberger, G., 1974: The structure and function of the primitive angiosperm flower – a discussion. – Acta. Bot. Neerl. **23**: 461–471.

Grant, V., 1950: The pollination of *Calycanthus occidentalis*. – Amer. J. Bot. **37**: 294–296.

Herendeen, P. S., Crepet, W. L., Nixon, K. C., 1993: *Chloranthus*-like stamens from the Upper Cretaceous of New Jersey. – Amer. J. Bot. **80**: 865–871.

– – – 1994: Fossil flowers and pollen of *Lauraceae* from the Upper Cretaceous of New Jersey. – Pl. Syst. Evol. **189**: 29–40.

Hesse, M., Kubitzki, K., 1983: The sporoderm ultrastructure in *Persea*, *Nectandra*, *Hernandia*, *Gomortega* and some other lauralean genera. – Pl. Syst. Evol. **141**: 299–311.

Hughes, N. F., Drewry, C. E., Lang, J. F., 1979: Barremian earliest angiosperm pollen. – Palaeontology **22**: 513–535.

Kemp, E. M., 1968: Probable angiosperm pollen from Biritish Barremian to Albian strata. – Palaeontology **11**: 421–434.

Kubitzki, K., 1981: The tubular exine of *Lauraceae* and *Hernandiaceae*, a novel type of exine structure in seed plants. – Pl. Syst. Evol. **138**: 139–146.

– Rohwer, J. G., Bittrich, V. (Eds), 1993: The families and genera of flowering plants, **II**. – Berlin: Springer.

McGinnes, E. A., Jr., Kandeel, S. A., Szopa, P. S., 1971: Some structural changes observed in the transformation of wood into charcoal. – WOOD Fiber **3**: 77–83.

– Szopa, P. S., Phelps, J. E., 1974: Use of scanning electron microscopy in studies of wood charcoal formation. – Scanning Electron Microscopy **1974**: 469–476.

Nixon, K. C., Crepet, W. L., 1993: Late Cretaceous fossil flowers of ericalean affinity. – Amer. J. Bot. **80**: 616–623.

– Wheeler, Q. D., 1992a: Extinction and the origin of species. – In Novacek, M. J., Wheeler, Q. D., (Eds): Extinction and phylogeny, pp. 119–143. – New York: Columbia University Press.

– – 1992b: Measures of phylogenetic diversity. – In NOVACEK, M. J., WHEELER, Q. D., (Eds): Extinction and phylogeny, pp. 216–234. – New York: Columbia University Press.
– CREPET, W. L., STEVENSON, D. M., FRIIS, E. M., 1994: A reevaluation of seed plant phylogeny. – Ann. Missouri Bot. Gard. **81**: 484–533.
PEDERSEN, K. R., CRANE, P. R., DRINNAN, A. N., FRIIS, E. M., 1991: Fruits from the mid-Cretaceous of North America with pollen grains of the *Clavatipollenites* type. – Grana **30**: 577–590.
TAKHTAJAN, A. L., 1980: Outline of the classification of flowering plants (*Magnoliophyta*). – Bot. Rev. **46**: 225–359.
TAYLOR, D. W., HICKEY, L. J., 1990: An Aptian plant with attached leaves and flowers: implications for angiosperm origin. – Science **247**: 702–704.
THIEN, L. H., 1980: Patterns of pollination in primitive angiosperns. – Biotropica **12**: 1–13.
TIFFNEY, B. H., 1984: Seed size, dispersal syndromes, and the rise of the angiosperms: evidence and hypothesis. – Ann. Missouri Bot. Gard. **71**: 551–576.
UPCHURCH, G. R. Jr., 1984: Cuticle evolution in Early Cretaceous angiosperms from the Potomac Group of Virginia and Maryland. – Ann. Missouri Bot. Gard. **71**: 522–550.
– DILCHER, D. L., 1990: Cenomanian angiosperm leaf megafossils, Dakota Formation, Rose Creek Locality, Jefferson County, southeastern Nebraska. – US Geol. Surv. Bull. **1915**: 1–55.
WALKER, J. W., DOYLE, J. A., 1975: The bases of angiosperm phylogeny: palynology. – Ann. Missouri Bot. Gard. **62**: 664–723.
– WALKER, A. G., 1984: Ultrastructure of Lower Cretaceous angiosperm pollen and the origin and early evolution of flowering plants. – Ann. Missouri Bot. Gard. **71**: 464–521.
– BRENNER, G. J., WALKER, A. G., 1983: Winteraceous pollen in the Lower Cretaceous of Israel: early evidence of a magnolialean angiosperm family. – Science **220**: 1273–1275.
WARD, J. V., DOYLE, J. A., HOTTON, C. L., 1989: Probable granular magnoliid angiosperm pollen from the early Cretaceous. – Pollen Spores **33**: 101–120.

Address of the authors: WILLIAM L. CREPET and KEVIN C. NIXON, L. H. Bailey Hortorium, Cornell University, Ithaca, NY 14853, USA.

Accepted May 2, 1994 by P. K. ENDRESS

Pl. Syst. Evol. [Suppl.] 8: 93–122 (1994)

# Patterns of floral evolution in the early diversification of non-magnoliid dicotyledons (eudicots)

ANDREW N. DRINNAN, PETER R. CRANE, and SARA B. HOOT

Received April 18, 1994

**Key words:** Eudicots, *Hamamelididae*, *Ranunculidae*. – Floral evolution, floral structure, fossil record, molecular phylogenetics.

**Abstract:** Recent cladistic analyses of angiosperms based on both morphological and molecular sequence data recognize a major clade of dicotyledons defined by triaperturate or triaperturate-derived pollen (non-magnoliids/eudicots). Evidence from morphology, as well as the *atp*B and *rbc*L genes (cpDNA), indicates that extant *Ranunculidae* (e.g., *Papaverales*, *Lardizabalaceae*, *Berberidaceae*, *Menispermaceae*, *Ranunculaceae*) as well as "lower" *Hamamelididae* [e.g., *Eupteleaceae* (allied to *Ranunculidae*), *Hamamelidaceae*, *Myrothamnaceae*, *Platanaceae*, *Trochodendraceae*] and several other families (e.g., *Gunneraceae*, *Nelumbonaceae*, *Proteaceae*, *Sabiaceae*) are basal in this group. The earliest records of diagnostic eudicot pollen are of mid-late Barremian age (c. 126 myr BP) and by around the latest Albian (c. 97 myr BP) several basal eudicot groups (e.g., *Trochodendrales*, *Platanaceae*, *Buxaceae*, and perhaps *Circaeasteraceae*, *Myrothamnaceae*, and *Nelumbonaceae*) are recognizable in the fossil record. Possible *Hamamelidaceae* and perhaps *Proteaceae* are present by the Turonian (c. 90 myr BP). Among basal eudicots, flowers are generally bisexual although unisexual flowers are also common. In some groups (e.g., *Myrothamnaceae*, *Buxaceae*, certain *Berberidaceae*), delimitation of the flower is not always clear and there is a more or less gradual transition between tepals and inflorescence bracts. Plasticity in floral form at this level of angiosperm evolution is predominantly encompassed by dimerous and trimerous cyclic floral organization and transitions from one to the other are common. Spiral floral phyllotaxis of numerous stamens and carpels is more or less restricted to the *Ranunculaceae*. The basic condition of the perianth in eudicots appears to lack differentiation into sepals and petals, and petals appear to have arisen independently numerous times from stamens. Based on the generality of its systematic distribution, cyclic floral architecture is probably basic for eudicots as a whole, and at this level of angiosperm evolution flowers with numerous, helically-arranged stamens and/or carpels (e.g., many *Ranunculaceae*) almost certainly reflect processes of secondary multiplication that have occurred independently many times.

A significant result from recent cladistic analyses of basal angiosperms has been the recognition that the traditional distinction between monocotyledons and dicotyledons does not accurately reflect probable phylogenetic patterns. Monocots appear monophyletic, and can be defined by the single cotyledon, and perhaps the

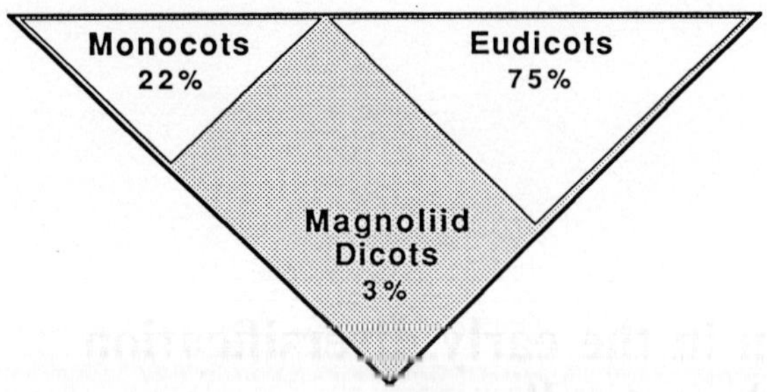


Fig. 1. Model of higher-level relationships among angiosperms, based on recent cladistic analyses of morphological and molecular data. See text for additional details. Percentages correspond to the proportion of extant angiosperm species in the three major groups. Data on species diversity based on MABBERLEY (1987)

triangular proteinaceous inclusions in their sieve tube plastids, as well as other features (DAHLGREN & al. 1985, BHARATHAN & ZIMMER 1994). Dicots, however, appear to lack unique characters and some dicot families (e.g., *Aristolochiaceae*, *Nymphaeaceae*) may be more closely related to monocots (DONOGHUE & DOYLE 1989a, b; DOYLE & DONOGHUE 1992, 1993; LOCONTE & STEVENSON 1991; CHASE & al. 1993; BHARATHAN & ZIMMER 1994). Most dicots, however, appear to comprise a major clade (eudicots or non-magnoliid dicots; CRANE 1989; DONOGHUE & DOYLE 1989a, b; DOYLE & HOTTON 1991) that includes the subclasses *Ranunculidae*, *Caryophyllidae*, *Hamamelididae*, *Rosidae*, *Dilleniidae*, *Lamiidae*, and *Asteridae* sensu TAKHATAJAN (1987). This clade is supported by analyses of molecular data (CHASE & al. 1993; HOOT & CRANE, unpubl.), and is defined morphologically by the presence of triaperturate (or triaperturate-derived) pollen (see DOYLE & HOTTON 1991 for discussion of the situation in *Illicium*). In this context, a useful model of angiosperm relationships interprets magnoliids (*Magnoliidae* sensu TAKHTAJAN 1987) as a depauperate basal grade of angiosperms that are variously related to two major clades: monocots and eudicots (Fig. 1).

Considerable recent effort has been directed toward resolving relationships at the magnoliid grade (DONOGHUE & DOYLE 1989a, b; LOCONTE & STEVENSON 1991; DOYLE & DONOGHUE 1992, 1993; DOYLE & al. 1994), but phylogenetic patterns at this level remain uncertain because of apparently conflicting results from different morphological and molecular data sets, as well as incomplete integration of existing morphological and molecular information. It has also proved difficult to root the angiosperm tree because of uncertainties over the appropriate outgroup, and the absence of critical features of angiosperms in potential outgroup taxa (DONOGHUE & DOYLE 1989b, DOYLE & DONOGHUE 1993, DOYLE & al. 1994).

A further problem has been how to represent certain very diverse groups of angiosperms (e.g., *Annonaceae*, *Monimiaceae*) in analyses at the magnoliid level, and this issue is particularly acute for the eudicots as a whole, which account for approximately 75% of extant angiosperm species. In this paper we briefly review recent attempts to understand relationships at the base of the eudicot clade as a basis for incorporating this group into phylogenetic analyses focused on magnoliids, and also to provide a starting point for resolving phylogenetic patterns among the bulk of extant dicotyledons. We present preliminary results from phylogenetic analyses of "basal" eudicots based on combined *rbc*L and *atp*B sequence data (chloroplast genome). We then use this phylogenetic model to review floral structure in extant and fossil taxa at this level of angiosperm evolution and offer preliminary ideas on the evolution of floral form among these groups.

## Relationships at the base of eudicots

**Hypotheses based on morphological data.** The genera *Trochodendron* and *Tetracentron* have figured prominently in ideas on the early evolution of eudicots because they appear to retain a variety of unmodified (plesiomorphic) features (e.g., ethereal oil cells, chloranthoid leaf teeth, dorsal carpel nectaries, valvate anther dehiscence) from putative precursors at the magnoliid grade (ENDRESS 1986a, CRANE 1989). In a preliminary phylogenetic diagram, CRANE (1989) placed *Trochodendron* and *Tetracentron* (*Trochodendraceae*) as sister taxa, and together these formed the sister group to all other non-magnoliid dicotyledons. Similarly, in an analysis of relationships among 85 eudicots based on morphological and chemical data, HUFFORD (1992) placed *Trochodendraceae* as the sister group to all other non-magnoliid dicots (see also CRANE 1989, HUFFORD & CRANE 1989).

The *Ranunculidae* sensu TAKHTAJAN (1987) have received less consideration than 'lower' hamamelidids in discussions of eudicot phylogeny, despite the fact that they uniformly produce triaperturate or triaperturate-derived pollen, and have often been placed in the *Magnoliidae* based on their perceived retention of primitive floral features (e.g., CRONQUIST 1981). Frequently, the *Ranunculidae* have been connected to magnoliids sensu stricto via the *Illiciales* (CRONQUIST 1981), and clearly need to be accounted for in analyses of relationships at the base of the eudicot clade.

**Hypotheses based on molecular data.** Despite the practical difficulties of analyzing such a large data set, recent comparative studies of nucleotide sequence data from the *rbc*L gene (chloroplast genome) for approximately 500 taxa currently provide the most comprehensive hypothesis of possible relationships among basal eudicots (CHASE & al. 1993). These cladograms recognize a ranunculid clade (including *Papaveraceae*, *Fumariaceae*, *Eupteleaceae*, *Lardizabalaceae*, *Menispermaceae*, *Ranunculaceae*, and *Berberidaceae*), which is resolved as the sister group to all other eudicots. A clade consisting of *Sabiaceae*, *Proteaceae*, *Nelumbonaceae*, and *Platanaceae* is basal among this diverse assemblage of non-ranunculid eudicots. *Trochodendraceae* plus *Buxaceae* are resolved as the sister group to *Gunneraceae* plus "higher eudicots" sensu CHASE & al. (1993). Despite previous suggestions of a relationship between caryophyllids and ranunculids (e.g., TAKHTAJAN 1987), caryophyllid taxa are placed well within the eudicot clade in these *rbc*L analyses. The pattern of relationships based on *rbc*L data is also broadly consistent with recent cladistic analysis of restriction site variation in the inverted repeat region of the chloroplast genome (MANOS & al. 1993). Despite limited sampling of key taxa, the restriction site analysis resolved *Euptelea* as basal, and *Platanus* as the sister group to *Trochodendron* plus "higher eudicots".

**Preliminary hypothesis based on combined *rbc*L and *atp*B cpDNA sequence data.** Over the last two years, using a sampling strategy based on results from the inclusive *rbc*L analysis as well as previous morphologically-based phylogenetic

Table 1. Comparison of taxa of basal eudicots sampled in the analysis of *rbc*L sequence data by Chase & al. (1993) with species sequenced for both *rbc*L and *atp*B genes in this analysis. **rbc*L data accessed from GenBank.

| Family | *rbc*L (Chase & al. 1993) | *rbc*L and *atp*B (this study) |
|---|---|---|
| *Eupteleaceae* | *Euptelea* | *Euptelea polyandra** |
| *Papaveraceae* | *Papaver* | *Eschscholzia californica* |
| | *Sanguinaria* | *Dendromecon rigida* |
| | | *Platystemon californicus* |
| | | *Sanguinaria canadensis** |
| *Fumariaceae* | *Dicentra* | *Corydalis nobilis* |
| | | *Dicentra eximia* |
| *Hypecoaceae* | – | *Hypecoum imberbe* |
| *Pteridophyllaceae* | – | *Pteridophyllum racemosum* |
| *Circaeasteraceae* | – | *Circaeaster agrestis* |
| *Kingdoniaceae* | – | *Kingdonia uniflora* |
| *Lardizabalaceae* | *Akebia* | *Akebia quinata** |
| | | *Decaisnea fargesii* |
| | | *Sinofranchetia chinensis* |
| *Berberidaceae* | *Caulophyllum* | *Caulophyllum thalictroides** |
| | *Mahonia* | *Nandina domestica* |
| *Menispermaceae* | *Cocculus* | *Menispermum canadensis* |
| | | *Tinospora caffra* |
| *Glaucidiaceae* | – | *Glaucidium palmatum* |
| *Hydrastidaceae* | – | *Hydrastis canadensis* |
| *Ranunculaceae* | *Caltha* | *Caltha palustris* |
| | *Ranunculus* | *Coptis trifolia* |
| | *Xanthorhiza* | *Ranunculus trichophyllus** |
| | | *Ranunculus hispidus* |
| | | *Xanthorhiza simplicissima** |
| *Platanaceae* | *Platanus* | *Platanus occidentalis** |
| *Sabiaceae* | *Sabia* | *Sabia japonica** |
| | | *Sabia swinhoei* |
| *Nelumbonaceae* | *Nelumbo* | *Nelumbo lutea** |
| *Proteaceae* | *Lambertia* | *Placospermum coriaceum* |
| | | *Roupala macrophylla* |
| *Trochodendraceae* | *Trochodendron* | *Trochodendron aralioides** |
| | *Tetracentron* | *Tetracentron sinensis** |
| *Cercidiphyllaceae* | *Cercidiphyllum* | *Cercidiphyllum japonicum** |
| *Buxaceae* | *Pachysandra* | *Buxus sempervirens* |
| | | *Pachysandra terminalis* |
| *Myrothamnaceae* | – | *Myrothamnus flabellifolia* |
| *Daphniphyllaceae* | *Daphniphyllum* | *Daphniphyllum* sp.* |
| *Hamamelidaceae* | *Altingia* | *Hamamelis mollis** |
| | *Hamamelis* | *Hamamelis virginiana* |
| | *Liquidambar* | |
| | *Rhodoleia* | |
| *Gunneraceae* | *Gunnera* | *Gunnera hamiltonii* |

and paleobotanical studies, we have obtained *rbc*L and *atp*B sequence data for approximately 100 species representing a broad range of basal eudicots. These data continue to be extended and a parallel data set for the 18S rDNA gene is also being compiled (HOOT & CRANE, unpubl.). Initial comparisons of results from the three different genes in the ranunculid family *Lardizabalaceae*, show that the 18S gene is the most conserved, the *atp*B gene is intermediate, and that the *rbc*L gene is the least conserved (HOOT & al. 1994).

For purposes of this analysis we use *Illicium* as a magnoliid outgroup (see below for experiments with other outgroups). The ingroup comprises less than half of our current data set, and includes species from 38 genera of putatively basal eudicots representing 24 families. By comparison, the recent analysis of *rbc*L data alone (CHASE & al. 1993) included species from 25 genera representing 17 families at this level of angiosperm evolution (Table 1). We use *Hamamelis*, *Cercidiphyllum*, *Daphniphyllum*, *Myrothamnus*, and *Gunnera* as "placeholders" for the bulk of the eudicots ("higher" eudicots).

The only families of *Ranunculidae* sensu TAKHTAJAN (1987) that are not represented in this preliminary analysis are *Sargentodoxaceae*, because of unavailability of adequate material, and *Paeoniaceae* because of their relatively derived position close to *Crassulaceae* in current *rbc*L analyses (CHASE & al. 1993). Other than segregate families of *Hamamelidaceae* (e.g., *Rhodoleiaceae*, *Altingiaceae*), only four families of potential "lower" *Hamamelididae* (based on TAKHTAJAN 1987) are not represented in this preliminary analysis. *Balanopaceae* and *Didymelaceae* were not considered because of lack of material, *Styloceratаceae* were excluded here because sequencing is not yet complete, and *Simmondsiaceae* and *Eucommiaceae* were not incorporated because of their relatively derived position in previous analyses (CHASE & al. 1993, HOOT & CRANE, unpubl.).

Technical details of DNA extraction, amplification and sequencing, as well as quality control of sequencing data, are provided in HOOT & al. (1994). The length of the combined *rbc*L (1,397bp) and *atp*B (1,468bp) sequences is approximately 2,865 base pairs, with 892 variable sites (532 potentially informative characters) in this data.

Analyses were performed on a Macintosh Quadra 840 using the heuristic search option of PAUP 3.1.1, using 100 random order replicate searches with the TBR search and MULPARS options (SWOFFORD 1993). To assess support for the various clades, bootstrap analysis with 100 replicates and decay analysis (examining trees up to three steps longer than the shortest tree) were conducted. Parsimony analysis of the combined *atp*B and *rbc*L data resulted in 53 equally parsimonious trees of 2,275 steps (CI excluding autapomorphies = 0.41, RI = 0.51). One of these most parsimonious trees is illustrated in Fig. 2 showing the number of synapomorphies and decay indices for each node. The strict consensus of the 53 trees is illustrated in Fig. 3. Experiments using *Houttuynia* (*Saururaceae*) and *Asarum* (*Aristolochiaceae*) as outgroups (either separately or in combination) in place of *Illicium*, yielded trees with topologies identical to one or more of the most parsimonious trees.

These phylogenetic results should be regarded as preliminary, pending more intensive systematic sampling and more detailed analysis (HOOT & CRANE, unpubl.). Nevertheless, the resulting consensus tree provides a useful model with which to examine patterns of floral evolution. The primary structure of the consensus

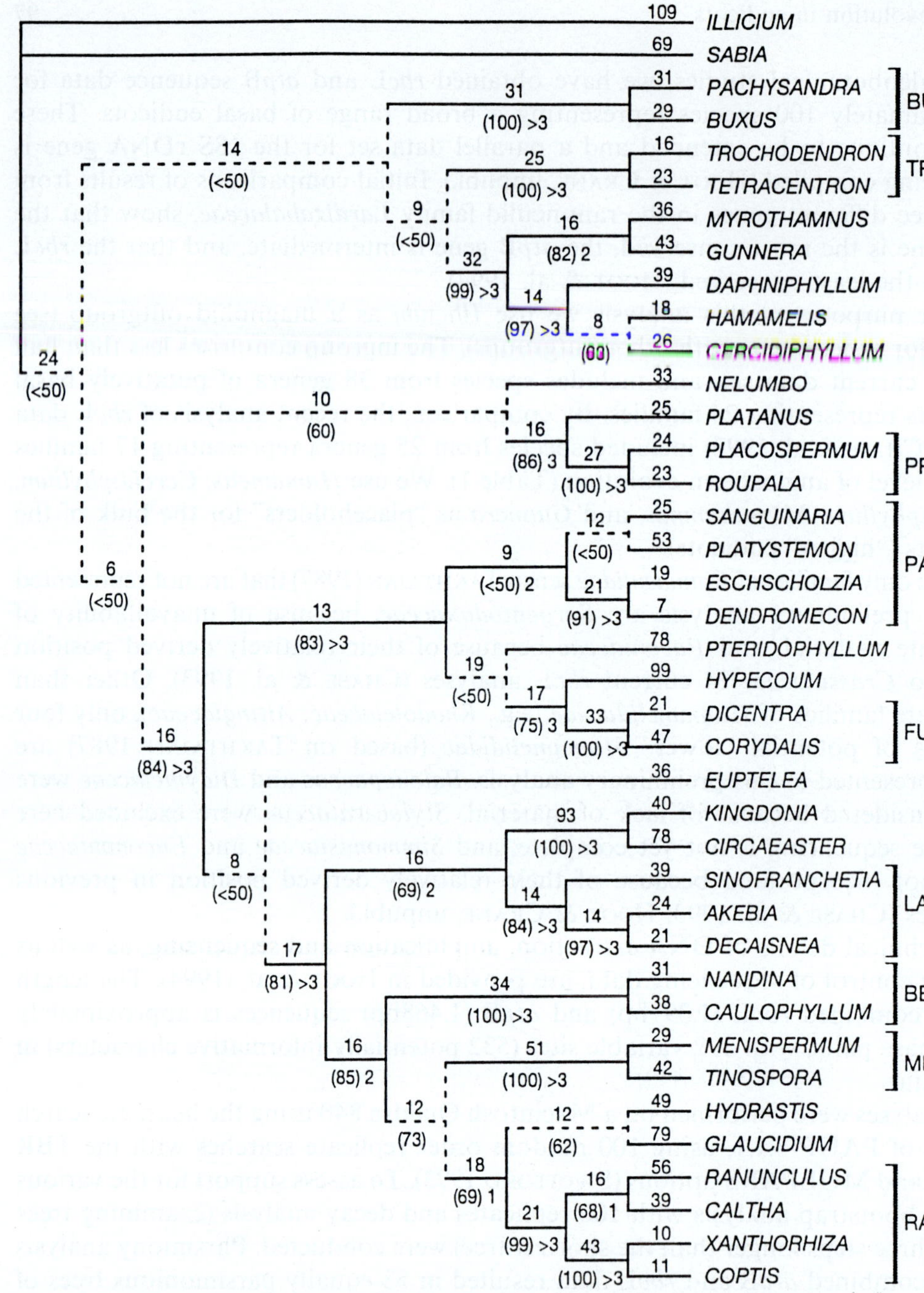


Fig. 2. One of the most parsimonious trees resulting from a preliminary cladistic analysis based on a combination of *rbc*L and *atp*B sequence data for families and taxa indicated in Table 1. Dotted lines indicate branches which collapse in the strict consensus tree derived from the 53 shortest trees. Tree length = 2275 steps, CI excluding autapomorphies = 0.41, RI = 0.51. Numerals above branches indicate the number of nucleotide changes supporting each branch. Numerals below in parentheses indicate the percentage of times that the branch was recovered in 100 bootstrap replications. Numerals below and to the right of the bootstrap values are decay indices. Acronyms after brackets indicate families with more than one genus represented in this study (BUXA *Buxaceae*, TROC *Trochodendraceae*, PROT *Proteaceae*, PAPA *Papaveraceae*, FUMA *Fumariaceae*, LARD *Lardizabalaceae*, BERB *Berberidaceae*, MENI *Menispermaceae*, RANU *Ranunculaceae*)

cladogram separates the eudicots into one major clade and six other lineages in a polychotomy (Figs. 2, 3). This is broadly consistent with previous results based on *rbc*L data (Chase & al. 1993). One clade includes most of the *Ranuculidae* sensu Takhtajan (1987) as well as *Euptelea*, which is often treated in the *Hamamelididae* allied to "lower" hamamelidids such as *Platanus*, *Cercidiphyllum* and *Trochodendron* (Cronquist 1981, Endress 1986a, Takhtajan 1987). Among ranunculids, three groups can be recognized; *Papaveraceae*, *Fumariaceae*, *Hypecoaceae*, and *Pteridophyllaceae*; *Euptelea* alone; and all other ranunculids. In the latter group, *Circaeasteraceae*, *Kingdoniaceae*, and *Lardizabalaceae* are sister to *Berberidaceae*, *Menispermaceae*, *Glaucidiaceae*, *Hydrastidaceae*, and *Ranunculaceae* (Figs. 2, 3). The six groups united with the ranunculids in the polychotomy include a variety of "lower" hamamelidids and other taxa. All the "placeholders" included for the "higher" eudicots (*Cercidiphyllum*, *Daphniphyllum*, *Gunnera*, *Hamamelis*, *Myrothamnus*) form a single clade. Another clade comprises *Platanaceae* and *Proteaceae*. The remaining four taxa in the basal polychotomy are *Buxaceae*, *Nelumbonaceae*, *Trochodendraceae* and *Sabiaceae* (Figs. 2, 3).

## Floral structure: *Ranunculidae*

***Eupteleaceae.*** The *Eupteleaceae* comprise a single genus *Euptelea* with two species, *E. pleiosperma* (northeastern India to China) and *E. polyandra* (Japan) (Smith 1946). The family is usually placed close to *Trochodendraceae*, *Cercidiphyllaceae*, and *Platanaceae* (Endress 1986a, 1993, Cronquist 1981), although similarities in carpel structure to *Schisandraceae* and *Winteraceae* have also been noted (Endress 1969, 1993). Our preliminary molecular phylogenetic analyses, and those of Chase & al. (1993), place *Euptelea* as more closely related to the *Ranunculidae* than other putative "lower" hamamelidids (Figs. 2, 3). The *Eupteleaceae* currently have no Cretaceous fossil record, although there are apparently reliable reports of leaves and wood from the Eocene (Collinson & al. 1993).

Flowers of *Euptelea* are bisexual and floral structure is similar in both species. Flowers lack a perianth, although there may be two small prophylls at the base of the lower flowers in the inflorescence (Endress 1969, 1986a). The number of stamens (six to 19) and stalked carpels (eight to 31) is variable and their arrangement is irregular. The gynoecium is apocarpous. The floral primordium is bilaterally symmetrical during development (Endress 1986a).

***Papaveraceae.*** The *Papaveraceae* are almost exclusively northern hemispheric in their distribution and comprise approximately 240 species divided among 23 genera (Kadereit 1993). Recent treatments recognize four subfamilies: *Chelidonioideae*, *Eschscholzioideae*, *Platystemonoideae*, and *Papaveroideae* (Kadereit 1993). The *Eschscholzioideae* and *Platystemonoideae* occur exclusively in the New World, and predominantly in western North America, while the *Chelidonioideae* and *Papaveroideae* are present in both the Old and New World. It is widely agreed that the *Papaveraceae* are closely related to *Fumariaceae* (e.g., Cronquist 1981, Takhatajan 1987) and this is also supported by current phylogenetic hypotheses based on *rbc*L data (Chase & al. 1993). Opinions differ, however, on the position of *Papaverales* (*Papaveraceae*, *Fumariaceae*) among the *Ranunculidae*. Recent cladistic analysis based on morphological data resolves the *Papaverales* as sister group to the *Ranunculaceae* (LOCONTE & STEVENSON 1991), in a relatively derived position within the *Ranunculidae*. In contrast, cladistic analyses based on *rbc*L data (CHASE & al. 1993) resolve *Papaverales* as the basal clade in the ranunculids. In our preliminary molecular phylogenetic analyses based on combined *rbc*L and *atp*B sequences (Figs. 2, 3) the *Papaverales* sensu lato are part of a trichotomy with *Euptelea* and other ranunculids. The *Papaveraceae* have no reliable fossil record, and flowers of *Princetonia allenbyensis* from the Middle Eocene of British Columbia, Canada, that have been compared with *Papaveraceae* are unlikely to be related to this family (axile rather than parietal placentation, STOCKEY & PIGG 1991).

Flowers of *Papaveraceae* are bisexual and cyclic. Typically there is one dimerous whorl of sepals, and two dimerous whorls of petals. Trimerous flowers are less common, but are characteristic of *Platystemonoideae* and also occur in *Papaveroideae* (KADEREIT 1993). Frequently dimerous and trimerous flowers are present in the same genus (e.g., *Meconella*, *Dendromecon*, many *Papaveroideae*, MUNZ 1959, KADEREIT 1993). Petals are lacking in *Bocconia* and *Macleaya*, while *Sanguinaria* is unique in being polypetalous with additional petals occuring in the position of stamens (KADEREIT 1993, LEHMANN & SATTLER 1993). Stamens are usually numerous, and arise centripetally. Studies of floral development (e.g., PAYER 1857, MURBECK 1912, SATTLER 1973, ENDRESS 1987, RONSE DECRAENE & SMETS 1990, KARRER 1991, LEHMANN & SATTLER 1993) show that the centripetal sequence frequently arises in a regular, whorled pattern that reflects the dimerous or trimerous arrangement of the petals. The gynoecium in *Papaveraceae* shows significant variation. In the *Papaveroideae* it is superficially ("internally") multi-carpellate and unilocular. In *Platystemonoideae* both tricarpellate (*Meconella*, *Hespermecon*) and superficially multicarpellate (six to 25 in *Platystemon*) but unilocular gynoecia occur. In *Meconella* there are typically three sepals, six petals, six to 12 stamens and a unilocular, tripartite ovary. In the *Chelidonioideae*, *Eschscholzioideae* the gynoecium is internally bicarpellate, except for *Stylophorum diphyllum* in which the gynoecium may be up to "five-carpellate" (KADEREIT 1993).

Floral architecture in most *Papaveraceae* is therefore cyclic, dimerous and organized on an opposite and decussate plan. However, cyclic, trimerous flowers are also an important feature of the family (especially *Platystemonoideae*) and there is variation at low taxonomic levels between dimerous and trimerous forms. Polypetaly in *Sanguinaria* is apparently a specialized condition that has arisen by homeotic modification of stamen primordia (LEHMANN & SATTLER 1993). The occurrence of numerous stamens (by increasing the number of stamen whorls per androecium, and the number of stamens per whorl) and numerous "carpels" (by increasing the number of placentae in a single ovary) is almost certainly secondary within the group.

***Fumariaceae.*** The *Fumariaceae* (excluding *Hypecoaceae*) comprise approximately 500 species in 16 genera that are divided among two tribes; *Fumarieae* with three subtribes (*Sarcocapninae*, *Fumariinae*, *Discocapninae*) and *Corydaleae* (LIDÉN 1993). The genus *Cysticapnos* is of uncertain affinity (LIDÉN 1993). The family is predominantly Northern Hemispheric in distribution with major centers of diversity in the Mediterranean and eastern Asia. Frequently the family is treated as the subfamily *Fumarioideae* of the *Papaveraceae*, to which it is clearly closely related. It is also often taken to include the *Hypecoaceae* (LIDÉN 1993). Our preliminary molecular phylogenetic studies, as well as those of CHASE & al. (1993), place the *Fumariaceae* sensu stricto as the sister group to *Hypecoaceae*. This clade (*Fumariaceae* sensu lato) is then placed as part of a trichotomy with *Pteridophyllum* and *Papaveraceae* sensu stricto (Figs. 2, 3). The fossil record of the family is poor with only a few scattered reports from the Tertiary (COLLINSON & al. 1993).

Flowers in the *Fumariaceae* are bisexual, cyclic and dimerous, with two petaloid sepals in the anterior-posterior positions and four petals in two opposite and decussate pairs. There are two stamen "bundles" opposite each other in the lateral positions. Each bundle consists of a central dithecate part and two lateral monothecate parts, the homology of which has been widely discussed, particularly with reference to the situation in *Hypecoum* (e.g., MURBECK 1912, ROHWEDER & ENDRESS 1983, RONSE DECRAENE & SMETS 1992). One interpretation (e.g., ROHWEDER & ENDRESS 1983) is that there are four stamens; the stamens in the lateral positions are bithecate (tetrasporangiate), and their lateral monothecate components are formed from half of each of the two stamens in the anterior-posterior position. Thus the two stamens in the anterior-posterior positions each appear to be deeply divided into two monothecate (bisporangiate) halves that fuse to the lateral stamens. An alternative interpretation (e.g., ARBER 1931, RONSE DECRAENE & SMETS 1992) is that the androecium is comprised of six stamens, and that the monothecate condition has arisen as a result of spatial factors during development (rather than splitting of a single stamen). The ovary is unilocular with a bifid style and two placentae. It is generally interpreted as bicarpellate. Despite its complex structure, floral architecture in *Fumariaceae* is clearly cyclic, dimerous and organized on an opposite and decussate plan.

*Hypecoaceae*. The family *Hypecoaceae* includes the single genus *Hypecoum* with about 15 living species that range from the Mediterranean through central Asia to northern China (MABBERLEY 1987). *Hypecoum* is thought to be closely related to the *Papaveraceae* and is frequently placed within the *Fumariaceae* (e.g., CRONQUIST 1981, LIDÉN 1993). Our preliminary molecular phylogenetic studies place the genus as sister group to the *Fumariaceae* (Figs. 2, 3). We are unaware of any fossil records of *Hypecoaceae*.

Flowers in *Hypecoum* have two small caducous anterior-posterior outer sepals, and four petals in two opposite and decussate pairs. There are apparently four stamens in two opposite and decussate pairs, and two carpels in the lateral positions. All stamens are dithecate, but in the anterior-posterior stamens the two thecae are separated and these stamens are supplied by two vascular bundles rather than one. Combined with evidence from floral ontogeny, this has led some authors to suggest that each anterior-posterior "stamen" is bipartite, and thus that the androecium of *Hypecoum* is derived from six free stamens (ARBER 1932, RONSE DECRAENE & SMETS 1992). The ovary is unilocular, has two placentae and two stigmatic arms, and is interpreted as bicarpellate.

Whatever the correct interpretation of the androecium in *Fumariaceae* and *Hypecoum*, floral architecture in the two families is clearly similar and organized on a cyclic, dimerous and opposite and decussate plan.

***Pteridophyllaceae*.** The *Pteridophyllaceae* include the monotypic genus *Pteridophyllum*, which is endemic to Japan. Like *Hypecoum*, *Pteridophyllum* is thought to be closely related to the *Papaveraceae* and is frequently placed either in this family (e.g., Takhajan 1987) or within the *Fumariaceae* (e.g., Cronquist, 1981). *Pteridophyllum* is also sometimes included with *Hypecoum* in the *Hypecoaceae*. Evidence from internal rDNA spacer sequences (Lidén 1993) suggests that the family is closer to the *Ranunculaceae* than the *Papaverales*. However, our preliminary molecular phylogenetic studies place the genus as remote from *Ranunculaceae*, in a polychotomy with *Papaveraceae* sensu stricto and *Hypecoaceae* plus *Fumariaceae* (Figs. 2, 3). We are unaware of any fossil records of *Pteridophyllaceae*.

Flowers in *Pteridophyllum racemosum* have two short, deciduous outer sepals, four petals and four stamens (Murbeck 1912). The ovary is unilocular with two placentae and a bifid style, and is interpreted as bicarpellate. Floral architecture in *Pteridophyllum* is cyclic, dimerous and organized on an opposite and decussate plan.

***Circaeasteraceae.*** The *Circaeasteraceae* contain the monotypic genus *Circaeaster*, which is endemic to southern and southeastern Asia. *Circaeaster* has often been allied with the *Ranunculaceae* (Oliver 1895), *Berberidales* (Hutchinson 1959), or even placed within *Chloranthaceae* (Bentham & Hooker 1883). Other authors (e.g., Foster 1963, Wu & Kubitzki 1993a) have emphasized its isolated systematic position. Our preliminary analyses, based on combined *rbc*L and *atp*B data, place the family as the sister group to *Kingdonia*, with *Circaeaster* and *Kingdonia* together forming the sister group to the *Lardizabalaceae* (Figs. 2, 3). Fossil fruits somewhat similar to those of *Circaeaster*, but associated with monocolpate pollen, are known from the mid-Albian of Virginia, USA (Crane & al. 1994).

Flowers in *Circaeaster agrestis* are typically bisexual, occasionally unisexual, and have a variable number of floral parts (Foster 1963). There is no clear differentiation into sepals and petals. Frequently each flower consists of three tepals, one stamen and one carpel but the number of tepals varies from two to three, and the numbers of stamens and carpels vary from one to three (Foster 1963). Despite the meristic variability, floral architecture in *Circaeaster* is clearly cyclic, dimerous and based on an opposite and decussate plan.

***Kingdoniaceae.*** The *Kingdoniaceae* contain the single genus *Kingdonia*, endemic to eastern Asia. *Kingdonia* has often been allied with the *Ranunculaceae* (Balfour & Smith 1914, Hutchinson 1959), but other authors have favored a more isolated position with *Circaeaster* in the *Circaeasteraceae* (e.g., Cronquist 1981, Foster 1961, Takhtajan 1987). One recent treatment places *Kingdonia* within the *Ranunculaceae* in the tribe *Anemoneae* (Tamura 1993). Our own preliminary analyses, based on combined *rbc*L and *atp*B data, place the family as the sister group to *Circaeaster*, with *Kingdonia* and *Circaeaster* together being the sister group to *Lardizabalaceae* (Figs. 2, 3). We are unaware of any fossil records of *Kingdonia*.

Flowers of *Kingdonia uniflora* are bisexual with a variable number of floral parts. Tepals range from five to seven, and there is no clear differentiation into sepals and petals. Tepals surround about eight to 12 (probably glandular) staminodia that themselves surround three to six fertile stamens. The center of the flower is occupied by five to eight helically-arranged carpels (Foster 1961). Floral architecture is clearly different from that in *Circaeaster* and is apparently based on a helical plan.

***Lardizabalaceae.*** The *Lardizabalaceae* (excluding *Sargentodoxaceae*) consist of seven to nine genera and about 40 species of climbers and shrubs from temperate regions (Taylor 1967, Qin 1989). The greatest species diversity is in eastern Asia, but there are two genera (*Boquila, Lardizabala*) in southwestern South America. The family is usually considered closely related to the monotypic family *Sargentodoxaceae*, with strong affinities to the *Menispermaceae* and/or *Berberidaceae* TAYLOR 1967, CRONQUIST 1981, TAKHTAJAN 1987, WU & KUBITZKI 1993b). A morphologically-based cladistic analysis (LOCONTE & STEVENSON 1991) resolves the family as basal among ranunculids. However, our preliminary analyses based on molecular data (not including *Sargentodoxa*) suggest that the *Lardizabalaceae* are most closely allied to the *Circaeasteraceae* and *Kingdoniaceae* (Fig. 2). Within the *Lardizabalaceae* sensu stricto, phylogenetic analyses based on molecular data (*atp*B, *rbc*L and 18S sequences; HOOT & al. 1994) place *Decaisnea* and *Sinofranchetia* as basal in the family. Fossil seeds similar to those of *Decaisnea* are recorded from the Miocene of Germany (MAI 1980).

Flowers in the *Lardizabalaceae* are typically unisexual, and species may be monoecious or dioecious. Bisexual flowers occur in *Decaisnea*. Floral structure is consistently trimerous, with most taxa having six overlapping or valvate, petaloid sepals (three in *Akebia*) in two whorls. When present, petals of probable staminodial origin are small and six in number. There are six stamens that may be free or monadelphous. Typically there are three free carpels (sometimes up to 12 in *Akebia*).

Floral architecture in the *Lardizabalaceae* is based on a cyclic, trimerous plan. The floral morphology of the two most basal members of the family, *Sinofranchetia* and *Decaisnea*, is typical of the *Lardizabalaceae* with the exception of the occasional presence of bisexual flowers in *Decaisnea*.

***Menispermaceae.*** The *Menispermaceae* consist of approximately 71 genera and 450 species, distributed throughout the world in tropical to warm temperate regions (CRONQUIST 1981, KESSLER 1993). The *Menispermaceae* is generally considered most closely related to the *Lardizabalaceae*, *Sargentodoxaceae*, and the *Berberidaceae* (KESSLER 1993). Our preliminary cladogram based on molecular data places the *Menispermaceae* as part of a trichotomy with the *Berberidaceae* and *Ranunculaceae* sensu lato (Figs. 2, 3). The *Menispermaceae* have an extensive fossil record based predominantly on their distinctive endocarps that are first recorded during the Maastrichtian and Paleocene (COLLINSON & al. 1993). From the Eocene onward, fossil endocarps of *Menispermaceae* are common and diverse (e.g., CHANDLER 1964, COLLINSON 1983, MANCHESTER 1994).

The *Menispermaceae* are generally dioecious and flowers are therefore regularly unisexual. Sepals and petals are commonly cyclic, in trimerous whorls. Most taxa have six imbricate or valvate sepals, although variants from one to 12 occur in some genera. There are typically six petals, but again variation is common. Stamens are free, and generally six in number, but range from one in *Odontocarya monandra* to up to 40 in *Hypserpa*. As in the *Lardizabalaceae*, floral architecture in the *Menispermaceae* is clearly based on a cyclic, trimerous plan that has been modified substantially in some members of the family.

***Berberidaceae.*** The *Berberidaceae* comprise about 650 species distributed among 15 genera (LOCONTE 1993). Despite its diversity, the family is usually regarded as monophyletic, and is often placed close to the *Ranunculaceae* (CRONQUIST 1981). A morphologically-based cladistic analysis of *Ranunculidae* (LOCONTE & STEVENSON 1991) places the family in an unresolved trichotomy with the *Menispermaceae* and the *Ranunculaceae* plus *Papaveraceae*. Our preliminary analyses based on molecular data place the *Berberidaceae* in a trichotomy with *Menispermaceae* and the *Ranunculaceae* sensu lato, including *Glaucidium* and *Hydrastis* (Figs. 2, 3). Phylogenetic analyses within the family include those of MEACHAM (1980) and LOCONTE & ESTES (1989). The *Berberidaceae* have a sparse fossil record, although there are records of *Mahonia*-like leaves from the Middle Eocene onward (COLLINSON & al. 1993).

Flowers in *Berberidaceae* are bisexual and regular. The perianth is commonly bicyclic and usually trimerous, but occasionally dimerous (*Epimedium*, possibly *Jeffersonia*). The perianth may be either tepaloid (*Nandina*); with sepals and nectariferous staminodia (*Caulophyllum, Gymnospermium, Leontice, Ranzania*); or may be totally absent (*Achlys*). Stamen number is commonly six (four in *Epimedium*, eight in some *Jeffersonia*, more than eight in *Achlys* and *Podophyllum*) in two whorls. The gynoecium is apparently composed of a single carpel, but is often interpreted as pseudomonomerous, representing two or three carpels (CRONQUIST 1981). Floral architecture in most *Berberidaceae* is cyclic and trimerous (e.g., *Berberis, Caulophyllum*). Opposite and decussate, dimerous forms (e.g., *Epimedium*) and irregular arrangements (e.g., *Achlys*, ENDRESS 1989d) occur only sporadically in a few genera and are probably derived within the family.

***Glaucidiaceae.*** The *Glaucidiaceae* consist of the monotypic genus *Glaucidium* endemic to Japan. *Glaucidium* is often placed either within the *Ranunculaceae* (e.g., CRONQUIST 1981) or together with *Hydrastis* in *Glaucidiaceae* (TAKHTAJAN 1987). Our preliminary analyses of molecular sequence data place *Glaucidium* in a trichotomy with *Hydrastis* and *Ranunculaceae* sensu stricto (Figs. 2, 3). We are unaware of any fossil records of *Glaucidiaceae*.

In *Glaucidium*, there are four petaloid tepals in two pairs, numerous stamens and two carpels. Stamens in *Glaucidium* are reportedly initiated centrifugally (TAMURA 1972), which has led to the suggestion that the genus may be more appropriately allied with *Paeonia* (*Dilleniidae*). Such a position is not favored by CRONQUIST (1981) or by our preliminary *rbc*L data (HOOT & CRANE unpublished). Floral architecture in *Glaucidium*, although variable (TAMURA 1972), appears to be basically on a cyclic, dimerous, opposite and decussate plan, which is expressed in both the perianth and carpels. The basic arrangement appears to have been modified by the secondary production of additional stamens.

***Hydrastidaceae.*** The *Hydrastidaceae* include the monotypic genus *Hydrastis* of central and eastern Northern America, which is often placed either within the *Ranunculaceae* (e.g., CRONQUIST 1981) or with *Glaucidium* in *Glaucidiaceae* (TAKHTAJAN 1987). Other authors (e.g., TOBE & KEATING 1985) have favored a more isolated systematic position. Our preliminary analysis based on sequence data places *Hydrastis* along with *Glaucidium* in a trichotomy with *Ranunculaceae* sensu stricto (Figs. 2, 3). We are unaware of any fossil records of *Hydrastidaceae*.

In *Hydrastis*, there are two to four tepals arranged in two opposite and decussate pairs, numerous stamens, and eight to 15 carpels. Floral architecture is possibly on a cyclic, dimerous, opposite and decussate plan, which appears to have been modified by the production of additional stamens and carpels.

***Ranunculaceae.*** The *Ranunculaceae* are a very diverse family, consisting of approximately 59 genera and 2,500 species (CRONQUIST 1981, TAMURA 1993). The family is widespread in both the Old and New World, especially in temperate and boreal regions. Phylogenetically, the *Ranunculaceae* are closely allied to the *Hydra-*

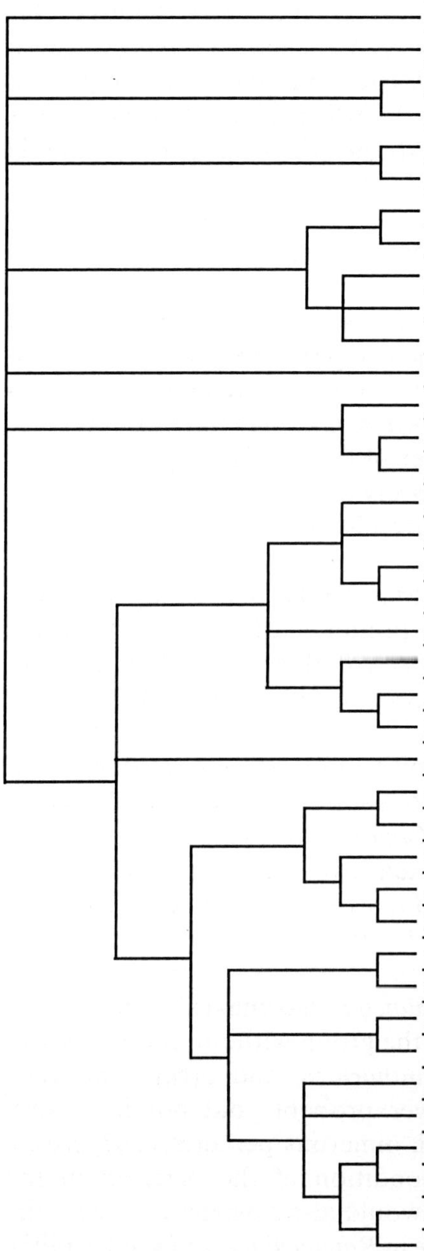

| | | P | A | G |
|---|---|---|---|---|
| *ILLICIUM* | | | | |
| *SABIA* | ⚥(♂♀) | 5, 5 | 5 | 2 |
| *PACHYSANDRA* | ♂/♀ | 6 | 4 | 2-3 |
| *BUXUS* | ♂♀ | 4 | 4 | 3 |
| *TROCHODENDRON* | ⚥ | 0 | ∞ | ∞ |
| *TETRACENTRON* | ⚥ | 4 | 4 | 4 |
| *MYROTHAMNUS* | ⚥ | 0 | 4 | 3-4 |
| *GUNNERA* | ♂♀(⚥) | 2, 2 | 1-2 | 2 |
| *DAPHNIPHYLLUM* | ♂♀ | 2-6 | 5-12 | 2-4 |
| *HAMAMELIS* | ⚥ | 4, 4 | 4, 4 | 2 |
| *CERCIDIPHYLLUM* | ♂♀ | 0 | 1-13 | 1 |
| *NELUMBO* | ⚥ | 2-5, 20-30 | ∞ | 2-30 |
| *PLATANUS* | ♂♀ | 3-4, 4 | 3-5 | 5-8 |
| *PLACOSPERMUM* | ♂♀(⚥) | 4 | 4 | 1 |
| *ROUPALA* | ⚥ | 4 | 4 | 1 |
| *SANGUINARIA* | ⚥ | 2, 6-12 | ∞ | 2 |
| *PLATYSTEMON* | ⚥ | 3, 6 | ∞ | 6-25 |
| *ESCHSCHOLZIA* | ⚥ | 2, 4 | ∞ | 2 |
| *DENDROMECON* | ⚥ | 2, 6 | ∞ | 2 |
| *PTERIDOPHYLLUM* | ⚥ | 2, 4 | 4 | 2 |
| *HYPECOUM* | ⚥ | 2, 4 | 4 | 2 |
| *DICENTRA* | ⚥ | 2, 4 | 4 | 2 |
| *CORYDALIS* | ⚥ | 2, 4 | 4 | 2 |
| *EUPTELEA* | ⚥ | 0 | 6-19 | 8-31 |
| *KINGDONIA* | ⚥ | 5-7 | 11-18 | 5-8 |
| *CIRCAEASTER* | ⚥(♂♀) | 2-3 | 1-3 | 1-3 |
| *SINOFRANCHETIA* | ♂♀ | 6, 6 | 6 | 3 |
| *AKEBIA* | ♂♀ | 3 | 6 | 3-12 |
| *DECAISNEA* | ♂♀(⚥) | 6 | 6 | 3 |
| *NANDINA* | ⚥ | 27-36 | 6 | 1 |
| *CAULOPHYLLUM* | ⚥ | 6, 6 | 6 | 1 |
| *MENISPERMUM* | ♂/♀ | 6, 6-19 | 12-18 | 3 |
| *TINOSPORA* | ♂/♀ | 6, 6(3) | 6 | 3 |
| *HYDRASTIS* | ⚥ | 2-4 | ∞ | 8-15 |
| *GLAUCIDIUM* | ⚥ | 4 | ∞ | 2 |
| *RANUNCULUS* | ⚥ | 3-5, 5+ | ∞ | ∞ |
| *CALTHA* | ⚥ | 5+ | ∞ | 3-∞ |
| *XANTHORHIZA* | ⚥ | 5, 5 | 5-10 | 5-10 |
| *COPTIS* | ⚥ | 5-8, 5-10 | ∞ | 5-15 |

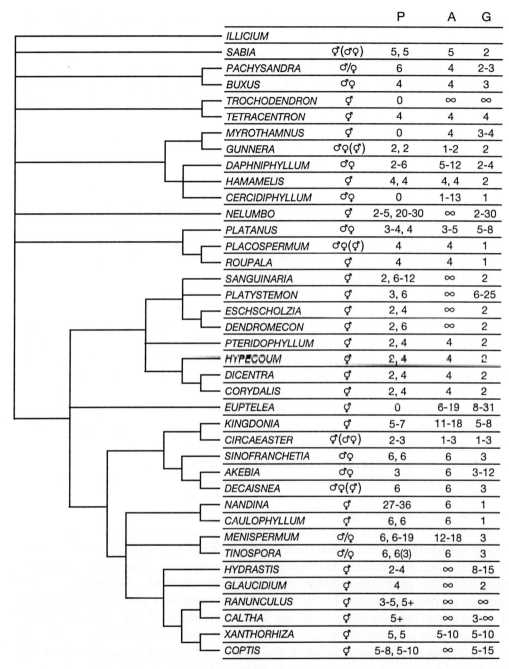

Fig. 3. Consensus tree of the 53 shortest trees derived from a preliminary cladistic analysis based on a combination of *rbc*L and *atp*B (as in Fig. 2) showing the distribution of floral types. The columns summarize floral structure as follows: ⚥ bisexual flowers (monoecious), ♂♀ unisexual flowers (monoecious), ♂/♀ unisexual flowers (dioecious); *P* perianth (first numeral represents number of sepals, second numeral the number of staminodia/petals; if only one numeral is present it represents the number of tepals); *A* androecium (number of stamens); *G* gynoecium (number of carpels). *Hypecoum*, *Dicentra*, and *Corydalis* are interpreted as having four stamens: alternative interpretations (e.g., Arber 1931, 1932; Ronse Decraene & Smets 1993a) consider the androecium of these taxa as composed of six stamens. The androecium in *Kingdonia* includes 8–12 staminodes and 3–6 fertile stamens

*stidaceae* and *Glaucidiaceae*, which are often considered as intermediates between the *Ranunculaceae* and the *Berberidaceae* (Cronquist 1981, Hoot 1991). Our preliminary molecular tree supports the monophyly of the *Ranunculaceae* sensu stricto and resolves it in a trichotomy with *Hydrastis* and *Glaucidium* (Figs. 2, 3) Analysis of our total data set, which includes more than 20 genera of *Ranunculaceae*, also supports the monophyly of the family sensu stricto (Hoot & Crane, unpubl.) and excludes *Glaucidium*, *Hydrastis* and *Kingdonia*. As far as we are aware, other than the *Thalictrum*-like fruits described from the Early Cretaceous of Portugal (Friis & al. 1994) there are no records of fossil *Ranunculaceae* prior to the Miocene (Collinson & al. 1993).

Within the *Ranunculaceae*, two major groups are commonly recognized based on chromosome size and other characters: *Ranunculus*-type (R-type) and *Thalictrum*-type (T-type) (Langlet 1932, Hoot 1991, Tamura 1993). Our preliminary analyses using molecular data confirm this dichotomy (Hoot & Crane, unpubl.), and place *Coptis* and *Xanthorhiza* as basal to the remainder of the family.

Flowers in the *Ranunculaceae* exhibit great variety of form and are often highly specialized. Typically, flowers in *Ranunculaceae* are bisexual, and rarely unisexual (*Thalictrum*). Sepals are commonly five to eight or more in number (rarely three to four), and are often petaloid and/or caducous. Staminodia/petals may be absent or present; when present, there is great variation in number, five to seven is most common. Staminodia and petals have clearly been derived several times from stamens in separate lineages (Janchen 1949; Tamura 1965, 1993; Hoot 1991). Stamens are usually numerous and free; carpels are usually numerous and free. According to Tamura (1965), the putatively primitive ranunculaceous flower had numerous sepals (monochlamydeous) with numerous stamens and carpels arranged spirally on an expanded receptacle. Flowers of *Coptis* have five to eight petaloid sepals, five to ten staminodia, numerous stamens, and five to 15 carpels. Flowers of *Xanthorhiza* have five small sepals, five staminodia, five to ten stamens in whorls, five or ten carpels. In both *Coptis* and *Xanthorhiza* the basic organization of the flower appears to be on a pentamerous plan.

The diversity of floral structure in the *Ranunculaceae* makes it difficult to interpret the likely patterns of floral evolution in the group without more intensive study. However, the flowers of *Coptis* and *Xanthorhiza*, and other ranuculids suggest that the primitive ranunculaceous flower probably did not have very numerous carpels or tepals. The undifferentiated, dimerous perianth of *Hydrastis* and *Glaucidium* may indicate that the basic condition of the perianth in the *Ranunculaceae* is few-parted and dimerous. This would also be consistent with the view that a monochlamydeous perianth is primitive in *Ranunculaceae* (Tamura 1993).

### Floral structure: "Basal" *Hamamelididae*, *Rosidae* and other eudicots

***Platanaceae.*** The only extant genus, *Platanus*, consists of about nine extant species with the greatest diversity in Mexico. Two subgenera have been recognized; subgenus *Platanus* contains all New World species, as well as *Platanus orientalis* from the eastern Mediterranean; subgenus *Castaneophyllum* contains the single extant species *Platanus kerrii* from south-eastern Asia. *Platanus kerrii* differs from all other extant *Platanus* species in having elliptical, pinnately-veined leaves and infructescence axes with numerous heads. The *Platanaceae* have been placed traditionally in the *Hamamelidales* (CRONQUIST 1981, TAKHTAJAN 1987, SCHWARZWALDER & DILCHER 1991), but recent phylogenetic analyses based on *rbc*L data (CHASE & al. 1993) place the family as the sister taxon to *Nelumbonaceae* in a clade with *Proteaceae* and *Sabiaceae*. Our preliminary analysis of combined *rbc*L and *atp*B data place the genus as sister group to the *Proteaceae* (Figs. 2, 3). *Platanaceae* have an extensive fossil record extending back to the mid-Cretaceous, and several thoroughly reconstructed fossil platanoids are known from the Cretaceous and Tertiary (e.g., MANCHESTER 1986, PIGG & STOCKEY 1991, CRANE & al. 1993).

Flowers in all *Platanus* species are unisexual and aggregated into compact, spherical heads. Interpretation of the perianth is not completely clear, but is often described as consisting of three to four (or up to seven) poorly developed "sepals" alternating with small "petals" (CRONQUIST 1981). In staminate flowers, stamens are opposite to, and equal in number to, the "sepals" (usually three, four or five). Staminate flowers occasionally contain vestigial carpels. Pistillate flowers commonly have three or four tepals and staminodes, and a variable number of carpels (usually five to eight, but ranging from three to nineteen).

Several late Albian-early Cenomanian platanoids are well-documented from North America. Flowers in these plants are unisexual and tightly clustered into spherical heads that are sessile along long infructescence axes as in *P. kerrii*. However, these mid-Cretaceous fossils differ from extant *Platanus* in several respects. Most notably, the number of carpels per flower is more tightly constrained to five or occasionally four. For example, pistillate flowers of *Platanocarpus brookensis* from the early-middle Albian of eastern North America are consistently pentamerous with a prominent perianth (CRANE & al. 1993). The number of tepals appears variable, and there is no obvious differentiation into sepals and petals. Carpels are situated opposite the innermost tepals. It is currently uncertain how to reconcile the floral structure in these Cretaceous platanoids with the flowers of Recent taxa. It is interesting however, that floral structure in well-preserved platanoid inflorescences from the early Tertiary (e.g., MANCHESTER 1986) is more similar to the floral structure of Cretaceous forms, than that of extant species.

***Proteaceae.*** The *Proteaceae* have over 1,500 species in 73 genera distributed predominantly in the Southern Hemisphere. The family is traditionally placed within the *Rosidae* allied to groups such as *Elaeagnaceae* or *Santalales* (CRONQUIST 1981, TAKHTAJAN 1987). However, a surprising result of recent phylogenetic studies based on *rbc*L data is that the family is placed in a clade with *Sabiaceae*, *Platanaceae* and *Nelumbo* (CHASE & al. 1993) and this is consistent with the relatively basal position suggested for the family by JOHNSON & BRIGGS (1975). Our preliminary analyses of combined *rbc*L and *atp*B data also broadly support this conclusion and place the family as sister group to the *Platanaceae* (Figs. 2, 3). The disjunct austral distribution of the family suggests that it has an ancient origin and this is supported by the occurrence of *Proteaceae*-like pollen grains in the late Cenomanian of Africa (WARD & DOYLE 1994). Reliably determined macrofossils of *Proteaceae* are not recorded until the Middle Eocene (e.g., CHRISTOPHEL 1984; see also COLLINSON & al. 1993).

Flowers of *Proteaceae* are bisexual. The perianth consists of four petaloid tepals that may be free or fused. The four stamens are opposite the tepals. Filaments are often lacking and the anthers are sessile and epitepalous. The ovary is superior and consists of a single carpel in the center of the flower. Within the family there are numerous modifications of floral morphology apparently associated with different animal pollination syndromes (zygomorphy, sessile anthers, epitepaly, syntepaly, stylar pollen presentation), but the basic floral architecture is regular and tetramerous.

***Trochodendraceae.*** The *Trochodendraceae* include the two monotypic genera *Trochodendron* and *Tetracentron*, both restricted to China and Japan. Both genera have vesselless wood, which was long thought to be a primitive feature but is now interpreted as almost certainly derived within flowering plants. Traditionally the *Trochodendrales* have been placed as the first group in the *Hamamelididae* (CRONQUIST 1981, TAKHATAJAN 1987) and the implied basal position is supported by recent phylogenetic analyses based on *rbc*L data (CHASE & al. 1993). Our preliminary analysis of combined *rbc*L and *atp*B data place the family in a polychotomy at the base of the eudicots (Figs. 2, 3).

Unequivocal fossils of *Trochodendron* are first recorded from the Miocene of Japan and North America (MANCHESTER & al. 1991). The fossil genus *Nordenskioldia*, known from the Maastrichtian to Miocene, is almost certainly a fossil member of this family (CRANE & al. 1990) and dispersed fruitlets similar to those of *Nordenskioldia* have been recorded from the early Cenomanian of North America (CRANE, unpubl.).

Flowers in *Tetracentron sinense* are bisexual, and have four tepals, four stamens and four carpels. ENDRESS (1986a) has shown that the floral organs develop as opposite and decussate pairs. The orientation of the four carpels suggests doubling in each of the two lateral positions from an originally bicarpellate condition (ENDRESS 1987).

Flowers in *Trochodendron aralioides* are bisexual with numerous stamens and carpels. Obvious tepals are lacking but between the prophylls and stamens on lateral flowers ENDRESS (1986a) recognized several very small scales that may be interpreted as perianth parts.

Floral architecture in *Tetracentron* is dimerous and clearly based on an opposite and decussate plan (ENDRESS 1986a). In *Trochodendron* this is less obvious, and there has apparently been an increase of both stamen and carpel number. The developing floral primordium, however, retains bilateral symmetry and this is also visible in the mature fruits of *Nordenskioldia* (CRANE & al. 1991, MANCHESTER & al. 1991).

***Sabiaceae.*** The *Sabiaceae* include three extant genera *Meliosma*, *Sabia* and *Ophiocaryon* and approximately 60 species. *Meliosma* is present in both southeastern Asia and tropical America. *Ophiocaryon* is restricted to tropical South America and *Sabia* is restricted to southeast Asia. The phylogenetic position of the *Sabiaceae* is uncertain, and while some place the family in the *Ranunculales* (e.g., CRONQUIST 1981) associated with *Menispermaceae* (ERDTMAN 1952), others place it in the *Sapindales* (HEIMSCH 1942). Phylogenetic analyses based on *rbc*L data have so far only sampled *Sabia*, which is placed as sister group to *Proteaceae* in a clade that also includes *Nelumbo* and *Platanus* (CHASE & al. 1993). Our own analysis based on a combination of *atp*B and *rbc*L data places *Sabia* in an unresolved polychotomy at the base of the eudicots (Figs. 2, 3). Fossil endocarps resembling those of extant

*Meliosma* provide the earliest records of the family and are first recorded from the Maastrichtian and Paleocene (VAN BEUSEKOM 1971, CRANE & al. 1990, COLLINSON & al. 1993).

Flowers in the *Sabiaceae* are usually bisexual (sometimes unisexual) and their structure is difficult to interpret. The perianth consists of three or more, typically five, unequal sepals, and four or five alternate petals. The two inner petals are often distinctly smaller than the others. Stamens (including staminodia) are as many as and opposite the petals. The gynoecium consists of two (sometimes three) carpels with two (or three) united locules. According to VAN BEUSEKOM (1971) the flowers of *Sabiaceae* (and *Meliosma* in particular) are essentially pentamerous with a series of reductions that give the appearance of trimery.

***Nelumbonaceae.*** The *Nelumbonaceae* consist of the single genus *Nelumbo* with two species ranging from eastern North America, Asia, and Australia. The traditional placement of the family within or near the *Nymphaeaceae* (e.g., CRONQUIST 1981, TAKHTAJAN 1987) is not supported by cladistic analyses based on morphological, anatomical, phytochemical, and sequence data (WILLIAMSON & SCHNEIDER 1993). Previous analyses based on *rbc*L data resolve the *Nelumbonaceae* as sister taxon to *Platanaceae* (CHASE & al. 1993). Our preliminary consensus tree based on combined *rbc*L and *atp*B data places the family in an unresolved polychotomy with "lower" hamamelidids, with some support (bootstrap value of 74%) for a closer alliance with a clade consisting of *Platanus* and the *Proteaceae* (Figs. 2, 3). Leaves assigned to *Nelumbo* occur from the Eocene onwards in Europe (COLLINSON & al. 1993). Possible leaves and reproductive structures of *Nelumbo* are also known from late Albian of eastern North America (UPCHURCH & al. 1994).

Flowers of *Nelumbonaceae* are bisexual. While they have often been interpreted as having spiral floral phyllotaxy (e.g., WILLIAMSON & SCHNEIDER 1993) this has not been studied critically and it is unknown whether the development is spiral, whorled or irregular (ENDRESS, pers. comm.). The perianth is caducous, consisting of two to five sepals and 20–30 petals. The androecium consists of 200–300 stamens, and the gynoecium of two to 30 free carpels embedded in a fleshy receptacle.

***Buxaceae.*** The *Buxaceae* are a family of three to five genera of tropical, subtropical and temperate plants. *Buxus*, with approximately 90 species in Central America, East Asia and North Africa-Mediterranean, is the largest genus. *Notobuxus*, which has been variously considered to be part of *Buxus*, has about eight species in Africa. *Sarcococca* has eleven species in Asia while *Pachysandra* has four species in North America and eastern Asia. *Styloceras*, with four species in South America has often been referred to its own family (*Stylocerataceae*), but preliminary molecular data place it within the *Buxaceae* (HOOT & CRANE unpublished). Ideas on the systematic affinities of the *Buxaceae* have varied considerably. CRONQUIST (1981) placed the family with the *Euphorbiaceae*, while TAKHTAJAN (1987) placed the family in the "lower" hamamelidids close to *Myrothamnaceae*. Phylogenetic analyses based on *rbc*L data resolve the *Buxaceae* as sister group to the *Trochodendraceae* (CHASE & al. 1993). Our own analysis based on a combination of *atp*B and *rbc*L data places the *Buxaceae* in an unresolved polychotomy at the base of the eudicots (Figs. 2, 3).

Fossil pollen very similar to that of *Pachysandra* and *Sarcococca* is known from the Campanian (Late Cretaceous) onwards and during the mid-Cretaceous pollen of the *Hexaporotricolpites*-type may also be related to the family (see Drinnan & al. 1991 for review). Two putatively buxaceous fossil inflorescences have been described from the late Albian and early Cenomanian of eastern North America (Drinnan & al. 1991). Inflorescences of *Spanomera mauldinensis* have a central pistillate flower with four tepals (two decussate, dimerous whorls) and two carpels. Staminate lateral flowers consist of five tepals and five stamens but it is clear that this pseudo-pentamerous arrangement is basically opposite and decussate, with the anterior-posterior tepal and stamen whorl being trimerous rather than dimerous. Staminate flowers of the other species (*Spanomera marylandensis*) are comprised of only four tepals and stamens (four dimerous whorls).

All extant *Buxaceae* have unisexual flowers in which the tepals are only weakly differentiated from the bracts of the inflorescence axis. Staminate flowers of *Buxus*, *Notobuxus*, *Sarcococca* and *Pachysandra* are constructed on a dimerous, decussate plan, and consist of four or six tepals (in two or three pairs), four or six stamens, and a central sterile pistillode. In *Pachysandra* and *Sarcococca*, the first pair of structures has been described either as two prophylls or the first two of six tepals. Stamens are positioned opposite the tepals as a consequence of the dimerous floral architecture. Pistillate flowers are usually, but not always, spirally constructed, with six to 20 tepals and two or three carpels. Ontogenetic data suggest that this spiral arrangement is probably secondary. In *Buxus*, the inflorescence (an axillary spike) starts with an opposite/decussate architecture and staminate flowers are formed in the axils of proximal inflorescence bracts. However, close to the initiation of the terminal pistillate flower the phyllotaxy changes into a spiral via several asymmetric bract pairs. Following initiation of the last staminate flower, six empty bracts are produced in a 2/5 spiral, and are followed by three carpel primordia which continue the phyllotactic spiral. 'Tepals' of the pistillate flowers are essentially bracts of the inflorescence axis that lack an axillary structure (staminate flower). They have an identical morphology to the inflorescence bracts and the "tepals" of the staminate flowers.

If the spiral phyllotaxy seen in the pistillate flowers is secondary, the basic architecture of flowers and inflorescences in extant *Buxaceae* is cyclic, dimerous, and opposite and decussate. The same pattern is seen in the putatively related mid-Cretaceous fossils.

***Cercidiphyllaceae.*** The *Cercidiphyllaceae* contain the single genus *Cercidiphyllum* with two species, *C. japonicum* and *C. magnificum*, both restricted to China and Japan. The family has generally been allied with *Trochodendraceae* and *Hamamelidales*. Cladistic analyses based on morphological data have placed *Cercidiphyllum* as sister group to *Myrothamnus* (Hufford & Crane 1989, Hufford 1992). Recent analyses of molecular sequence data (Chase & al. 1993) place the family as close to *Daphniphyllum* and *Hamamelidaceae* with the precise relationships among these groups depending on the search procedure used and the taxa included. Our preliminary phylogenetic analyses based on combined *rbc*L and *atp*B data place *Cercidiphyllum* in a polychotomy with the *Hamamelidaceae* (represented by *Hamamelis*) and *Daphniphyllaceae*.

Unequivocal fossil *Cercidiphyllum* is recorded from the Early Oligocene of North America, but closely related fossil plants (e.g., *Joffrea speirsii*) are well known from leaves, shoots, infructescences, seeds and staminate inflorescences from the early Tertiary (e.g., Crane 1984; Crane & Stockey 1985, 1986).

Floral structure is identical in both species of *Cercidiphyllum*. Flowers are unisexual. Staminate flowers consist of one to 13 stamens in the axil of a bract, but because they lack both floral prophylls and perianth they are often difficult to delimit (ENDRESS 1986a) and the arrangement of stamens is apparently irregular. Pistillate flowers are borne in one or two decussate pairs. Each is delimited by a bract but the flower lacks both floral prophylls and perianth. In the extant species there is a single carpel per flower, but in fossil *Cercidiphyllaceae* (e.g., *Joffrea*) there may be one or two carpels per flower.

The reduced reproductive structures of *Cercidiphyllum* makes the floral architecture difficult to determine. However, the phyllotaxy of the plant, the occurrence of paired carpels in fossil forms, and the paired flowers in the inflorescence suggest a dimerous organization.

***Daphniphyllaceae.*** The *Daphniphyllaceae* include the single genus *Daphniphyllum* with about nine to 40 species restricted to eastern and southeastern Asia (HUANG 1965, 1966; ROSENTHAL 1916; SUTTON 1989). The relationships of *Daphniphyllum* have been controversial and in many respects ideas have paralleled changing views on the relationships of the *Buxaceae* (SUTTON 1989). Frequently the family has been allied with the *Euphorbiaceae* (e.g., BENTHAM 1880, STEBBINS 1974) but a range of alternative views have placed it close to various *Rosidae* (e.g., *Pittosporales*, THORNE 1983), or "lower" *Hamamelididae*, including the *Cercidiphyllaceae* and *Eucommiaceae* (CROIZAT & METCALFE 1941, CROIZAT 1941, BARABÉ & al. 1987), *Hamamelidaceae* (HALLIER 1904), or *Hamamelidales/Didymelales* (CRONQUIST 1981, TAKHTAJAN 1987). Recent analyses of molecular sequence data (CHASE & al. 1993) place the family close to *Cercidiphyllum* and *Hamamelidaceae* with the precise relationships among these groups depending on the search procedure used and the taxa included. Our preliminary molecular phylogenetic analyses place *Daphniphyllum* in a trichotomy with *Cercidiphyllum* and *Hamamelidaceae* (Figs. 2, 3). There is no reliable fossil record of the *Daphniphyllaceae*.

Flowers are unisexual, typically with two to six sepals (rarely absent) and no petals. The androecium in staminate flowers consists of five to 12 stamens. In pistillate flowers the gynoecium consists of two, or occasionally four (rarely three), carpels that are more or less united to form an incompletely septate ovary. Both staminate and pistillate flowers may have staminodes. Floral architecture is irregular, perhaps reflecting the variable and irregular form of the calyx.

***Hamamelidaceae.*** The *Hamamelidaceae* comprise four subfamilies (*Hamamelidoideae, Exbucklandioideae, Rhodoleioideae, Altingioideae*) with a total of 30 genera and approximately 100 species (ENDRESS 1989a, b 1993). The family is generally placed close to the "lower" *Hamamelididae* (e.g., *Trochodendrales, Platanaceae*, CRONQUIST 1981), "higher" *Hamamelididae* (e.g., *Betulaceae, Fagaceae*, ENDRESS 1967, 1977) or *Rosidae* (e.g., *Cunoniaceae, Escalloniaceae*, ENDRESS 1993). Recent cladistic analyses of morphological and molecular sequence data, as well as restriction sites (chloroplast genome), place the family in a broadly transitional position between the "lower" *Hamamelididae* and a large group of eudicots including "higher" *Hamamelididae* (e.g., *Fagaceae, Betulaceae*) and many *Rosidae* (e.g., *Cunoniaceae, Fabaceae*: HUFFORD & CRANE 1989, HUFFORD 1992, CHASE & al. 1993, MANOS & al. 1993). Evidence from *rbc*L sequence data resolves the family as the sister group to *Cercidiphyllum* and/or *Daphniphyllum* (CHASE & al. 1993), and also suggest that *Rhodoleia* may be more closely related to various rosids than other

*Hamamelidaceae*. Our own preliminary analyses based on combined *rbc*L and *atp*B data, and using *Hamamelis* as a "placeholder" (Figs. 2, 3), place the family in a polychotomy with *Cercidiphyllaceae* and *Daphniphyllaceae*. Several fossil flowers similar to those of extant *Hamamelidaceae* are known from the Late Cretaceous (Friis & Crane 1989, Endress & Friis 1991, Crepet & al. 1992, Herendeen & al., unpubl.). Many modern genera were clearly differentiated by the Early Tertiary (Friis & Crane 1989).

Flowers in the *Hamamelidaceae* vary considerably in size and in the development of the perianth and androecium. Flowers are typically bisexual and only rarely unisexual (e.g., *Altingia, Liquidambar, Sinowilsonia, Fortunearia*). The perianth is usually differentiated into calyx and corolla. Floral structure is generally pentamerous, or more rarely tetramerous (e.g., *Hamamelis*, Endress 1993). Stamen number is typically four or five, and sometimes there are four or five staminodes. The number of stamens in the androecium is increased in genera in which the perianth is reduced (e.g., *Fothergilla, Parrotia*). Interestingly, in these cases, initiation of the androecium may be either centrifugal (e.g., *Fothergilla*) or centripetal (e.g., *Matudaea*, Endress 1993). The gynoecium is consistently bicarpellate, although rare tricarpellate forms also occur.

In *Archamamelis* from the late Santonian/early Campanian of southern Sweden, the flowers are bisexual and six- to seven-merous, probably with a perianth of two whorls, an androecium of one whorl of six to seven stamens, and a gynoecium of three (or occasionally two?) carpels (Endress & Friis 1991). Flowers of hamamelidaceous affinity from the Turonian of New Jersey, USA, are unisexual and tetramerous. Staminate flowers have a four-lobed calyx, alternating with a whorl of four staminodes, which then alternate with a whorl of four stamens (Crepet & al. 1992). Corresponding pistillate flowers also have a four-lobed calyx and bicarpellate gynoecium. Recently discovered fossil flowers from the Campanian of Georgia, U.S.A., resemble those of *Embolanthera* and related genera. The perianth is unknown but the androecium appears to have two whorls of five stamens and the gynoecium is bicarpellate (Herendeen & al., unpubl.).

Based on extant and putatively related fossil taxa, floral architecture in the *Hamamelidaceae* is most commonly pentamerous in the perianth and androecium and dimerous in the gynoecium. Tetramerous forms also occur (e.g., *Hamamelis*, New Jersey fossils) and there is also some indication of a trimerous plan both from the fossil *Archamamelis* and occasional deviant flowers in extant taxa (Endress & Friis 1991).

***Myrothamnaceae***. The *Myrothamnaceae* contain the single genus *Myrothamnus* with two species; *Myrothamnus flabellifolia* (southern and eastern Africa) and *M. moschata* (Madagascar) (Jäger-Zürn 1966, Endress 1989c). The family is frequently allied with the *Hamamelidaceae* (Cronquist 1981), but Endress (1989c) emphasized its similarities to *Cercidiphyllaceae*, and recent cladistic analyses of morphological data have resolved *Cercidiphyllaceae* and *Myrothamnaceae* as sister taxa (Hufford & Crane 1989, Hufford 1992). Our preliminary molecular phylogenetic analyses place *Myrothamnus* as the sister group to *Gunneraceae* (Figs. 2, 3). Fossil pollen grain tetrads from the mid-late Albian and early Cenomanian of North America have been compared to those of extant *Myrothamnus* (Doyle 1969, Ward & Doyle 1988).

Floral structure is the same in both species (Jäger-Zürn 1966). Flowers are unisexual. Staminate flowers typically include four stamens, but stamen number varies from three to eight (Cronquist 1981). Pistillate flowers consist of one or more pairs of opposite and decussate bracts and three or four carpels that are fused at the base. Floral architecture is based on a cyclic, dimerous, opposite/decussate plan, but some pistillate flowers exhibit an apparent radial symmetrical arrangement of three carpels similar to that seen in the gynoecia of the *Buxaceae*.

***Gunneraceae.*** The family *Gunneraceae* include the single genus *Gunnera* with approximately 40–50 species, occurring in tropical and southern Africa, South America, South-east Asia, Tasmania, New Zealand, and the Pacific. *Gunnera* has often been included with the *Halorhagaceae* but this relationship is not secure (Cronquist 1981). Takhtajan (1987) places *Gunneraceae* close to the *Rosales*. Recent cladistic analyses based on molecular sequence data separate *Gunnera* from the *Halorhagaceae* and place it as the sister taxon to a major clade that includes all eudicots except ranunculids and "lower" hamamelidids (Chase & al. 1993). Our own analyses, based on combined *rbc*L and *atp*B data (Figs. 2, 3), place the family as the sister group to *Myrothamnaceae*. Fossil pollen grains assigned to the *Gunneraceae* have been reported from the Turonian of Peru (Brenner 1968) and the Campanian of North America, the Antarctic, Australasia and Africa (Jarzen 1980, Jarzen & Dettmann 1989).

Flowers are typically unisexual but occasionally bisexual. Often, staminate and pistillate flowers occur in different parts of the inflorescence with bisexual flowers in the intermediate zone (Cronquist 1981). Staminate flowers typically have two (rarely three) sepals, two petals and one or two stamens. In pistillate flowers the gynoecium is unilocular but bicarpellate with two styles. Floral architecture in *Gunnera* appears based on a dimerous plan.

## Patterns of floral evolution in "basal" eudicots

The above summary of floral structure among "basal" eudicots permits several generalizations concerning patterns of floral evolution in the early radiation of this major clade of angiosperms.

**Floral phyllotaxy and merosity.** Flowers constructed on a cyclic plan are common, and are unquestionably basic for eudicots as a whole. Among the taxa considered here, well-developed helical floral phyllotaxy is only known to occur in *Ranunculaceae*, and is almost certainly secondary. Among the cyclic forms, dimerous (opposite/decussate) and trimerous (ternate) arrangements are widespread and the transition from one to the other appears to have occurred numerous times (e.g., *Buxaceae*, *Myrothamnaceae*, *Papaveraceae*). Pentamerous flowers (e.g., *Hamamelidaceae*) are more restricted than dimerous or trimerous forms and are also clearly secondary under current interpretations of phylogenetic patterns. Tendencies to pentamery also occur in certain *Ranunculaceae* and perhaps *Sabiaceae*.

An important conclusion from reviewing floral structure in 'basal' eudicots is that detailed ontogenetic studies are crucial for discriminating among dimerous, trimerous and pentamerous flowers that have apparently arisen in different ways. For example, the tetramerous condition in *Buxaceae* (Drinnan & Crane, unpubl.) clearly arises as two decussate dimerous tepal whorls, and the position of the stamens (opposite the tepals) arises through the subsequent production of two decussate pairs of stamens. In contrast, in the tetramerous flowers of the *Proteaceae*, in which the stamens are also opposite the tepals, the floral parts are initiated helically with short plastochrons between successive primordia (Douglas, pers. comm). Similarly, trimerous whorls may arise in a variety of ways. In some groups the initiation of primordia is more or less simultaneous with a divergence angle of approximately 120°, in other groups initiation is helical with a divergence angle of 137.5°, while in other taxa trimery may arise through apparent duplication of a primordium in one sector of a dimerous whorl (e.g., *Spanomera mauldinensis*). Similar considerations apply to the ontogenetic basis of pentamery. In addition to "normal" helical initiation, in the mid-Cretaceous fossil *Spanomera mauldinensis*, and perhaps fossil *Platanaceae*, the "pentamerous" condition appears to have arisen by the superposition of a dimerous and trimerous whorl. In this case pentamery results from an apparent duplication of a stamen and tepal in one sector of the floral apex. A similar origin for pentamerous whorls has been recorded in other groups (e.g., early development of leaf whorls in *Casuarina* seedlings, Johnson & Wilson 1989: fig. 9.1).

**Transitions in the apical meristem:** While there is considerable variation in merosity of the flower across various groups of "basal" eudicots it is also interesting that in some groups similar variation occurs during the ontogeny of a single flower (i.e., through the development of a single apical meristem). In many *Papaveraceae* the floral apex passes from the production of few primordia (two plus two tepals) to the production of many primordia per whorl (many stamens) and back to the production of a single (internally "doubled") carpel. However, in general the basic underlying dimerous pattern is retained. In the *Ranunculaceae* there is a common transition from dimery or trimery in the perianth to helices in the androecium and gynoecium. Similarly, in *Pachysandra* there is clear transition from an opposite and decussate to helical floral phyllotaxy associated with the development of the pistillate flowers. Interestingly, this transition appears irreversible and the helical arrangement is retained even where the gynoecium consists of only two carpels (*Pachysandra terminalis*). The nature of these transitions in a single floral apex and associated phenomena (e.g., reversibility of phyllotactic transitions, presence or absence of prophylls, intergradation of inflorescence bracts and tepals, floral ontogeny in the context of inflorescence architecture) requires more detailed investigation. Comparison with changes in phyllotaxis during the development of a single plant would also be interesting. Frequently the arrangement of the first primordia initiated on the apical meristem is dimerous (two cotyledons) but subsequently there is a transition to helical phyllotaxis which often reverts to a dimerous whorl (two floral prophylls) with the transition to the floral apex.

**Increase in number of floral parts.** Based on current ideas of phylogenetic relationships, flowers with relatively few floral parts are almost certainly basic in the eudicot clade. Even in families where polymerous flowers are common (e.g., *Ranunculaceae*, *Papaveraceae*), genera that appear phylogenetically basal tend to have flowers with relatively few floral parts (especially perianth and carpels). This conclusion conflicts with traditional interpretations of floral evolution among basal angiosperms in which the possession of numerous floral organs is generally assumed to be primitive and the overriding trend is thought to be toward reduction (Bessey

1915, CRONQUIST 1981). Based on phylogenetic analysis of combined *rbc*L and *atp*B data the polymerous androecium and gynoecium of many *Papaveraceae* and *Ranunculaceae* almost certainly reflect secondary multiplication. It is also important that such multiplication may be achieved developmentally in a variety of ways, as discussed for the androecium by RONSE DECRAENE & SMETS (1993a, b). For example in the *Ranunculaceae* the polymerous androecium is associated with helical initiation of stamen primordia. In contrast, in the *Papaveraceae* the increase in the number of parts is achieved by an increase in the number of androecial whorls and by an increase in the number of primordia per whorl (e.g., ENDRESS 1987, KARRER 1991). Such examples emphasize the need for developmental data in evaluating potential homologies among different floral architectures.

**Differentiation of perianth and nature of petals.** An obvious and common feature of the flowers of 'basal' eudicots is differentiation of the perianth into calyx and corolla, perhaps associated with specialization for protection during development (sepals) and attraction for pollination (petals). In ranunculids this feature is particularly common. Among "lower" hamamelidids differentiation of calyx and corolla occurs only in *Hamamelidaceae* and perhaps (more equivocally) in *Platanus*. However, among 'basal' eudicots as a whole there are sufficient flowers in which sepals and petals are not differentiated (e.g., *Hydrastis*, *Glaucidium*, *Euptelea*, *Circaeaster*, certain *Lardizabalaceae*) to suggest that petals have arisen independently from staminodes several times (cf. *Kingdonia*, *Lardizabala*, *Sanguinaria*). This is also consistent with the view (HOOT 1991, TAMURA 1993) that the perianth of *Ranunculaceae* is primitively undifferentiated and that petals have arisen numerous times within the family.

**Unisexual versus bisexual flowers.** Bisexual flowers predominate among the 'basal' eudicots reviewed here, but it is perhaps surprising that unisexual flowers are relatively common among *Magnoliidae* (ENDRESS 1989b, 1990). On current evidence, the bisexual condition is basic but unisexual flowers appear to have arisen separately several times in different lineages. In most *Lardizabalaceae* and *Menispermaceae* the occurrence of unisexual flowers is more or less constant, and frequently associated with dioecy.

## Conclusions

Review of patterns of floral evolution among "basal" eudicots highlights significant similarities and differences compared to floral structure in magnoliids, and reinforces the view that *Ranunculidae* and "lower" *Hamamelididae* are in some sense "transitional" between magnoliids on the one hand and "higher" eudicots on the other. This critical phylogenetic position in the diversification of angiosperms is also supported by the long fossil history of several families in the eudicot alliance (Fig. 4). The earliest records of diagnostic eudicot pollen are of mid-late Barremian age (c. 126 myr BP) and by around the latest Albian (c. 97 myr BP) several groups (e.g., *Trochodendrales*, *Platanaceae*, *Buxaceae*, and perhaps *Circaeasteraceae*, *Myrothamnaceae*, and *Nelumbonaceae*) are recognizable in the fossil record. Possible *Hamamelidaceae* and *Proteaceae* are recorded by the Turonian (c. 90 myr BP). However, a striking feature of the fossil record as it is currently known is the relative paucity of putative ranunculids in the Late Cretaceous. Attempts to expand

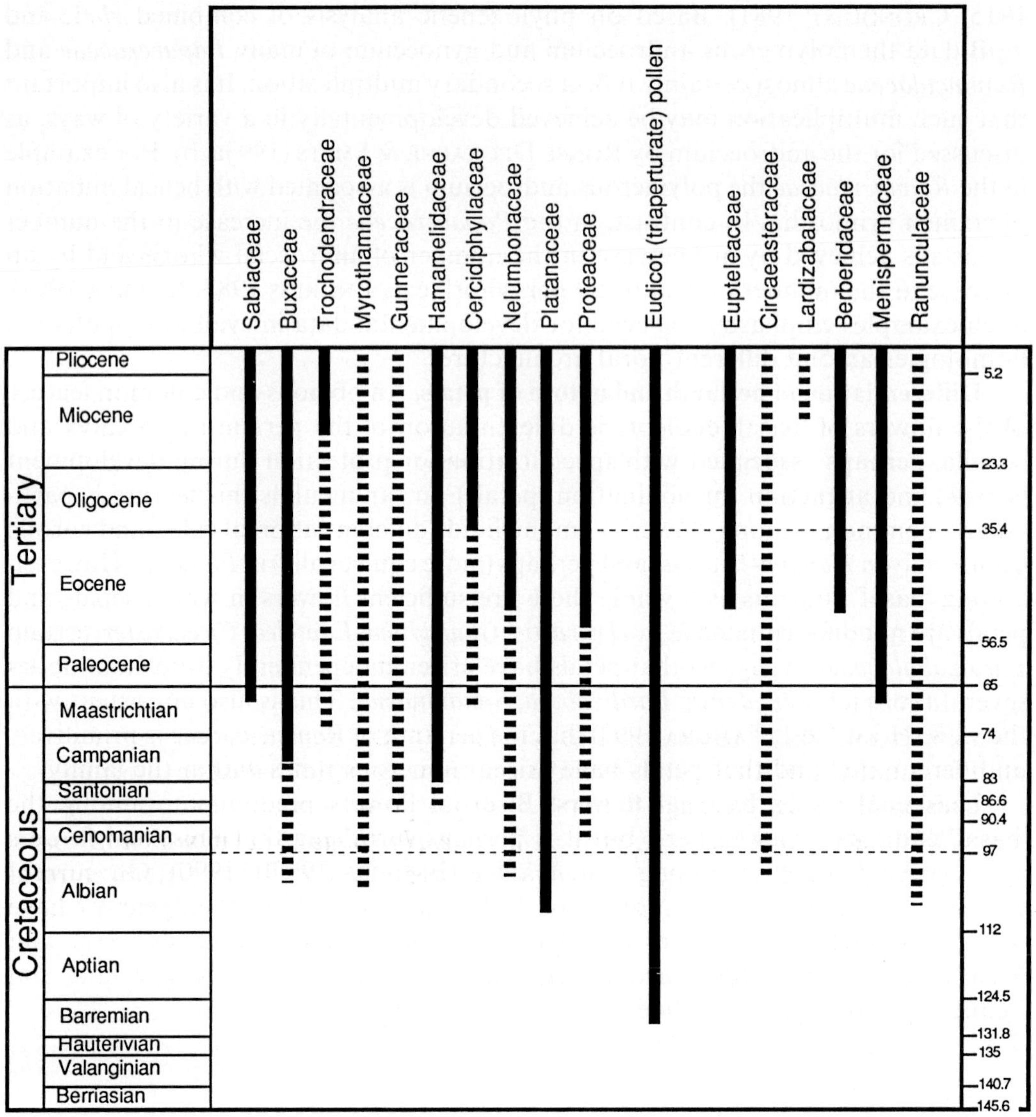


Fig. 4. First appearance in the fossil record of selected extant families of "basal" eudicots. Time scale based on HARLAND & al. (1989). Solid bars indicate unequivocal records of extant families. Dotted bars indicate records of extinct genera, dispersed pollen, or other dispersed organs, that are not securely referable to the family

the known fossil record of such groups are currently a high priority of ongoing studies of Late Cretaceous fossil floras.

For interpretation of the fossil record, a particularly intriguing result of the molecular phylogenetic studies is the possible close relationship between *Proteaceae* and *Platanaceae*. *Platanaceae* are among the most common fossils in mid-Cretaceous floras and the molecular results suggest that the *Proteaceae* may be a southern hemisphere vicariant of the platanoids that diverged very early in angiosperm evolution. A close relationship of these two families could potentially explain the platanoid aspect of many leaf taxa in Late Cretaceous floras from the Southern Hemisphere (K. JOHNSON, pers. comm.), and highlights the need for more detailed comparison of the architecture of pinnate/pinnatifid foliage in many extant *Proteaceae*, with similar foliage of the *Sapindopsis* type that is known to have been produced by certain early platanoids.

In terms of the overall evolution of floral form in the early diversification of angiosperms, the most striking feature of "basal" eudicots is that cyclic flowers with dimerous or trimerous floral architecture predominate. Indeed, based on our preliminary assessment of relationship (Figs. 2, 3), flowers with a cyclic arrangement of relatively few floral parts are probably primitive for eudicots as a whole. Flowers with numerous floral parts occur only in relatively derived groups (e.g., *Ranuculaceae*) or in isolated genera such as *Nelumbo* and *Trochodendron*. In *Papaveraceae*, the whorled ontogeny of the androecium indicates that the polymerous condition is clearly not homologous with that in magnoliids or ranunculids where helical initiation of primordia predominates (spiral polyandry sensu RONSE DECRAENE & SMETS 1993a, b).

If the basic condition in eudicot flowers is dimery or trimery with relatively few parts, then these flowers are very different from those flowers of magnoliid angiosperms that have traditionally been regarded as primitive (e.g., *Magnoliales*) (BESSEY 1915, CRONQUIST 1981). However, there is increasing evidence that the basic angiosperm flower type is small, few-parted and with cyclic floral phyllotaxy. Flowers of this kind predominate in the Early Cretaceous fossil record (FRIIS & al. 1994, CRANE & al. 1994), and several recent phylogenetic analyses of morphological and molecular data suggest that the angiosperm tree may be rooted in the vicinity of the paleoherbs (e.g., *Aristolochiaceae*, *Nymphaeaceae*) in which the basic floral type is oligomerous. Such analyses suggest that the floral architecture of "basal" eudicots could be close to the basic condition for angiosperms as a whole, and that polymerous magnolialean flowers of the classic "primitive" type may themselves be the result of secondary increase in the number of floral parts in a manner analogous to the situation in *Papaveraceae* and *Ranunculaceae*. Further analysis of phylogenetic and structural patterns at the magnoliid grade, as well as among "basal" eudicots, will be needed for further resolution of this issue.

We are grateful to Professors ELSE MARIE FRIIS and PETER K. ENDRESS for stimulating discussion and the opportunity to participate in this symposium at the XV International Botanical Congress, Yokohama. This work was supported in part by National Science Foundation Grants DEB-9020237 and INT-9015123 to PETER R. CRANE, DEB-9306533 to SARA B. HOOT, and Australia Research Council grant A190131108 to ANDREW N. DRINNAN.

## References

ARBER, A., 1931: Studies in floral morphology. III. On the *Fumarioideae* with special reference to the androecium. – New Phytol. **30**: 317–354.

– 1932: Studies in floral morphology. IV. On the *Hypecooideae* with special reference to the androecium. – New Phytol. **31**: 145–173.

BALFOUR, I. B., SMITH, W. W., 1914: *Kingdonia uniflora*. – Notes Roy. Bot. Gard. Edinburgh **8**: 191–192.

Barabé, D., Bergeron, Y., Vincent, G., 1987: La répartition des caractères dans la classification des *Hamamelididae* (*Angiospermae*). – Canad. J. Bot. **65**; 1756–1767.

Bentham, G., 1880: Ordo CLI. *Euphorbiaceae*. – In Bentham, G., Hooker, J. D. (Eds): Genera Plantarum **3**, pp. 239–340. – London: Reeve & Co.

– Hooker, J. D., 1883: Genera plantarum ad exemplaria imprimis in herbariis Kewensibus servata definita. – London: Reeve & Co.

Bessey, C. E., 1915: The phylogenetic taxonomy of flowering plants. – Ann. Missouri Bot. Gard **2**: 109–164.

Bharathan, G., Zimmer, E. A., 1994: Early branching events in monocotyledons – partial 18S ribosomal DNA sequence analysis. (in press).

Brenner, G. J., 1968: Middle Cretaceous spores and pollen from northeastern Peru. – Pollen Spores **10**: 341–383.

Chandler, M. E. J., 1964: The Lower Tertiary floras of Southern England. IV. A summary and survey of findings in the light of recent botanical observations. – London: British Museum (Natural History).

Chase, M. W., Soltis, D. E., Olmstead, R. G., Morgan, D., Les, D. H., Mishler, B. D., Duvall, M. R., Price, R. A., Hills, H. G., Qiu, Y.-L., Kron, K. A., Retting, J. H., Conti, E., Palmer, J. D., Manhart, J. R., Sytsma, K. J., Michaels, H. J., Kress, W. J., Karol, K. G., Clark, W. D., Hedrén, M., Gaut, B. S., Jansen, R. K., Kim, K.-J., Wimpee, C. F., Smith, J. F., Furnier, G. R., Strauss, S. H., Xiang, Q.-Y., Plunkett, G. M., Soltis, P. S., Swensen, S. M., Williams, S. E., Gadek, P. A., Quinn C. J., Eguiarte, L. E., Golenberg, E., Learn, G. H., Graham, S. W., Barrett, S. C. H., Dayanandan, S., Albert, V. A., 1993: Phylogenetics of seed plants: an analysis of nucleotide sequences from the plastid gene *rbc*L. – Ann. Missouri Bot. Gard. **80**: 528–580.

Christophel, D. C., 1984: Early Tertiary *Proteaceae*: The first floral evidence of *Musgraveinae*. – Austral. J. Bot. **32**: 177–186.

Collinson, M. E., 1983: Fossil plants of the London Clay. – London: The Palaeontological Association.

– Boulter, M. C., Holmes, P. L., 1993: *Magnoliophyta* ("*Angiospermae*"). – In Benton, M. J., (Ed): The fossil record **2**, pp. 809–841. – London: Chapman and Hall.

Crane, P. R., 1984: A re-evaluation of *Cercidiphyllum*-like plant fossils from the British early Tertiary. – Bot. J. Linn. Soc. **89**: 199–230.

– 1989: Paleobotanical evidence on the radiation of non-magnoliid dicotyledons. – Pl. Syst. Evol. **162**: 165–191.

– Stockey, R. A., 1985: Growth and reproductive biology of *Joffrea speirsii* gen. et sp. nov., a *Cercidiphyllum*-like plant from the late Paleocene of Alberta, Canada. – Canad. J. Bot. **63**: 340–364.

– – 1986: Morphology and development of pistillate inflorescences in extant and fossil *Cercidiphyllum*. – Ann. Missouri Bot. Gard. **73**: 382–393.

– Manchester, S. R., Dilcher, D. L., 1990: A preliminary survey of fossil leaves and well-preserved reproductive structures from the Sentinel Butte Formation (Paleocene) near Almont, North Dakota. – Fieldiana **20**: 1–63.

– – – 1991: Reproductive and vegetative structure of *Nordenskioldia* (*Trochodendraceae*), a vesselless dicotyledon from the Early Tertiary of the Northern Hemisphere. – Amer. J. Bot. **78**: 1311–1334.

– Pedersen, K. R., Friis, E. M., Drinnan, A. N., 1993: Early Cretaceous (early to middle Albian) platanoid inflorescences associated with *Sapindopsis* leaves from the Potomac Group of eastern North America. – Syst. Bot. **18**: 328–344.

– Friis, E. M., Pedersen, K. R., 1994: Paleobotanical evidence on the early radiation of magnoliid angiosperms. – Pl. Syst. Evol [Suppl.] 8: 51–72.

Crepet, W. L., Nixon, K. C., Friis, E. M., Freudenstein, J. V., 1992: Oldest fossil flowers of hamamelidaceous affinity, from the Late Cretaceous of New Jersey. – Proc. Natl. Acad. Sci. **89**: 8986–8989.

Croizat, L., 1941: On the systematic position of *Daphniphyllum* and its allies. – Lingnan Sci. J. **20**: 79–103.

– Metcalfe, F. P., 1941: The Chinese and Japanese species of *Daphniphyllum*. – Lingnan Sci. J. **20**: 104–127.

Cronquist, A., 1981: An integrated system of classification of flowering plants. – New York: Columbia University Press.

Dahlgren, R. M. T., Clifford, H. T., Yeo, P. F., 1985: The families of the monocotyledons. – Berlin: Springer.

Donoghue, M. J., Doyle, J. A., 1989a: Phylogenetic studies of seed plants and angiosperms based on morphological characters. – In Fernholm, B., Bremer, K., Jornvall, H., (Eds): The hierarchy of life, pp. 181–193. – Amsterdam: Elsevier Excerpta Medica.

– – 1989b: Phylogenetic analysis of the angiosperms and the relationships of "*Hamamelidae*". – In Crane, P. R., Blackmore, S., (Eds): Evolution, systematics and fossil history of the *Hamamelidae*, pp. 17–45. – Oxford: Clarendon Press.

Doyle, J. A., 1969: Cretaceous angiosperm pollen of the Atlantic Coastal Plain and its evolutionary significance. – J. Arnold Arb. **50**: 1–35.

–, Donoghue, M. J., 1992: Fossils and seed plant phylogeny reanalyzed. – Brittonia **44**: 89–106.

– – 1993: Phylogenies and angiosperm diversification. – Paleobiology **19**: 141–167.

– – Zimmer, E. A., 1994: Integration of morphological and ribosomal RNA data on the origin of angiosperms. – Ann. Missouri Bot. Gard. **81**: 405–418.

– Hotton, C. L., 1991: Diversification of early angiosperm pollen in a cladistic context. – In Blackmore, S., Barnes, S. H., (Eds): Pollen and spores, pp. 169–196. – Oxford: Clarendon Press.

Drinnan, A. N., Crane, P. R., Friis, E. M., Pedersen, K. R., 1991: Angiosperm flowers and tricolpate pollen of buxaceous affinity from the Potomac Group (mid-Cretaceous) of eastern North America. – Amer. J. Bot. **78**: 153–176.

Endress, P. K., 1967: Systematische Studie über die verwandtschaftlichen Beziehungen zwischen den Hamamelidaceen und Betulaceen. – Bot. Jahrb. Syst. **87**: 431–527.

– 1969: Gesichtspunkte zur systematischen Stellung der Eupteleaceen (*Magnoliales*). – Ber. Schweiz. Bot. Ges. **79**: 229–278.

– 1977: Evolutionary trends in the *Hamamelidales-Fagales*-group. – Pl. Syst. Evol. [Suppl.] **1**: 321–347.

– 1986a: Floral structure, systematics and phylogeny in *Trochodendrales*. – Ann. Missouri Bot. Garden **73**: 297–324.

– 1986b: Reproductive structures and phylogenetic significance of extant primitive angiosperms. – Pl. Syst. Evol. **152**: 1–28.

– 1987: Floral phyllotaxis and floral evolution. – Bot. Jahrb. Syst. **108**: 417–438.

– 1989a: Aspects of evolutionary differentiation of the *Hamamelidaceae* and the Lower *Hamamelididae*. – Pl. Syst. Evol. **162**: 193–211.

– 1989b: A suprageneric taxonomic classification of the *Hamamelidaceae*. – Taxon **38**: 371–376.

– 1989c: The systematic position of the *Myrothamnaceae*. – In Crane, P. R., Blackmore, S. (Eds): Evolution, systematics, and fossil history of the *Hamamelidae* 1, pp. 193–200. – Oxford: Clarendon Press.

– 1989d: Chaotic floral phyllotaxis and reduced perianth in *Achlys* (*Berberidaceae*). – Bot. Acta **102**: 159–163.

– 1990: Evolution of reproductive structures and functions in primitive angiosperms (*Magnoliidae*). – Mem. New York Bot. Gard. **55**: 5–34.

– 1993: *Eupteleaceace/Hamamelidaceae.* – In Kubitzki, K., Rohwer, J. G., Bittrich, V., (Eds): The families and genera of vascular plants, II, pp. 299–301/322–331. – Berlin: Springer.
– Friis, E. M., 1991: *Archamamelis*, hamamelidalean flowers from the Upper Cretaceous of Sweden. – Pl. Syst. Evol. **175**: 101–114.
Erdtman, G., 1952: Pollen morphology and plant taxonomy. An introduction to palynology. I. Angiosperms. – Stockholm: Almqvist & Wiksell.
Foster, A. S., 1961: The floral morphology and relationships of *Kingdonia uniflora.* – J. Arnold Arbor. **42**: 397–416.
– 1963: The morphology and relationships of *Circaeaster.* – J. Arnold Arbor. **44**: 299–321.
Friis, E. M., Crane, P. R., 1989: Reproductive structures of Cretaceous *Hamamelidae.* – In Crane, P. R., Blackmore, S., (Eds): Evolution, systematics and fossil history of the *Hamamelidae*, **II**, pp. 155–174. – Oxford: Clarendon Press.
– Pedersen, K. R., Crane, P. R., 1994: Angiosperm floral structures from the Early Cretaceous of Portugal. – Pl. Syst. Evol. [Suppl.] **8**: 31–49.
Hallier, H., 1904: Ueber die Gattung *Daphniphyllum*, ein Uebergangsglied von den Magnoliaceen und Hamamelidaceen zu den Kätzchenblüthlern. – Bot. Mag. (Tokyo) **18**: 55–69.
Harland, W. B., Armstrong, R. L., Cox, A. V., Craig, L. E., Smith, A. G., Smith, D. G., 1989: A geologic time scale 1989. – Cambridge: Cambridge University Press.
Heimsch, C., 1942: Comparative anatomy of the secondary xylem of the '*Gruinales*' and '*Terebinthales*' of Wettstein with reference to taxonomic groupings. – Lilloa **8**: 83–198.
Hoot, S. B., 1991: Phylogeny of the *Ranunculaceae* based on epidermal microcharacters and macromorphology. – Syst. Bot. **16**: 741–755.
– Culham, A., Crane, P. R., 1994: The utility of *atp*B gene sequences in resolving relationship in the *Lardizabalaceae*, including comparisons with *rbc*L and 18S ribosomal DNA sequences. – Ann. Missouri Bot. Gard. (in press).
Huang, T.-C., 1965: Monograph of *Daphniphyllum* (1). – Taiwania **11**: 57–98.
– 1966: Monograph of *Daphniphyllum* (2). – Taiwania **12**: 137–234.
Hufford, L., 1992: *Rosidae* and their relationship to other nonmagnoliid dicotyledons: a phylogenetic analysis using morphological and chemical data. – Ann. Missouri Bot. Gard. **79**: 218–248.
– Crane, P. R., 1989: A preliminary cladistic analysis of the 'lower' Hamamelidae. – In Crane, P. R., Blackmore, S., (Eds): Evolution, systematics, and fossil history of the *Hamamelidae*, I, pp. 175–192. – Oxford: Clarendon Press.
Hutchinson, I., 1959: The families of flowering plants. 1. Dicotyledons, 2nd edn. – Oxford: Clarendon Press.
Jäger-Zürn, I., 1966: Infloreszenz- und blütenmorphologische, sowie embryologische Untersuchungen an *Myrothamnus* Welw. – Beitr. Biol. Pfl. **42**: 241–271.
Janchen, E., 1949: Die systematische Gliederung der Ranunculaceen und Berberidaceen. – Österr. Akad. Wiss., Math. – Naturwiss. Kl., Denkschr. **108**: 1–82.
Jarzen, D. M., 1980: The occurrence of *Gunnera* pollen in the fossil record. – Biotropica **12**: 117–123.
– Dettmann, M. E., 1989: Taxonomic revision of *Tricolpites reticulatus* Cookson ex Couper, 1953 with notes on the biogeography of *Gunnera* L. – Pollen Spores **31**: 97–112.
Johnson, L. A. S., Briggs, B. G., 1975: On the *Proteaceae* – the evolution and classification of a southern family. – Bot. J. Linn. Soc. **70**: 83–182.
– Wilson, K. L., 1989: *Casuarinaceae*: a synopsis. – In Crane, P. R., Blackmore, S., (Eds): Evolution, systematics and fossil history of the *Hamamelidae*, II, pp. 167–188. – Oxford: Clarendon Press.
Kadereit, J. W., 1993: *Papaveraceae.* – In Kubitzki, K., Rohwer, J. G., Bittrich, V., (Eds): The families and genera of vascular plants, II, pp. 494–506. – Berlin: Springer.

KARRER, A. B., 1991: Blütenentwicklung und systematische Stellung der *Papaveraceae* und *Capparaceae*. – Ph.D. Thesis, Universität Zürich.
KESSLER, P. J. A., 1993: *Menispermaceae*. – In KUBITZKI, K., ROHWER, J. G., BITTRICH, V., (Eds): The families and genera of vascular plants, II, pp. 402–418. – Berlin: Springer.
LANGLET, O., 1932: Über Chromosomenverhältnisse und Systematik der *Ranunculaceae*. – Svensk Bot. Tidskr. **26**: 381–401.
LEHMANN, N. L., SATTLER, R., 1993: Homeosis in floral development of *Sanguinaria canadensis* and *S. canadensis* 'multiplex' (*Papaveraceae*). – Amer. J. Bot. **80**: 1323–1335.
LIDÉN, M., 1993: *Fumariaceae*. – In KUBITZKI, K., ROHWER, J. G., BITTRICH, V., (Eds): The families and genera of vascular plants, II, pp. 310–318. – Berlin: Springer.
LOCONTE, H., 1993: *Berberidaceae*. – In KUBITZKI, K., ROHWER, J. G., BITTRICH, V., (Eds): The families and genera of vascular plants, II, pp. 147–152. – Berlin: Springer.
– ESTES, J. R., 1989: Phylogenetic systematics of *Berberidaceae* and *Ranunculales* (*Magnoliidae*). – Syst. Bot. **14**: 565–579.
– STEVENSON, D. W., 1991: Cladistics of the *Magnoliidae*. – Cladistics **7**: 267–296.
MABBERLEY, D. J., 1987: The plant book. – Cambridge: Cambridge University Press.
MAI, D. H., 1980: Zur Bedeutung von Relikten in der Florengeschichte. – In VENT, W., (Ed.): 100 Jahre Arboretum 1879–1979, pp. 281–307. – Berlin.
MANCHESTER, S. R., 1986: Vegetative and reproductive morphology of an extinct plane tree (*Platanaceae*) from the Eocene of Western North America. – Bot. Gaz. **147**: 200–226.
– 1994: The flora of the Clarno Chert. – Palaeontographica Americana **58**: 1–205.
– CRANE, P. R., DILCHER, D. L., 1991: *Nordenskioldia* and *Trochodendron* (*Trochodendraceae*) from the Miocene of Northwestern North America. – Bot. Gaz. **152**: 357–368.
MANOS, P. S., NIXON, K. C., DOYLE, J. J., 1993: Cladistic analysis of restriction site variation within the chloroplast DNA inverted repeat region of selected *Hamamelididae*. – Syst. Bot. **18**: 551–562.
MEACHAM, C. A., 1980: Phylogeny of the *Berberidaceae* with an evaluation of classifications. – Syst. Bot. **5**: 149–172.
MUNZ, P. A., 1959: A Californian flora. – Berkeley: University of California Press.
MURBECK, S., 1912: Untersuchungen über den Blütenbau der Papaveraceen. – Kungl. Svenska Vet. Akad. Handl. **50**: 1–168.
OLIVER, D., 1895: *Circaeaster agrestis* MAXIM – In HOOKER, J. D., (Ed.): Icones Plantarum, IV, p.4.
PAYER, J. B., 1857: Traité d'organogénie comparée de la fleur. – Paris: Masson.
PIGG, K. B., STOCKEY, R. A., 1991: Plantanaceous plants from the Paleocene of Alberta, Canada. – Rev. Palaeobot. Palynol. **70**: 125–146.
QIN, H.-N., 1989: An investigation of carpels of *Lardizabalaceae* in relation to taxonomy and phylogeny. – Cathaya **1**: 61–82.
ROHWEDER, O., ENDRESS, P. K., 1983: Samenpflanzen, Morphologie und Systematik der Angiospermen und Gymnospermen. – Stuttgart: Thieme.
RONSE DECRAENE, L. P., SMETS, E. F., 1990: The systematic relationship between the *Begoniaceae* and *Papaveraceae*: a comparative study of their floral development. – Bull. Jard. Bot. Nat. Belgique. **60**: 229–273.
– – 1992: An updated interpretation of the androecium of the *Fumariaceae*. – Canad. J. Bot. **70**: 1765–1776.
– – 1993a: Dédoublement revisited: towards a renewed interpretation of the androecium of the *Magnoliophytina*. – Bot. J. Linnean Soc. **113**: 103–124.
– – 1993b: The distribution and systematic relevance of the androecial character polymery – Bot. J. Linnean Soc. **113**: 285–350.
ROSENTHAL, K., 1916: Monographie der Gattung *Daphniphyllum*. – Dissertation, Breslau.
SATTLER, R., 1973: Organogenesis of flowers: a photographic text-atlas. – Toronto: University of Toronto Press.

Schwarzwalder, R. N., Dilcher, D. L., 1991: Systematic placement of the *Platanaceae* in the *Hamamelidae*. – Ann Missouri Bot. Gard. **78**: 962–969.

Smith, A. C., 1946: A taxonomic review of *Euptelea*. – J. Arnold Arbor. **27**: 215–226.

Stebbins, G. L., 1974: Flowering plants: evolution above the species level. – Cambridge, MA: Belknap Press of Harvard University Press.

Stockey, R. A., Pigg, K. B., 1991: Flowers and fruits of *Princetonia allenbyensis* (*Magnoliopsida*; family indet.) from the Middle Eocene Princeton chert of British Columbia. – Rev. Palaeobot. Palynol. **70**: 163–172.

Sutton, D. A., 1989: *Daphniphyllales*. – In Crane, P. R., Blackmore, S., (Eds): Evolution, systematics, and fossil history of the *Hamamelidae*, I, pp. 285–291. – Oxford: Clarendon Press.

Swofford, D. L., 1993: PAUP: Phylogenetic analysis using parsimony, version 3.1. Champaign, IL: Illinois Natural History Survey.

Takhtajan, A., 1987: Systema Magnoliophytorum. – Leningrad: Nauka.

Tamura, M., 1965: Morphology, ecology and phylogeny of the *Ranunculaceae*. – Sci. Rep. Osaka Univ. **14**: 52–71.

– 1972: Morphology and phyletic relationship of *Glaucidiaceae*. – Bot. Mag. (Tokyo) **85**: 29–41.

– 1993: *Ranunculaceae*. – In Kubitzki, K., Rohwer, J. G., Bittrich, V., (Eds): The families and genera of vascular plants, II, pp. 563–583. – Berlin: Springer.

Taylor, B. A. S., 1967: The comparative morphology and phylogeny of the *Lardizabalaceae*. – Ph.D. Thesis, Indiana University.

Thorne, R. F., 1983: Proposed new realignments in the angiosperms. – Nordic J. Bot. **3**: 85–117.

Tobe, H., Keating, R. C., 1985: The morphology and anatomy of *Hydrastis* (*Ranunculales*): Systematic reevaluation of the genus. – Bot. Mag. (Tokyo) **98**: 291–316.

Upchurch, G. R., Crane, P. R., Drinnan, A. N., 1994: The megaflora from the Quantico locality (Upper Albian), Lower Cretaceous Potomac Group of Virginia. – Virginia Museum of Natural History, Mem. **4**: 1–57.

Van Beusekom, C. F., 1971: Revision of *Meliosma* (*Sabiaceae*), section *Lorenzanea* excepted, living and fossil, geography and phylogeny. – Blumea **19**: 355–529.

Ward, J. V., Doyle, J. A., 1988: Possible affinities of two triporoidate tetrads from the mid-Cretaceous of Laurasia. – Abstracts, 7th Int. Palyn. Congress, Brisbane: 179.

– – 1994: Ultrastructure and relationships of mid-Cretaceous polyporate and triporate pollen from northern Gondwana. – In Kurmann, M. H., Doyle, J. A., (Eds): Ultrastructure of fossil spores and pollen, pp. 161–172. – Kew: Royal Botanic Gardens.

Williamson, P. S., Schneider, E. L., 1993: *Nelumbonaceae*. – In Kubitzki, K., Rohwer, J. G., Bittrich, V., (Eds): The families and genera of vascular plants, II, pp. 470–473. – Berlin: Springer.

Wu, C.-Y., Kubitzki, K., 1993a: *Circaeasteraceae*. – In Kubitzki, K., Rohwer, J. G., Bittrich, V., (Eds): The families and genera of vascular plants, II, pp. 288–289. – Berlin: Springer.

– – 1993b: *Lardizabalaceae*. – In Kubitzki, K., Rohwer, J. G., Bittrich, V., (Eds): The families and genera of vascular plants, II, pp. 361–365. – Berlin: Springer.

Addresses of the authors: Andrew N. Drinnan, School of Botany, University of Melbourne, Parkville 3052, Victoria, Australia. – Peter R. Crane and Sara B. Hoot, Department of Geology, The Field Museum, Roosevelt Road at Lake Shore Drive, Chicago, IL 60605, USA.

Accepted April 27, 1994 by P. K. Endress

Pl. Syst. Evol. [Suppl.] 8: 123–135 (1994)

# *Elsemaria*, a Late Cretaceous angiosperm fructification from Hokkaido, Japan*

HARUFUMI NISHIDA

Received March 21, 1994

**Key words:** Angiosperms, *Magnoliatae, Dilleniidae, Elsemaria.* – Paleobotany, morphology, permineralization, placentation, Cretaceous.

**Abstract:** *Elsemaria kokubunii* gen. et spec. nov. is a permineralized angiosperm fruit from the Upper Cretaceous of Hokkaido, Japan. The dry dehiscent capsule has axile placentation consisting of ten carpels that are laterally fused with each other. Intersecting the carpels are ten columnar structures that may be interpreted as reduced carpels. Dehiscence is loculicidal. Capsular fruits similar to *Elsemaria* occur in some extant groups of the *Dilleniidae*, but the lack of stamens and other floral parts prevent a detailed discussion of the systematic affinities of the fossil. The morphology of *Elsemaria* suggests that in some angiosperms axile placentation may have evolved through the fusion of conduplicate carpels.

Angiosperms are characterized by the presence of carpels forming a diverse array of gynoecia. Based mainly on comparative morphology of living taxa, an apocarpous gynoecium with conduplicate carpels that mature into follicles is considered by several authors to be the most primitive condition in angiosperms (BAILEY & SWAMY 1951, CRONQUIST 1968, WEBERLING 1989, TAKHTAJAN 1991). Findings of Cretaceous conduplicate carpels of simple morphology support the hypothesis that this condition evolved early in the history of angiosperms (DILCHER & CRANE 1984, NISHIDA & NISHIDA 1988). For a discussion of the derivation of the angiosperm carpel see also DOYLE (1994).

Neobotanical evidence suggests that the multilocular ovary with axile placentation has evolved from conduplicate carpels through union and lateral fusions (e.g., TAKHTAJAN 1991). The permineralized angiosperm capsule described here may represent the first direct fossil evidence supporting this hypothesis. The capsule shows axile placentation formed from carpels that still exhibit a primitive conduplicate nature. This morphology may represent one of phylogenetic intermediate states between free conduplicate and syncarpous gynoecia.

*Structure and affinities of petrified plants from northern Japan and Saghalien. XVII.

## Material and methods

The fruit was collected by Mr Hakuji Kokubun of Mikasa City, Hokkaido, in the river bed of the Obirashibe River near the town of Obira, Hokkaido (Fig. 1). The specimen is embedded in a calcium-carbonate nodule and was split into the vertical halves (Figs. 2, 3a). Similar plant-containing nodules occur abundantly in the Upper Cretaceous shallow marine sediments of the Upper Yezo Group which are exposed in the upper courses of the Obirashibe River. The geology of the area has been reviewed by Tsushima & al. (1958).

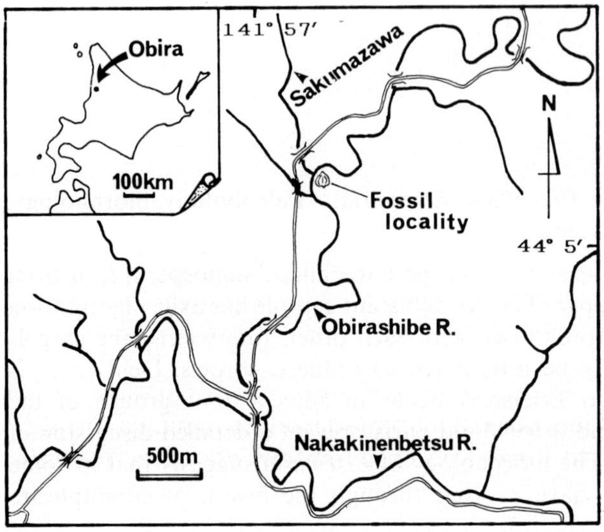


Fig. 1. Map showing fossil locality

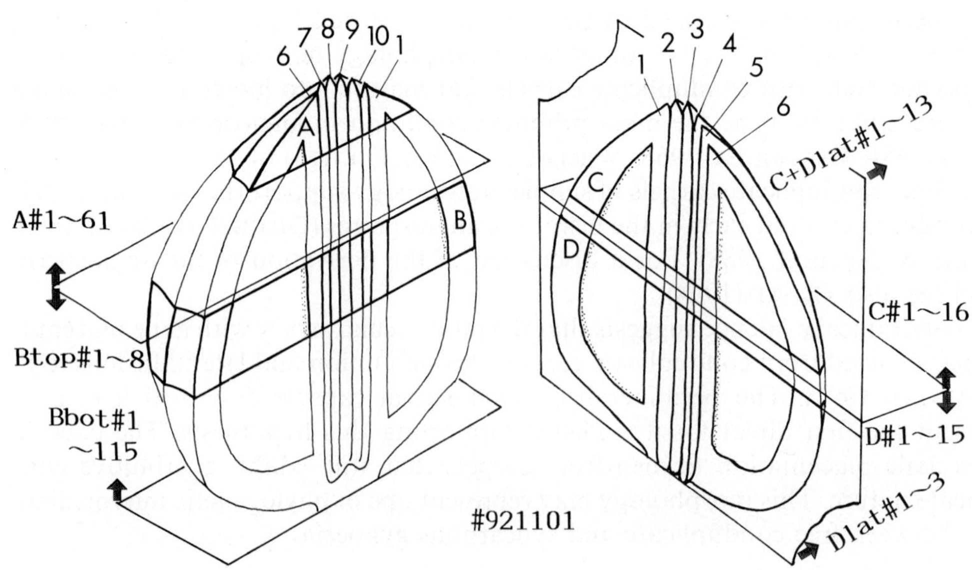


Fig. 2. Diagrammatic figure showing orientations of sections (A-D pieces of the specimen remained). Carpel numbers correspond to those in other Figures

The stratigraphy and global correlations are mainly based on marine molluscs associated with the permineralized plant debris (TANABE & al. 1977). Both lithological and paleontological information clearly indicate that the fossil belongs to the Upper Yezo Group. The age of this group in the upper courses of the Obirashibe River is estimated as Coniacian to Santonian.

The specimen is deposited in the Laboratory of Phylogenetic Botany, Faculty of Science, Chiba University (#921101). It was divided into four pieces to obtain sections for the anatomical study (Fig. 2). Micropreparations were made using the standard peel technique (JOY & al. 1956). The peels were mounted in Canada Balsam. Internal structures were reconstructed from composite camera lucida drawings of serial preparations.

## Results

### Systematics

Division *Angiospermae*
Class *Magnoliatae*
Subclass *Dilleniidae*
Order and family incertae sedis
***Elsemaria*** H. NISHIDA gen. nov.
Type: ***Elsemaria kokubunii*** H. NISHIDA spec. nov (Figs. 3–9).

**Diagnosis:** Multilocular angiosperm fruit or ovary composed of ten laterally fused carpels. Peduncle woody. Dehiscence loculicidal. Individual carpels basically conduplicate; triangular in cross section; adaxial margins deeply invaginated towards the locule and fused to each other to form a parenchymatous placenta. Placenta basally massive, apically bipartite in two longitudinal plates. More than 40 anatropous (?), bitegmic (?) ovules per locule. Each carpel supplied by two ventral, two placental and one dorsal bundle derived from a single bundle. Two major lateral bundles on two dorsal corners of carpel, forming a network with dorsal and ventral bundles. Ten sclerenchymatous columns each accompanied with massive vascular bundles of basically the same organization as carpel bundles surround a hollow cylinder in capsule center. Longitudinal slit between sclerenchymatous columns continuous with ventral suture of carpel and a middle split line of placenta. Seeds small, pyriform. Seed-coat thin; outer layer of seed coat composed of longitudinally-elongated, thick-walled cells; inner layer membranous.

**Derivation of generic name:** After ELSE MARIE FRIIS, Swedish Museum of Natural History, for her contributions to the studies of early angiosperms.

***Elsemaria kokubunii*** H. NISHIDA spec. nov (Figs. 3–9).

**Holotype:** Specimen no. 921101, housed in the Laboratory of Phylogenetic Botany, Faculty of Science, Chiba University.

**Diagnosis:** As for the genus.

**Derivation of species epithet:** After Mr HAKUJI KOKUBUN who collected the specimen.

**Description:** The fruit is 13.5 mm in diameter and 11.7 mm high (Fig. 3a, b). Because it is completely embedded in the carbonate nodule, the fruit morphology as well as internal structure can only be reconstructed from serial sections. Most of the external surface is abraded, and loss of stamens and other floral parts together with some fungal damage, indicate that the fossil was somewhat degraded before fossilization and was probably transported a considerable distance.

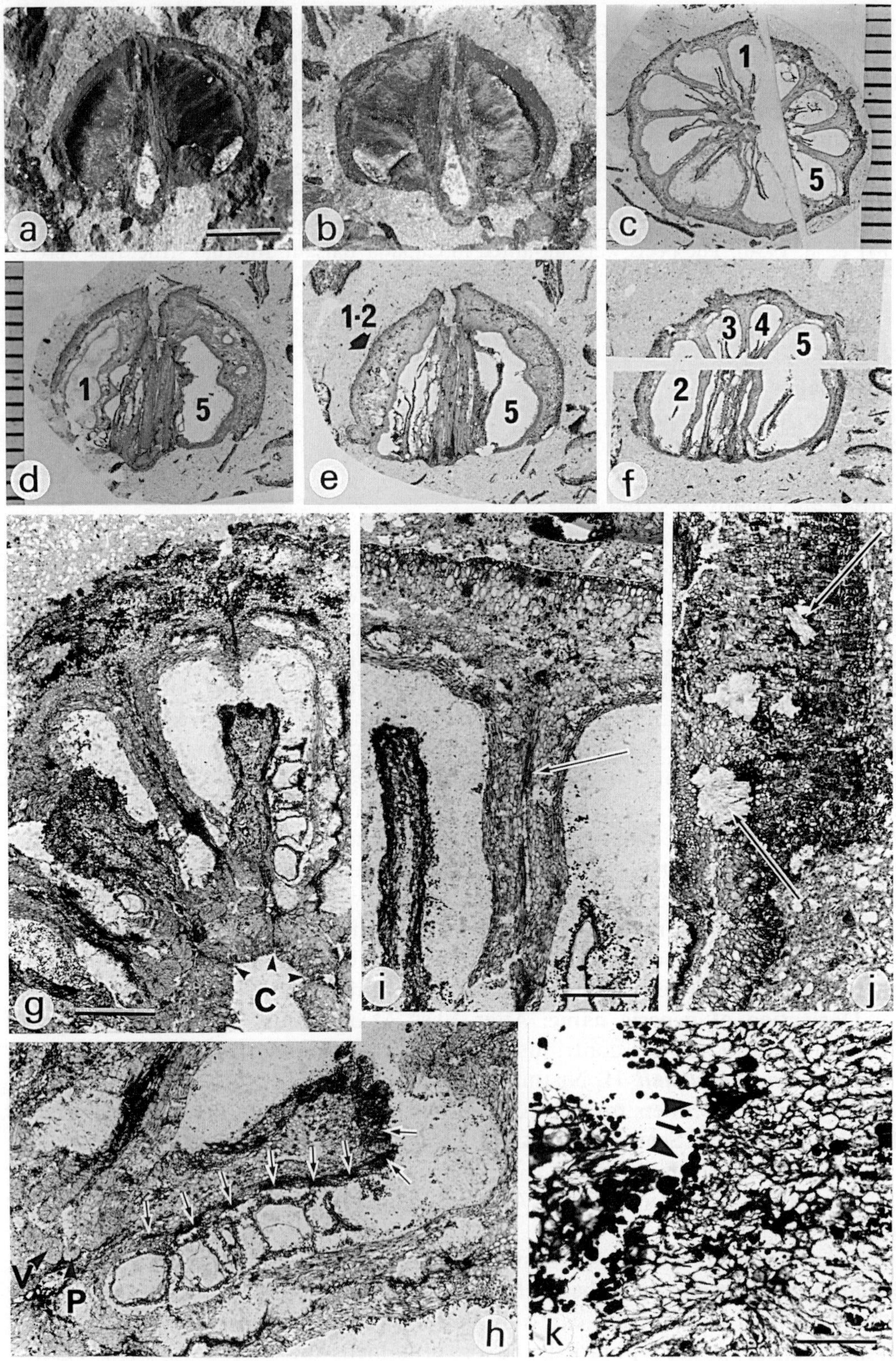
a
b
c
1
5
d
1
5
e
1·2
5
f
3
4
5
2
g
c
i
j
h
V
P
k

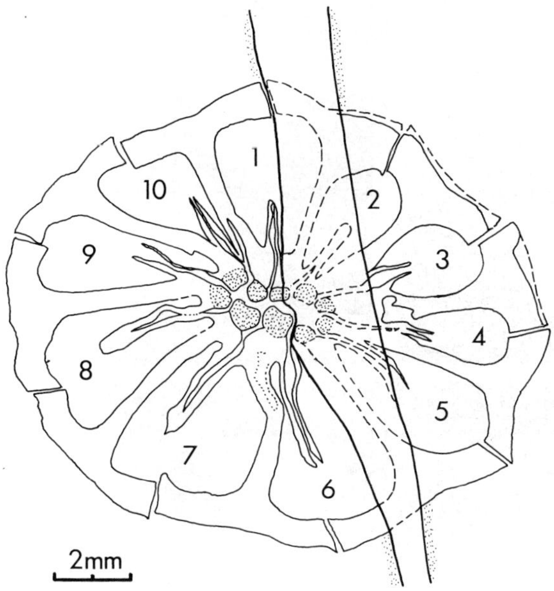


Fig. 4. *Elsemaria kokubunii* gen. et spec. nov. Illustrated reconstruction of a cross section based on slides shown in Fig. 3c

The fruit is a capsule consisting of a whorl of ten laterally fused carpels (Figs. 3c–f, 4). Dehiscence is loculicidal. In cross section at the widest part of the capsule, the carpel segments are triangular up to 6.8 mm deep (radially) and up to 4.5 mm wide (tangentially) (Fig. 3g). The placentae are parenchymatous extending radially from the ventral margins of the carpel. Basally they are massive (Fig. 3e, g, h, i), but towards the apex they separate into two longitudinal halves (Fig. 3c, f, i).

There is a hollow cylinder in the center of the capsule, surrounded by ten densely spaced sclerenchymatous columns that alternate with carpels (Figs. 3g, 4, 6c). The longitudinal slit between the sclerenchymatous columns is connected to the ventral suture of the carpel and further inside to the middle line of the massive placenta along which the placenta splits. Horizontally elongated tubular cells filled with black contents occur frequently along the middle split-line of the placenta (Fig. 5a). The external surface of the sclerenchymatous columns consists of cells also containing a dark substance (Fig. 5b). Styles have not been observed and may have broken off since the fossil is preserved in the fruiting or postanthetic stage.

◀

Fig. 3. *Elsemaria kokubunii* gen. et spec. nov. (C central hollow cylinder, P placenta bundle, V ventral bundle). *a*, *b* Two halves of original specimen. *c* Cross sections of two halves showing ten locules, slides Bbot#115 (left) and D#1, see Fig. 4. *d*, *e* Longitudinal sections C + Dlat#3 & #7 (1.2 wall between carpels 1 and 2). *f* Cross section D#13 and longitudinal section Dlat#2. *g* Cross section of capsule enlarged, arrowheads indicate the slit between sclerenchyma columns, Bbot#45. *h* Parenchymatous placenta and bundles to ovules (arrows), Bbot#45. *i* Lateral bundle in lateral wall (arrow), Bbot#66. *j* Cross section of dorsal parenchymatous projection of carpel showing radially aligned cells and vascular bundles (arrows), Bbot#48. *k* Cross section of carpel base showing a dorsal bundle being split into two (arrowheads) by dorsal dehiscence line (arrow), parenchyma of placenta base on the left, Bbot#13. – Bars: *a* 5 mm, *h* 0.5 mm, *k* 0.1 mm; *a–f*, *h–j* at same magnification; *c*, *d* scales in mm

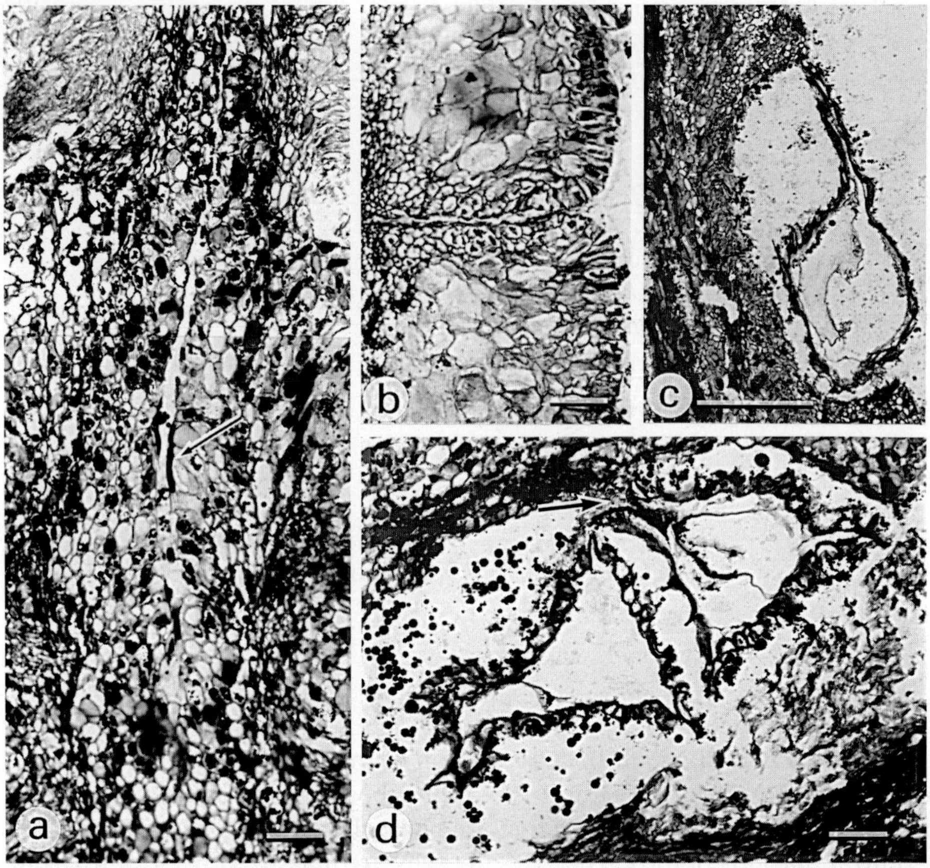


Fig. 5. *Elsemaria kokubunii* gen. et spec. nov. *a* Cross section of placenta showing dark tubelike cell (arrow) along a median split line, Bbot#16. *b* Cross section of two adjacent sclerenchymatous columns showing a slit and surface cells, Bbot#15. *c* Longitudinal section of ovule attached to the placenta, C + Dlat#2. *d* cross section of ovules showing wall structure and a possible micropylar end (arrow), Bbot#17. – Bars: *a*, *b*, *d* 10 μm, *c* 0.5 mm

Fig. 6. *Elsemaria kokubunii* gen. et spec. nov. (*C* central hollow cylinder, *D* dorsal bundle, *F* funicular bundle, *IC* intercarpellary structure, *PA* parenchyma of intercarpellary structure, *P* placenta bundle, *S* seed or ovule, *V* ventral bundle). *a–c* Serial cross sections of stipe and capsule base. *a* Top of stipe, Bbot#2. *b* Floral apex, vascular cylinder being divided, and slits between sclerenchyma columns (arrowheads), Bbot#9. *c* Basalmost section of a carpel, Bbot#12. *d* Fig. 3*c* enlarged, showing vascular bundles to the carpel and two intercarpellary structures besides, arrowheads indicate a set of bundles to intercarpellary structure consisting of a central (larger) and pair of sister (smaller) bundles. *e* Cross section of a carpel base showing pair of basalmost bundles branched from placenta bundles (white-margined arrows) providing funicular bundles, Bbot#27, a set of bundles to intercarpellary structure is enclosed by white-margined bars. *f* Triangular vascular bundle mass of the intercarpellary structure (enclosed by white-margined bars), Bbot#41. *g* The bundle mass in *f* dorsally departing a bundle inbetween the fused lateral walls of two adjacent carpels (arrow), Bbot#48. *h* Small bundle from intercarpellary structure splitting the lateral wall of fused carpels, upper side dorsal, Bbot#48. *i*, *j* Longitudinal section of placentae (shadow) in locules (single number) and fused lateral walls (double number), arrowheads in *i* indicate bundles in the placenta, in *j* indicate split lines in lateral wall, see Fig. 7*b*, C + Dlat#10. – Bars: *a*, *c*, *i* 1 mm, *d* 0.5 mm; *a*, *b* and *d–h* at same magnification

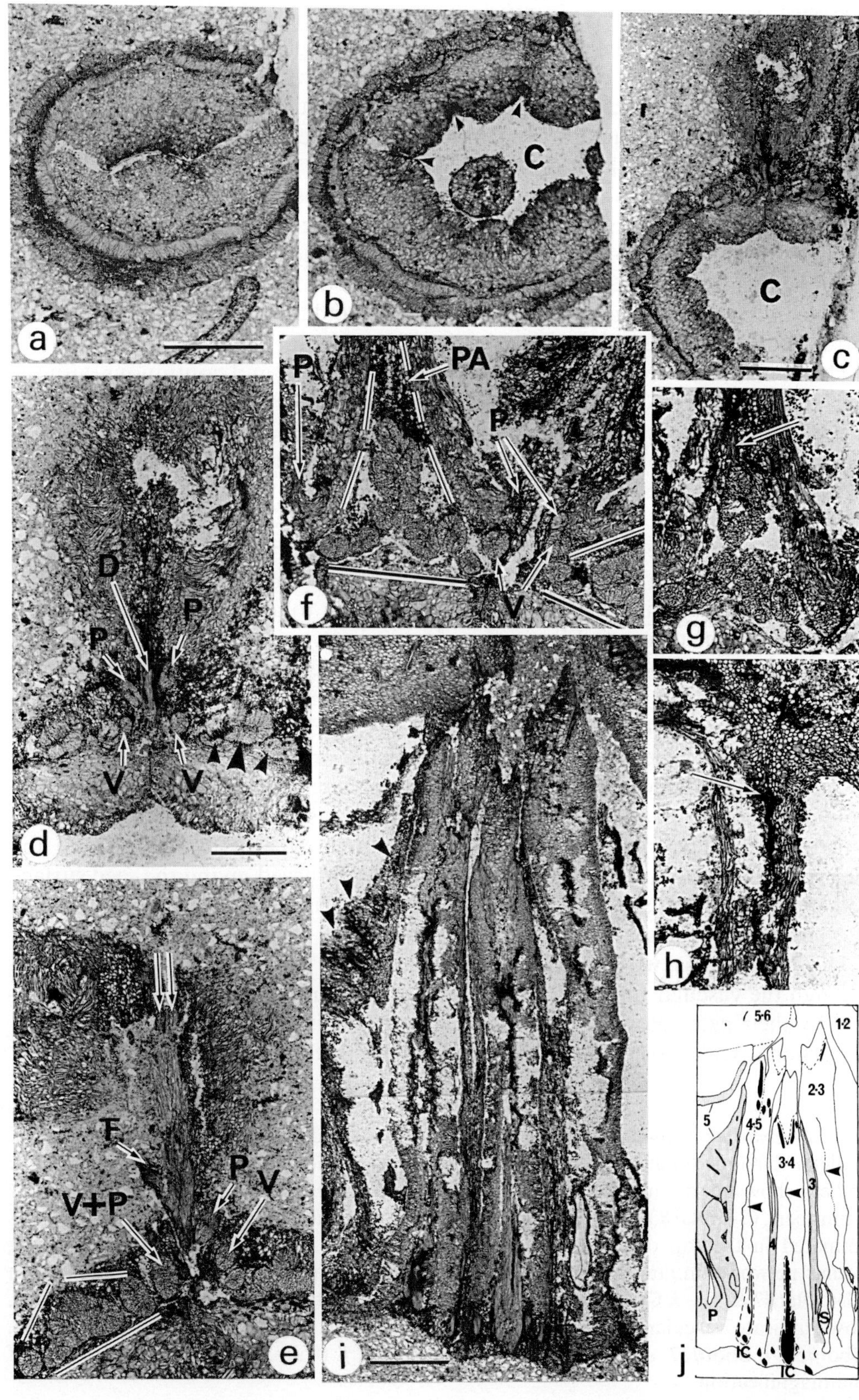
a
b
C
c
C
P
PA
P
f
V
g
D
P
P
V
V
d
h
T
P
V
V+P
e
i
5·6
1·2
2·3
5
4·5
3·4
3
4
P
IC
IC
j

Dark-coloured parenchyma fills the triangular space between each sclerenchymatous column and the walls of two adjacent carpels in cross section (Fig. 6f). The parenchyma encloses a central triangular mass of vascular tissue. The sclerenchymatous column and the parenchyma with its triangular mass of vascular tissues together constitute a structural unit intersecting the carpels. This unit may here be termed the intercarpellary structure.

The lateral walls of the carpels are thick and fused with the neighbouring carpels over their entire surfaces (Figs. 3e, g, i, 7d). The fused walls form a septum which is composed of a fibrous layer lining the locule and a parenchymatous middle layer. The locules are horizontally widened in four or five areas around the seeds and constricted inbetween (Fig. 3a,b,d). Lateral bundles occasionally traverse the ventral and dorsal sides of the carpel (Fig. 3i).

The dorsal walls are thicker but histologically similar and continuous with the septa, although their external layer is composed of larger, thick-walled cells (Fig. 3i). Small rounded protrusions consisting of a radial series of storied parenchyma cells occur dorsally at the base of the fruit in front of the septa (Fig. 3a, b, d, e, j). These protrusions are traversed by several vascular offshoots from the lateral bundles (Fig. 3j).

Ovules are attached to the placenta along four or five rows of obliquely to horizontally ascending pairs of vascular bundles from which the funicular bundles separate (Figs. 3h, 6e, i, j). The lowermost pair has a total of ten ovules with five on each side of the placenta. Eight pairs of funicular traces are found in the second series which corresponds to the widest space of the locule (Fig. 3h). Based on these numbers and the volume of each locule the total number of ovules per locule is estimated to be more than 40.

The ovules are pendulous from the placenta and probably anatropous (Fig. 5c). The seed wall is composed of two distinct layers, here interpreted as formed from an outer and inner integument. The outer integument is composed of longitudinally elongated cells. The inner periclinal and the anticlinal walls are thick, and the outer periclinal walls are membranous and rarely preserved giving the seed surface a reticulate pattern (Fig. 5c, d). The inner integument is thin and membranous. No other structural details are preserved in the ovules.

The peduncle is woody and has a continuous vascular cylinder (Fig. 6a). Although the vascular bundles are generally poorly preserved, they may be recognized by their pale colour and massive nature. This is confirmed by the presence of occasional scalariform tracheids. Towards the distal part of the axis a central

Fig. 7. *Elsemaria kokubunii* gen. et spec. nov. (*D* dorsal bundle, *P* placenta bundle, *V* ventral bundle). *a* Oblique longitudinal section of capsule base showing vascular courses, double numbers show bundle masses to intercarpellary structures between carpels indicated by the numbers, C + Dlat#3. *b* Fig 6i enlarged, showing vascular courses. *c* Two lateral bundles terminated in the top of carpel no. 5, C + Dlat#9. *d*, *e* Longitudinal sections showing vascular networks in lateral walls of carpels no. 2 and 5, one of major lateral bundles in *d*, C + Dlat#7 & 9. *f*–*h* Cross sections of dorsal walls at dehiscence line (in *h*, thick arrows), arrows indicate vascular bundles, dorsal bundles split by dehiscence in *g* and *h*; *f* D#4. *g* Bbot#54. *h* Bbot#27. – Bars: *a*, *d* 1 mm, *b*, *c* 0.5 mm, *f* 10 µm, *h* 0.2 mm; *d*, *e* and *g*, *h* of same magnification

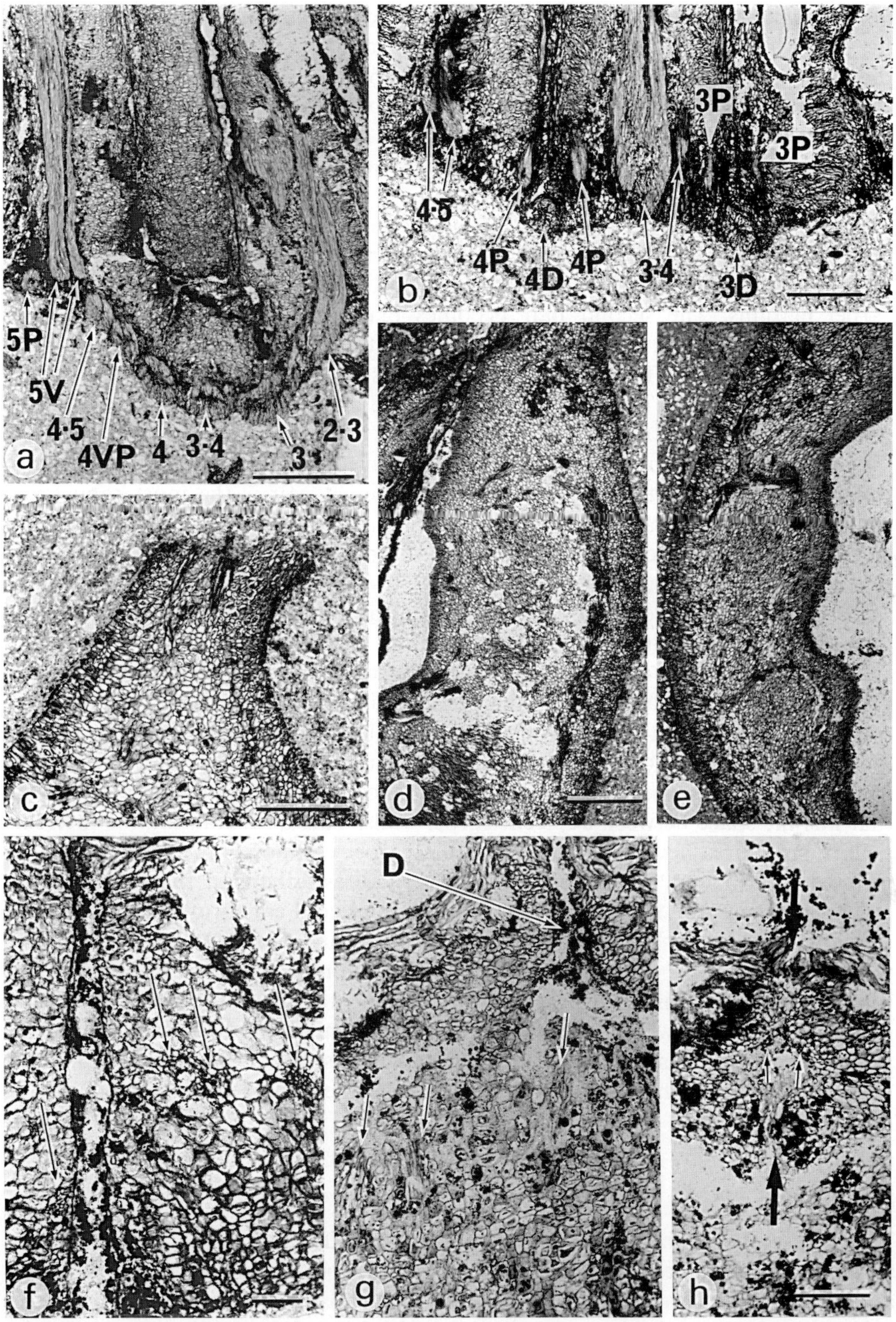
5P
5V
4·5
4VP
4
3·4
3
2·3
a
4·5
4P
4D
4P
3·4
3D
3P
3P
b
c
d
e
f
D
g
h

hollow cylinder appears in the pith. This cylinder is surrounded by the bases of the intercarpellary structures. The floral apex remains in the central cylinder at the gynoecium base but soon disappears (Fig. 6b, c).

Distally, the vascular cylinder of the main axis is separated into 20 bundles of almost equal size: ten corresponding to the ten carpels, and ten corresponding to the intercarpellary structures (Fig. 6b, c). Subsequently two smaller bundles depart from each of the 20 units, resulting in a bundle unit composed of one central and two lateral bundles (Fig. 6c–e). Further on, one pair of bundles departs from the central bundle to the inner side of the first lateral pair (Fig. 6d).

The first pair of lateral bundles in the carpel arises vertically along the ventral carpel margins. The second pair becomes placental bundles that also ascend vertically (Figs. 3h, 6d, e, 7a, b). The ventral and the placental bundles occasionally anastomose (Fig. 6e). The central bundle passes through the base of the placenta and becomes a dorsal bundle running along the line of dehiscence. This bundle is generally fragmented or decomposed due to the carpel splitting (Figs. 3k, 6e, 7a, b, f–h). Several lateral bundles depart from the ventral bundles to supply the lateral walls of the carpel. Of these bundles the basalmost pair is the most robust and runs up along the dorsal corners of the carpel anastomosing with the dorsal bundle and other lateral bundles (Fig. 7d, e). The ventral bundles terminate at the apex of the carpel as two bundles (Fig. 7c).

The placental bundles run along the ventral margins of the carpel and give rise to four or five pairs of obliquely ascending ovule traces (Figs. 6i, j, 7a, b).

It is interesting to note the course of bundles in the intercarpellary structure. Each set of bundles has basically the same organization as seen in the carpellary bundles, with one central and four lateral bundles comparable to the two ventral and two placental bundles (Fig. 6f). The central unit is further divided into smaller irregular subunits forming a triangular mass enclosed in a triangular parenchyma that fills the space between adjacent carpel walls. This unit, entering the intercarpellary structures never contributes to the carpel composition, remaining as an isolated conical unit. The intercarpellary structures adhere to the sclerenchyma columns, which are regarded as part of the former, and constitute structural units that are apparently distinguishable from the adjacent carpels. Only a small bundle which is comparable with the dorsal carpellary bundle traverses the base of the septum (Fig. 6g, h). Tissue disorganization in the septa is characterized by intrusion of a dark substance which follows the path of the small bundle from the intercarpellary structure. This may result in some septicidal splits in the capsule wall. Such tissue disorganization does not occur following the course of lateral bundles in the septa (Fig. 3i).

## Discussion

The structural organization of the fossil capsule is illustrated in Fig. 8 and the basic carpel structure is reconstructed in Fig. 9. Each carpel is supplied with one dorsal and two ventral strands, and the decamerous capsule indicates that *Elsemaria* belongs to the dicotyledonous angiosperms. Multilocular capsules expressing the loculicidal dehiscence occur in several separate orders of angiosperms. In the *Dilleniidae* a number of taxa show fruits morphologically comparable to that of

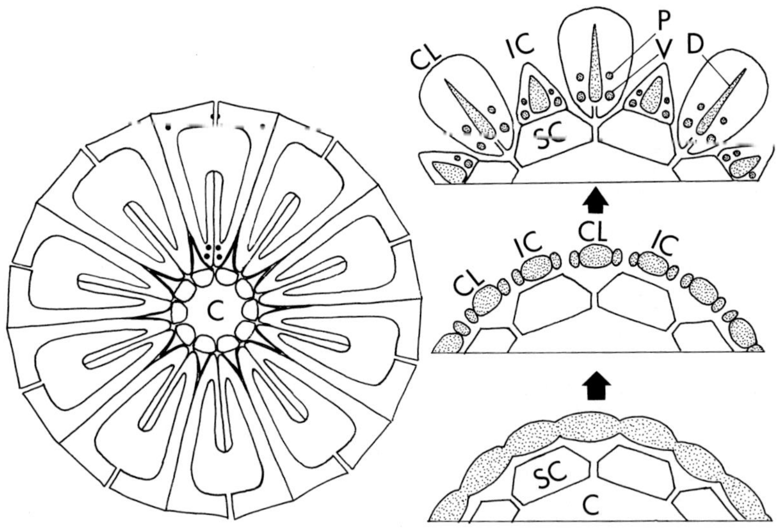


Fig. 8. Diagrammatic illustrations of capsule construction (left, small dots symbolize major vascular bundles) and a series of sections explaining vascular behaviour from stipe to capsule base (*C* central hollow cylinder, *CL* carpel, *D* dorsal bundle, *IC* intercarpellary structure, *P* placenta bundle, *SC* sclerenchymatous column, *V* ventral bundle)

*Elsemaria* with ten carpels and numerous small seeds. One example, is *Cistus ladaniferus* L. of the *Cistaceae* (GROSSER 1903), but comparable fruits also occur in the *Ericales* (P. K. ENDRESS, pers. comm.). However, the lack of other floral parts in *Elsemaria* and the possible polyphyletic nature of multilocular capsules make it difficult to attribute *Elsemaria* to a particular order or family.

Structurally preserved multilocular capsules with axile placentation have been reported from a few Cretaceous and Tertiary localities. The latest Cretaceous to Miocene fructification of *Nordenskioldia* assigned to the *Trochodendraceae* (CRANE & al. 1991) has about 15 locules each containing only a single seed and differs in fruit and in seed morphology from *Elsemaria*.

*Daberocarpon* CHITALEY et SHEIKH from the Deccan Intertrappean Series of India of Late Cretaceous or Early Tertiary age has ten fused locules (CHITALEY & SHEIKH 1973). However, *Daberocarpon* has a single seed per locule only and the dehiscence is septicidal. A variety of other multilocular capsules have also been reported from the same series (SAHNI 1933, 1943; JAIN 1964; CHITALEY & NAMBUDIRI 1968; CHITALEY & PATIL 1973). Comparison with *Elsemaria* is restricted to the multiocular nature of the capsules. Because there are no other similar fructifications in living or fossil plants, I propose a new genus *Elsemaria* for the fossil.

*Elsemaria* is important from a morphological perspective. It may demonstrate one possible origin of axile placentation in angiosperms. Although completely fused with each other, the individual carpels maintain features of the supposed primitive conduplicate carpel. The massive placentae of *Elsemaria* probably originated from the deeply invaginated ventral margins, and this is consistent with the observed vasculature. A possible ancestor for *Elsemaria* would comprise a fructification with spirally arranged conduplicate carpels like *Protomonimia* from the Santonian of

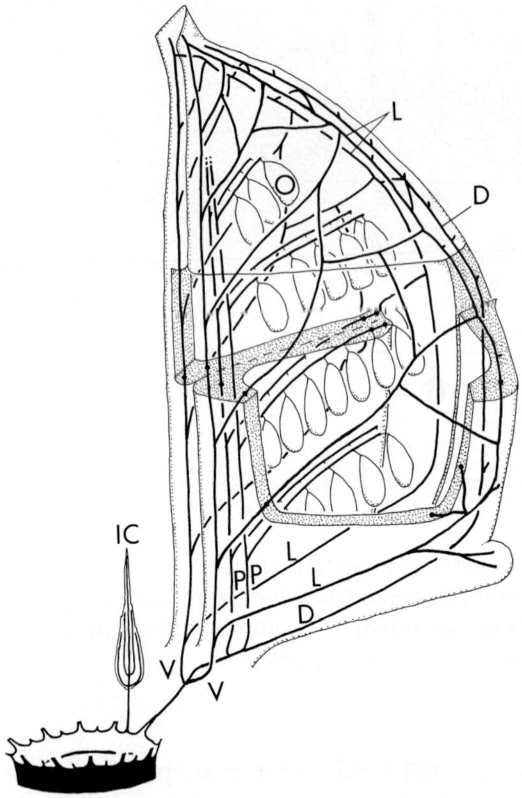


Fig. 9. Suggested reconstruction of single carpel illustrated diagramatically (*D* dorsal bundle, *IC* intercarpellary structure, *L* lateral bundle, *O* ovule, *P* placenta bundle, *V* ventral bundle)

Hokkaido (Nishida & Nishida 1968). *Protomonimia*, however, has marginal placentation, while in *Elsemaria* the placenta has bundles separate from the ventral bundles (Fig. 9). Such differences could represent extremes in variation in early carpel morphologies and would be better understood in relation to the origins of the carpel itself.

Most interesting is the presence of the intercarpellary structures. They are structurally independent from the carpels, but have a vascular pattern basically similar to that of the carpel. It is probable that these structures are reduced carpels. If so, *Elsemaria* can be interpreted as a compound structure consisting of 20 both functional and non-functional carpels. Some of the various placentation types of living angiosperms could be reinterpreted as descendant of such a compound structure.

I thank Dr Makoto Nishida, Prof. Emeritus, Chiba University, and Mr Takeshi Ohsawa, Chiba University, for various helps. Thanks are due to Mr Hakuji Kokubun who provided the specimen. I also thank Dr P. Kenrick and Prof. Else Marie Friis, Stockholm, and Prof. P. K. Endress, Zürich, for comments on the manuscript. I am, however, fully responsible for the content of this manuscript. This study was financially supported by the Ministry of Education, Science and Culture of Japan, no. 05304009 (represented by Dr Hiroshi Tobe, Kyoto University), and by Fujiwara Natural History Foundation for 1993.

**References**

BAILEY, I. W., SWAMY, B. G. L., 1951: The conduplicate carpel of dicotyledons and its initial trends of specialization. – Amer. J. Bot. **38**: 373–379.

CHITALEY, S. D., NAMBUDIRI, E. M. V., 1968: *Harrisocarpon sahnii* gen. et sp. nov., from the Deccan Intertrappean beds of Mohgaonkalan. – Proc. 55th Indian Sci. Congr., 31.

– PATIL, G. V., 1973: *Sahniocarpon harrisii* gen. et sp. nov. from the Mohgaonkalan beds of India. – Palaeobotanist **20**: 288–292.

– SHEIKH, M. T., 1973: A ten locular petrified fruit from the Deccan Intertrappean Series of India. – Palaeobotanist **20**: 297–299.

CRANE, P. R., MANCHESTER, S. R., DILCHER, D. L., 1991: Reproductive and vegetative structure of *Nordenskioldia* (*Trochodendraceae*), a vesselless dicotyledon from the early Tertiary of the northern hemisphere. – Amer. J. Bot. **78**: 1311–1344.

CRONQUIST, A., 1968: The evolution and classification of flowering plants. – London: Nelson.

DILCHER, D. L., CRANE, P. R., 1984: *Archaeanthus*: an early angiosperm from the Cenomanian of the western interior of North America. – Ann. Missouri Bot. Gard. **71**: 351–383.

DOYLE, J. A., 1994: Origin of the angiosperm flower: a phylogenetic perspective. – Pl. Syst. Evol. [Suppl.] **8**: 7–29.

GROSSER, W., 1903: *Cistaceae*. – In ENGLER, A. (Ed.), Das Pflanzenrcich, IV, p. 193. – Leipzig: Engelmann.

JAIN, R. K., 1964: *Indocarpa intertrappea* gen. et sp. nov. from the Deccan Intertrappean Series of India. – Ann. Bot. **125**: 26–33.

JOY, K. W., WILLICE, A. J., LACEY, W. S., 1956: A rapid cellulose peel technique in paleobotany. – Ann. Bot. **20**: 635–637.

NISHIDA, H., NISHIDA, M., 1988: *Protomonimia kasai-nakajhongii* gen. et sp. nov.: a permineralized magnolialean fructification from the mid-Cretaceous of Japan. – Bot. Mag. (Tokyo) **101**: 397–426.

SAHNI, B., 1933: A fossil pentalocular fruit from Pondicherry, South India. – Rec. Geol. Surv. India **56**: 430–437.

– 1943: Indian silicified plants. 2. *Enigmocarpon parijai*, a silcified fruit from the Deccan, with a review of the fossil history of the Lythraceae. – Proc. Indian Acad. Sci. **3**: 59–96.

TAKHTAJAN, A., 1991: Evolutionary trends in flowering plants. – New York: Columbia University Press.

TANABE, K., HIRANO, H., MATSUMOTO, T., MIYATA, Y., 1977: Stratigraphy of the Upper Cretaceous deposits in the Obira area, northwestern Hokkaido. – Sci. Repts. Dept. Geol., Kyushu Univ. **12**: 181–202. (In Japanese with English abstract).

TSUSHIMA, S., TANAKA, K., MATSUNO K., YAMAGUCHI, S., 1958: Tappu. Expl. Text, Geological Map of Japan, no. 38. – Geological Survey of Japan.

WEBERLING, F., 1989: Morphology of flowers and inflorescences. – Cambridge: Cambridge University Press.

Address of the author: Dr HARUFUMI NISHIDA, International Budo University, 841 Shinkan, Katsuura, Chiba, 299–52 Japan.

Accepted April 7, 1994 by P. K. ENDRESS

Pl. Syst. Evol. [Suppl.] 8: 137–158 (1994)

# Ontogeny of staminate and carpellate flowers of *Schisandra glabra* (*Schisandra*)

SHIRLEY C. TUCKER and J. ALLEN BOURLAND

Received November 17, 1993

**Key words:** *Schisandra*, *Schisandraceae*. – Flower, carpel, stamen, staminal shield, ontogeny, synandry, unisexuality.

**Abstract:** *Schisandra glabra* (*Schisandraceae*) is a rare monoecious liana in forests of the southeastern United States. Both types of flowers are solitary in leaf axils and radially symmetrical, with eight to 13 greenish yellow to red tepals in the flower. Each male flower has five (four to seven) spirally arranged stamens forming a red pentagonal synandrous shield. Transitions between tepals and stamens occur occasionally. Each carpellate flower contains 25–30 free, spirally arranged carpels on a conical receptacle. Histological study shows that the floral apical meristem is more highly convex than the vegetative apex; both have tunica-corpus configurations. All primordia, starting with tepals, are initiated acropetally, in a continuous 2/5 phyllotaxis. Apical diameter increases greatly after tepal initiation, more so in carpellate than in staminate flowers. The apical residuum in staminate flowers expands to form the center of the staminal shield, with the stamen primordia projecting as flattened marginal extensions. The connective region of each stamen broadens markedly, resulting in wide separation of the sporangial pairs of each stamen. Just before anthesis, each connective arches outward so that the sporangia appear lateral. In carpellate flowers, the carpel primordia are initiated helically; the apical residuum forms a narrow, spinelike structure. Floral development in *Schisandra* is compared with that of other primitive angiosperms such as *Illicium* and *Myristica*.

The genus *Schisandra* in the family *Schisandraceae* is of morphological and phylogenetic interest because it has many primitive character states (flowers with numerous free parts; radial symmetry, helical arrangement of floral organs, and free conduplicate carpels; WOOD 1958). *Schisandra* includes 25 species of eastern Asia and southeastern United States. SMITH (1947) divided *Schisandra* into four sections: *Euschisandra*, *Sphaerostema*, *Maximowiczia*, and *Pleiostema*, mostly distinguished on differences in the androecium of male flowers. In *Euschisandra*, the androecium is a sessile pentagonal flattened shield consisting of five radiating stamens, the thecae borne on the lower (dorsal) margins of anthers, and the perianth segments number seven to 13. It includes three species of southern Korea, Japan (*S. bicolor*, *S. repanda*) and southeastern United States (*S. glabra* = *S. coccinea*). *Schisandra glabra* (BRICKELL) REHDER, Bay Star Vine or Wild Sarsaparilla, is a rare monoecious woody vine in forests of the southeastern United States.

*Schisandra* has some relatively specialized features, such as the vineing habit, unisexuality, low ovule number, and the staminal shield, with cryptic homology. Because the structure of the gynoecium and androecium is so important in systematics of *Schisandra*, floral ontogeny holds the potential to show the developmental basis for systematic distinctions. This paper will examine development of the carpellate and staminate flowers, with the object of determining homologies between the two, particularly of the staminal shield in the staminate flower. The floral ontogeny of *Schisandra glabra* will also be compared with that of other primitive taxa such as *Illicium floridanum* Ellis (Robertson & Tucker 1979) in *Illiciaceae* which Cronquist (1981) placed in the same order as *Schisandraceae*, and with *Myristica fragrans* Houtt., *Myristicaceae* (Armstrong & Tucker 1986).

## Material and methods

Vegetative and floral buds of *Schisandra glabra* were collected weekly, monthly, or bimonthly as required during the growing season from February to June over two years. Collecting sites were in Rapides, St Helena and West Feliciana parishes, Louisiana. Additional material was collected in Shelby Country, Tennessee. Voucher specimens are deposited in the herbarium of Louisiana State University, Baton Rouge. Vegetative and floral buds were fixed in formalin – acetic acid – ethanol (90 cc 50% ETOH: 5 cc acetic acid: 5 cc 37% formaldehyde). Buds for sectioning were dehydrated in a tertiary butyl alcohol series, embedded in paraffin (Johansen 1940), and sectioned at 8–10 µm. The sections were stained with safranin – fast green (Sass 1951) and mounted in "Harleco" (Hartman-Leddon Co.) or "Permount" (Fisher Scientific Co.).

For studying the vasculature, clearings were made of older buds, flowers, and fruits. The buds were washed and incubated in 2% unbuffered papain activated with sodium sulfide (Rodin & Davis 1967), then bleached in 5–10% sodium hydroxide, "Chlorox", or Stockwell's Bleach. Cleared material was then dehydrated in an ethyl alcohol series, stained with basic fuchsin, and stored in xylene.

Sections and clearings were photographed with a Leitz Orthomat-Orthoplan Photomicroscope with bright-field illumination. Carpels too large for this method were photographed immersed in xylene with a Topcon camera attached to the phototube of an American Optical dissecting microscope.

For scanning electron microscopy, fixed floral buds were dissected in 95% alcohol. Some older buds were bisected longitudinally. The buds were further dehydrated in an ethanol series, placed in amyl acetate, and dried in a Denton critical point drying apparatus with liquid carbon dioxide. Buds were then mounted on aluminium stubs with silver paint, coated with gold approximately 20 nm thick or with a layer of carbon followed by gold. Micrographs were taken at 25 kV with a Jeol JSM-2 or a Hitachi S-500 scanning electron microscope on Tri-X Ortho film.

Height and width of vegetative and floral meristems were measured on longitudinal sections using an ocular micrometer. The apical meristem was defined from the base as the level just above the last-formed appendage or initiation site. The height of an appendage was measured from a line perpendicular to an imaginary line between the abaxial and adaxial base of the primordium.

## Results

**Vegetative organography.** *Schisandra glabra* is a high-climbing woody liana in the hardwood forests of the southeastern United States from western North Carolina to eastern Louisiana. In Louisiana it usually grows in rich woods on shaded, moist embankments or the sides and bottoms of ravines. Numerous sterile shoots of different ages occur, while larger branched trunks up to 2 inches in diameter clamber upward in trees and produce extensive canopies. Phyllotaxy is 2/5 clockwise or counterclockwise. Leaves are alternate, simple, entire, petiolate, glabrous, and deciduous (Fig. 1a). The leaf blades are oblong-elliptic to ovate-cordate or lanceolate, with acuminate tips and cuneate bases. The shoots are either

Fig. 1. Flowers of *Schisandra glabra*. *a* Flowering branch with leaves and both staminate and carpellate flowers in leaf axils; *b* staminate flower; *c* carpellate flower; *d* tepals of one flower in order of attachment from the base, at left; *e* fruits. – Bar: 1 cm

determinate or indeterminate, and have simple, ovate, glabrous leaves (Fig. 1a), with axillary racemose inflorescences. Vegetative growth is sympodial, with an axillary bud continuing next year's growth of a shoot.

**Floral organography.** The species is usually monoecious, with staminate and carpellate flowers in different inflorescences (rarely on the same shoot, Fig. 1a); some plants lack carpellate flowers. Male and female flowers usually bloom in May-June in southern Louisiana. Each shoot has zero to seven flower buds. Flowers form in all the leaf axils of determinate shoots; in indeterminate shoots, flower buds are produced at the older nodes, while the distal-most two to five axillary buds remain vegetative. Anthesis occurs approximately at the same time in both sexes, but some carpellate flowers open several days prior to the male flowers. Staminate flowers (Fig. 1b) are borne on pedicels 17–40 mm long, while carpellate flowers (Fig. 1c) have pedicels 31–55 mm long; both are in pendent racemes.

Each flower of *S. glabra* is borne in the axil of a leaf or fugacious bud scale, and is subtended by a single bracteole 0.8–1.4 mm long with ciliolate margins. Flowers are actinomorphic, with tepals below the reproductive organs. All organs of a flower are in the same phyllotactic spiral, clockwise or counterclockwise. In the flowers examined, phyllotaxy was 2/5 in staminate flowers, 3/8 in carpellate flowers. Each flower has tepals (Fig. 1d) plus one type of sexual organ, either carpels or stamens. No primordia are formed of the other sex in either staminate or carpellate flowers.

**Ethology of flower development.** In Louisiana, tepals of *S. glabra* are initiated during February and early March. By mid March, tepal initiation is complete and stamen and carpel initiation have begun. Meiosis is occurring in stamen primordia in April; mature pollen is present in the anthers by late May. In the carpellate flowers, ovule initiation occurs in early April. Differentiation of the primary sporogenous cells and their subsequent meiotic divisions occur in early May. Maturation of the embryo sac is complete in late May. Fruits (Fig. 1e) are ripe in late August and early September.

**Tepal organography.** Since staminate and carpellate flowers are nearly alike during tepal initiation, tepal form will be described first. Staminate flowers have

Table 1. Tepals from one flower grouped according to size, shape, color, and margins

| Order of development | Number | Color | Margin | Height in mm | Width in mm | Shape |
|---|---|---|---|---|---|---|
| Group 1 | 3–4 | green, greenish yellow | ciliolate | 1.1–8.2 | 1.0–6.4 | suborbicular to obovate |
| Group 2 | 3–5 | yellow, shading to crimson | weakly ciliolate | 4.8–8.9 | 3.0–6.8 | obovate |
| Group 3 | 2–4 | crimson | weakly ciliolate | 3.0–8.1 | 3.0–5.0 | obovate to elliptic |

about nine tepals (range: eight to 13), while carpellate ones have 11–13 tepals, which are larger than those of staminate flowers. The small outer tepals are 1.1–8.2 mm long, suborbicular to obovate in shape, and shade from green through yellow to crimson in the innermost, largest tepals. Maximum tepal size is realized among the middle tepals which are 4.8–8.9 mm long and obovate; the earliest and last tepals are smaller (Fig. 1d). Table 1 shows tepal transition for several characters: number, color, nature of the margin, size, and shape. The three groups are artificial in that there are not sharp boundaries. The tepals of a flower intergrade in size and shape (Fig. 1d), color, and nature of the margin. Each tepal has one to three vascular traces; if only one, it trifurcates at the base of the organ. The three bundles branch again, forming an open vasculature. The innermost tepals show less branching and anastomoses than the outer ones. Although the flowers lack nectaries, the bases of the inner tepals are thickened and contain tannin-filled parenchyma and ethereal oil cells.

**Tepal initiation.** Preceding tepal initiation, the floral apex (Fig. 2a) is 103–130 μm wide and 29–32 μm high. It is more highly convex than the vegetative apex (not shown). It has two tunica layers and a corpus of homogeneous meristematic cells. Near the summit, occasional periclinal divisions occur in the inner tunica layer and contribute to the corpus. Cells are differentiating in the pedicel below the floral apex, with tannin-containing cells and vacuolation evident.

As tepal initation proceeds, the floral apex enlarges and changes from a tunica-corpus configuration to a mantle-core configuration (Fig. 3c). The biseriate tunica and the outermost two or three corpus layers comprise a densely staining meristematic zone, the mantle (Figs. 2c, 3c). The core is the central area in which cell differentiation is occurring (Figs. 2c, 3c).

The tepals are initiated in acropetal sequence. Initiation begins with one or several periclinal divisions (at arrow, Fig. 2a) in the inner of the two tunica layers low on the flanks of the floral apex. Anticlinal divisions in the T1 (Fig. 2a) accompany the periclinal divisions. Cells in the inner mantle also divide periclinally adjacent to the initiatory site. Initiating cells divide both anticlinally and periclinally to produce a tepal primordium (Fig. 2c). A transection through a young floral bud (Fig. 2b) shows the floral apex with four tepal primordia and a bracteole. Procambium extends to the base of the tepal primordium (P, Fig. 2c).

**Tepal development.** The tepal primordium increases at first by general cell division and enlargement (Fig. 2d, e). A uniseriate protoderm is present. Apical and subapical initial cells appear when the primordium is 50–60 μm high, although cell divisions are also scattered throughout the primordium at this stage. The apical initial divides anticlinally only to perpetuate the protoderm, while the subapical initial divides both anticlinally and periclinally to produce four to six cell layers internally.

The young tepal primordium increases in height, first by general cell division and enlargement (Fig. 2d) and later by subapical initial activity (not shown). In transverse section of the tepal primordium (Fig. 2f) submarginal initials (at arrows) extend the margins. Broadening of the tepal margins is evident in Fig. 2e. The tepal primordia show early vacuolation of the abaxial protodermal and sub-protodermal layers. Tepals of varying sizes and degree of maturation are seen in Fig. 3a, b.

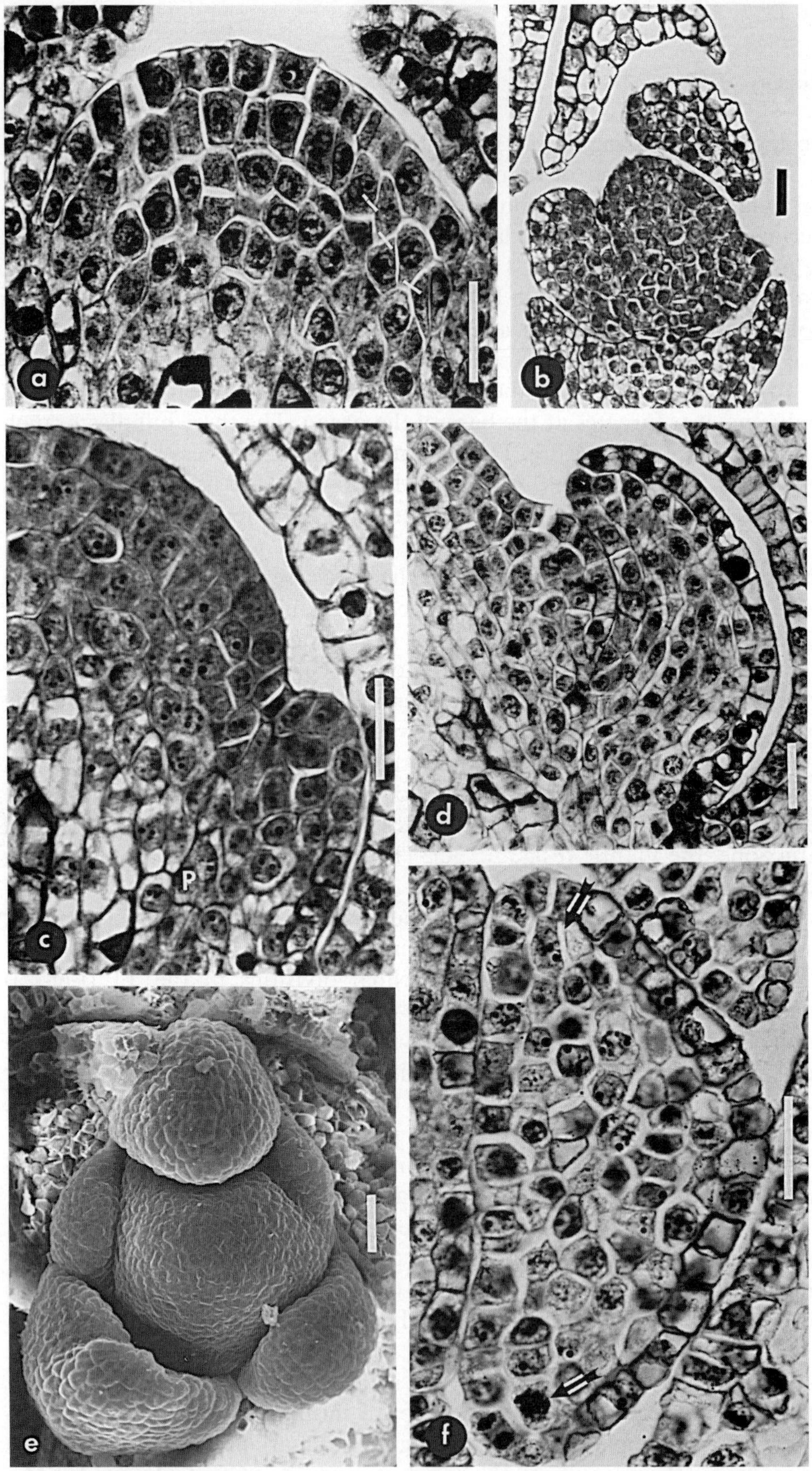


Fig. 2. *Schisandra glabra*: initiation of tepals. *a–d*, *f* Photomicrographs of sections, *e* SEM micrograph. *a* Floral apex at organ inception (at white arrows, at right; longitudinal section); *b* young flower (transverse section) with bracteole and four tepals; *c* young tepal primordium (longitudinal section), p procambium; *d* tepal primordium (longitudinal section); *e* floral apex (polar view) with first few tepals; *f* tepal (transverse section) with marginal and submarginal initials (at arrows). – Bar: 50 μm

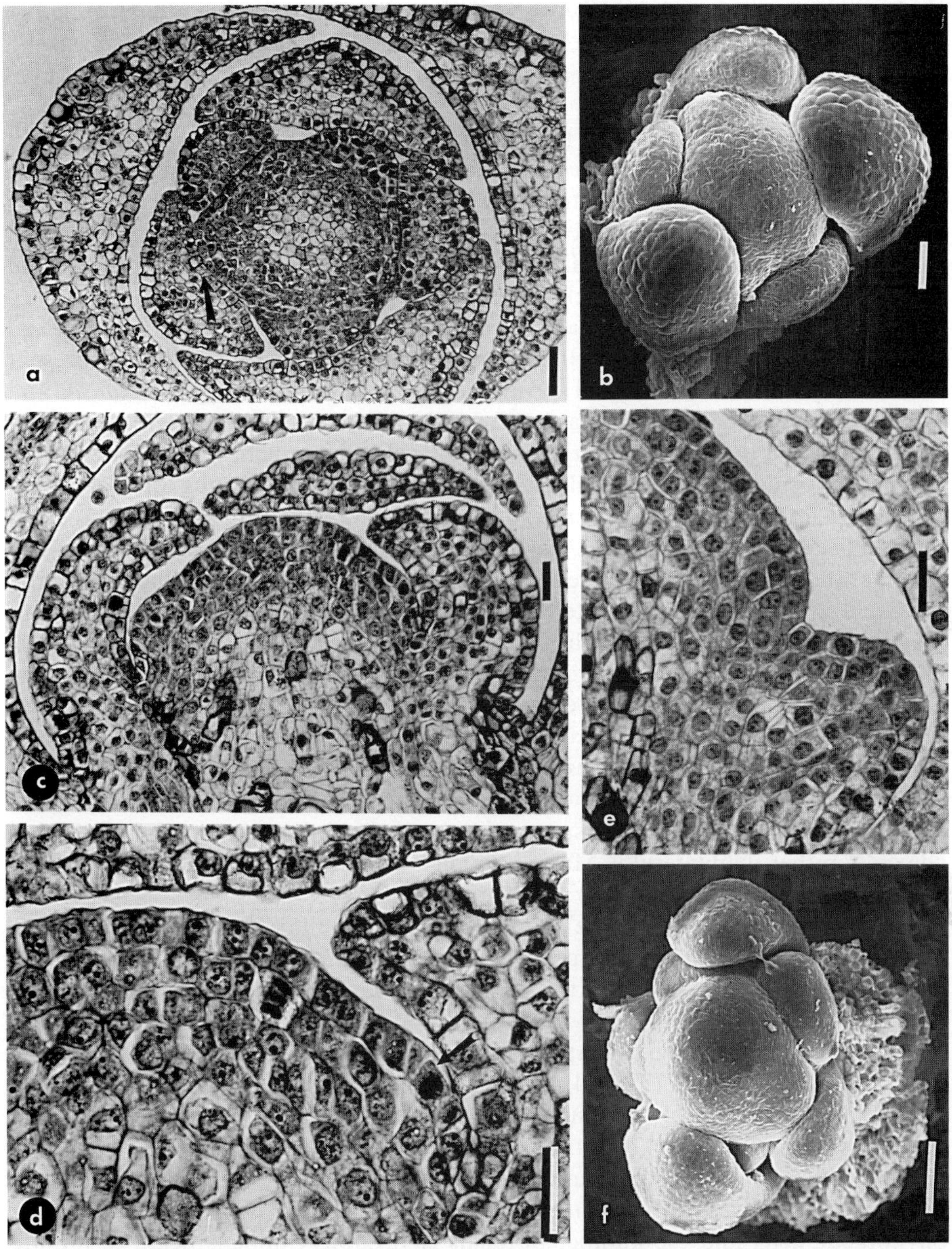


Fig. 3. *Schisandra glabra*, initiation of tepals and of early stamens in staminate flower. *a*, *c–e* photomicrographs of sections, *b*, *f* SEM micrographs. *a* Floral axis with several tepals of varying sizes round it, adaxial meristem is seen at arrow; *b* floral apex initiating late tepals and/or early stamen primordia; *c* floral apex (longitudinal section) of staminate flower at early stamen initiation; *d* stamen initiation at right (arrow) on flank of floral apex; *e* young stamen primordium (at right) on flank of floral apex; *f* floral apex with early stamen primordia (near-polar view). – Bars: *a*, *f* 100 μm, *b–e* 50 μm

Procambium is present below the periclinal division initiating a tepal (Fig. 2c). It extends acropetally from the receptacle, and will become the midrib strand (Fig. 2d). Lateral strands develop acropetally from the receptacle later in tepal maturation.

Marginal growth to produce the blades begins in the tepals (Fig. 2f) at a height of 100–136 µm. Marginal initial cells in linear files divide anticlinally to perpetuate the protoderm, both adaxially and abaxially. Linear files of submarginal initials divide both anticlinally and periclinally to produce the mesophyll and procambial strands. The tepals broaden and become arcuate (Fig. 3a, b) as a consequence of marginal growth plus general cell division.

The tepals undergo an increase in thickness due to the activity of an adaxial meristem (at arrow, Fig. 3a). Numerous periclinally dividing cells are present immediately below the adaxial protoderm in the midrib region. Apical and subapical growth cease in tepals at heights of 700–850 µm. Further increase in tepal size (Fig. 3a, b) is due to intercalary growth and cell enlargement. Cell differentiation within the tepals includes oil cells, tannin cells, and crystalliferous idioblasts.

**Stamen organography.** Staminate flowers become distinguishable from carpellate ones only after tepal initiation. Five (four to seven) stamens are found in the staminate flower, in crevices on the edge of a red, flattened shield. At anthesis (Fig. 1b), the shield is 3.0–4.8 mm in diameter and 1.0–1.5 mm high. Each anther is 1.0–2.9 mm long and tetrasporangiate, and the sporangia are separated by a broad connective. Dehiscence is latrorse and occurs by longitudinal slits. Approximately 18% of the flowers in one population had six stamens, and 1.5% had either four or seven. The sixth or seventh stamen, if present, may be sterile, bisporangiate, or rarely tetrasporangiate. In the instances where the sixth stamen is below the upper five, it shows transitional features between a stamen and tepal. Such transitional organs are either bisporangiate or tetrasporangiate, and the sporangia may be positioned adaxially rather than laterally. Mature pollen is hexacolpate, has a reticulate exine, and is released in the two-celled state.

**Stamen initiation.** Just before stamen initiation, the floral apex (Fig. 3b) has a mantle-core configuration. The apex increases at this time to a width of 140–170 µm and a height of 40–48 µm. Shortly after stamen initiation is completed (Fig. 4a), the floral apical residuum is 160–230 µm in diameter and 56–100 µm high. As the stamen primordia develop, the apical residuum is wide and low-convex, and may become flat at the center (Fig. 4d) as the surrounding tissue heightens. Cells of the outer tunica of the apical residuum continue to divide anticlinally. At anthesis the apical residuum is about 1100–1400 µm in diameter.

The last tepal primordium and the first stamen primordium are difficult to distinguish at stages close to initiation (Fig. 3b, c), since they lie on the same phyllotactic helix and the number of tepals is variable. Stamens begin to initiate quickly (Fig. 3d, e) after the tepals, since the last-formed tepals are only 40–104 µm high when the last stamen is initiated (Fig. 4a). Initiation of the stamen primordia is first seen as periclinal divisions in T2 and the outer corpus, plus clusters of T1 cells undergoing anticlinal divisions, low on the flanks of a convex floral apex (Fig. 3c, d). As a consequence of these divisions and subsequent cell enlargement and division, a protuberance is formed (Figs. 3f, 4a).

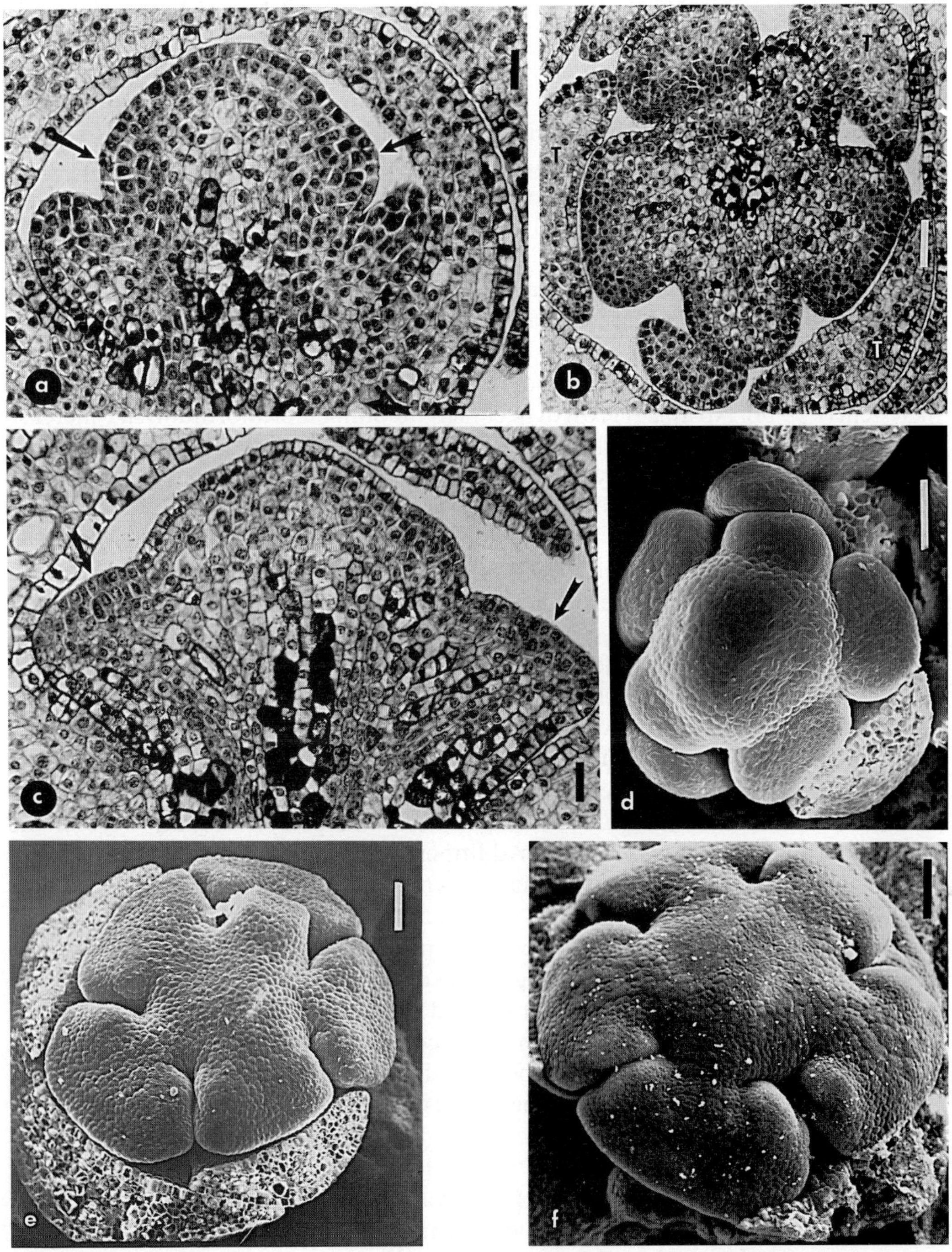


Fig. 4. *Schisandra glabra*, stamen development in staminate flower *a–c* Photomicrographs of sections, *d–f* SEM micrographs. *a* Late stamen initiation (at arrows) on floral apex (longitudinal section); *b* flower (transverse section) showing three broad stamen primordia and three tepals with tapered margins; *c* floral apex after stamen initiation, showing flattened stamen primordia (at arrows); *d–f* three stages of stamen initiation (polar or near-polar views); *d* three stamens have been initiated; *e* all five have been initiated, the order is indicated by stage of expansion of each stamen; *f* apical residuum has broadened, and sporogenous tissue is beginning to differentiate in stamen primordia. – Bars: *a*, *c* 50 μm *b*, *d–f* 100 μm

The floral apex in the staminate flower has a further role, after organogeny is completed. The apex and the stamen primordia, following stamen initiation, enlarge greatly in diameter to form the staminal shield (Fig. 4d–f). Since stamens are sessile in *S. glabra*, the term stamen will be used to include the anthers and the connective.

**Stamen development.** Stamen development will be described up to the time of sporogenesis, which will be included in a subsequent paper. The stamen primordia at first appear in longitudinal section as small acute-tipped structures (Fig. 3e), rather similar to the tepal primordia, but subsequently they expand greatly at the primordium base (Fig. 4b, c). In transverse section, each stamen primordium is wide and flat-topped (Figs. 3e, 4b, d). Unlike the tepals which begin to grow in height continuously, stamen primordia expand outward, parallel to the flat apical meristem (Fig. 3f). No apical growth occurs in the stamen primordium; size increase rather results from general cell division throughout the stamen (the protodermal cells divide to extend the layer over the surface as in Fig. 4c). Eventually the meristem is restricted to the surface layers, with a protoderm persisting (Fig. 5b). Submarginal initial activity does not occur. The stamen primordia vary in transverse outline from broadly ovoid in younger primordia to broadly obdeltoid in older ones (Fig. 4b, d). The helical order of stamen initiation remains evident from size of organs and degree of differentiation (Fig. 4d–f).

As sporogenous tissue differentiates, cell division expands the connective area of stamen primordia around the flattened apical residuum (Fig. 4f, 5a). Cells of the connective differentiate earlier than the sporogenous tissue (Fig. 5a–e). Cell enlargement and differentiation occur distally and abaxially in the stamen primordia (Fig. 5b–d), delimiting the meristematic tissue that will become sporogenous. At this time the connectives are 115–180 μm wide and project slightly beyond the developing microsporangia (Fig. 5d). Additional lateral expansion of the connective results in wide separation of sporangial pairs (Fig. 5c). At this stage, the connective is 240–260 μm long, 70–260 μm wide at the base, and 330–540 μm wide at the top. This huge connective constitutes the major part of each stamen (Fig. 6a–c). Just before anthesis, there is a slight ridge over each sporangial pair, and the distal edge of each connective is prolonged beyond the microsporangia (Figs. 5d, 6a–c). The distally prolonged connective apex or "connective appendage" of the stamen and the lateral position of the sporangial pairs are shown in Fig. 6c, d in longisection. At anthesis, the connective is 0.9–1.3 mm wide proximally and 1.3–3.1 mm wide distally.

**Procambialization of stamens.** Procambium is present below each stamen primordium at inception (Fig. 4c). Each stamen receives a solitary trace that bifurcates in the connective (Fig. 5d) and terminates distally below, or adjacent to, the lowermost sporangium of each pair. In the single bundle, procambium develops in the stamen acropetally and continuously. Procambium is first present within a stamen 60 μm long and 115 μm high. At anthesis, the bundle dichotomy is positioned in the mid-portion of the connective.

**Staminal shield and the fate of the floral apex.** Five radiating stamen primordia in helical succession surround the domelike apical residuum which continues to enlarge to 340–500 μm diameter (Fig. 5a) and begins to appear flattened. The stamen primordia, originally helical, become realigned so that they are approximately at one level. Internally in the apical residuum, the cells of the uniseriate

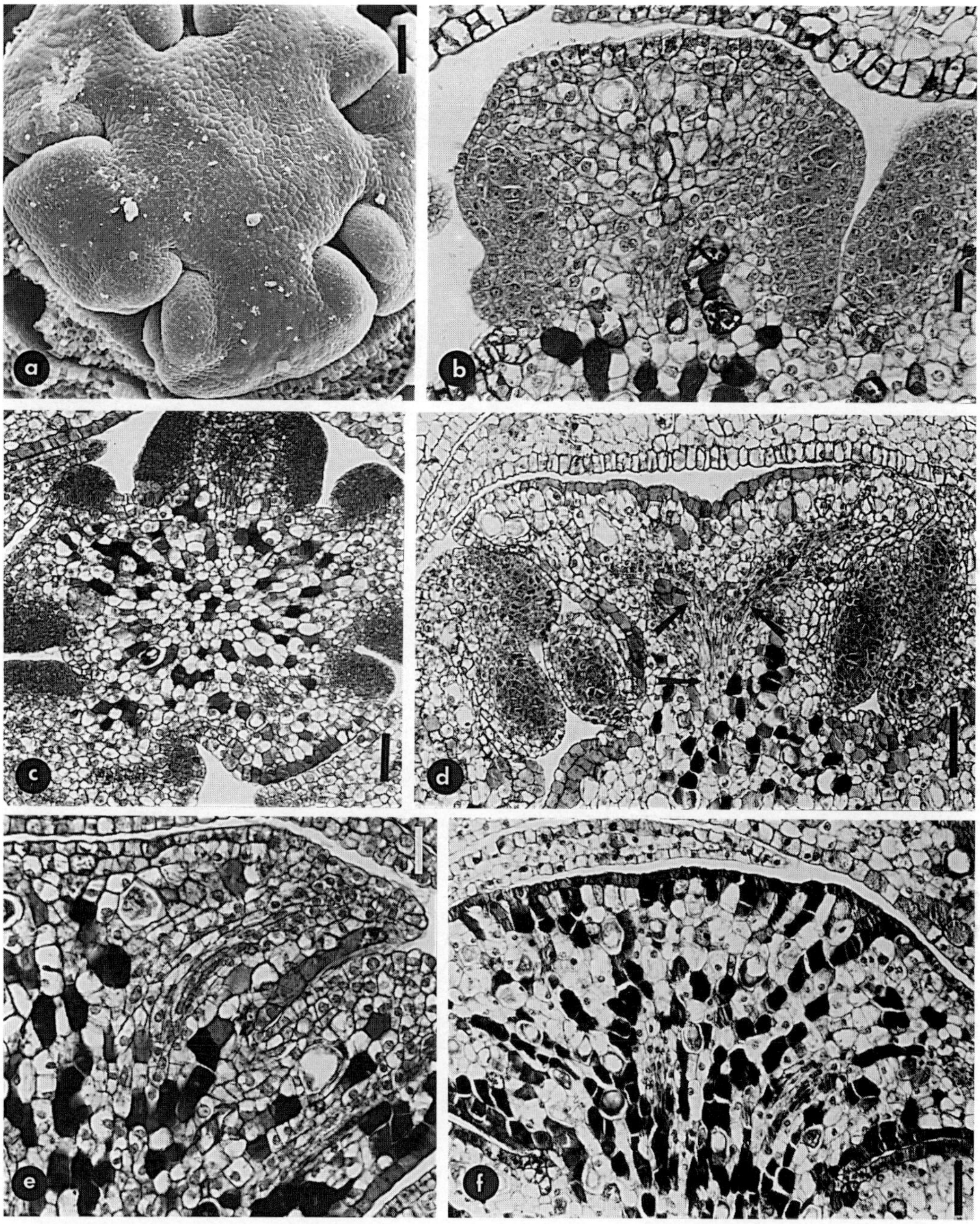


Fig. 5. *Schisandra glabra*, staminal shield development in staminate flower. *a* SEM micrograph, *b–f* photomicrographs of sections. *a* Staminal shield resulting from broadened apical residuum, and stamens developing sporogenous tissue; *b* stamen (tangential longitudinal section) showing sporogenous tissue at sides; *c* floral axis with stamens and their sporangia cut at slightly different levels (transverse section); *d* stamen showing bifurcating procambial bundle (at arrows), sporogenous tissue (dark-staining) and differentiated tissue at surface of connective; *e, f* staminal shield (longitudinal section) showing progression of cell differentiation. – Bars: *a, c, d* 100 µm, *b, e, f* 50 µm

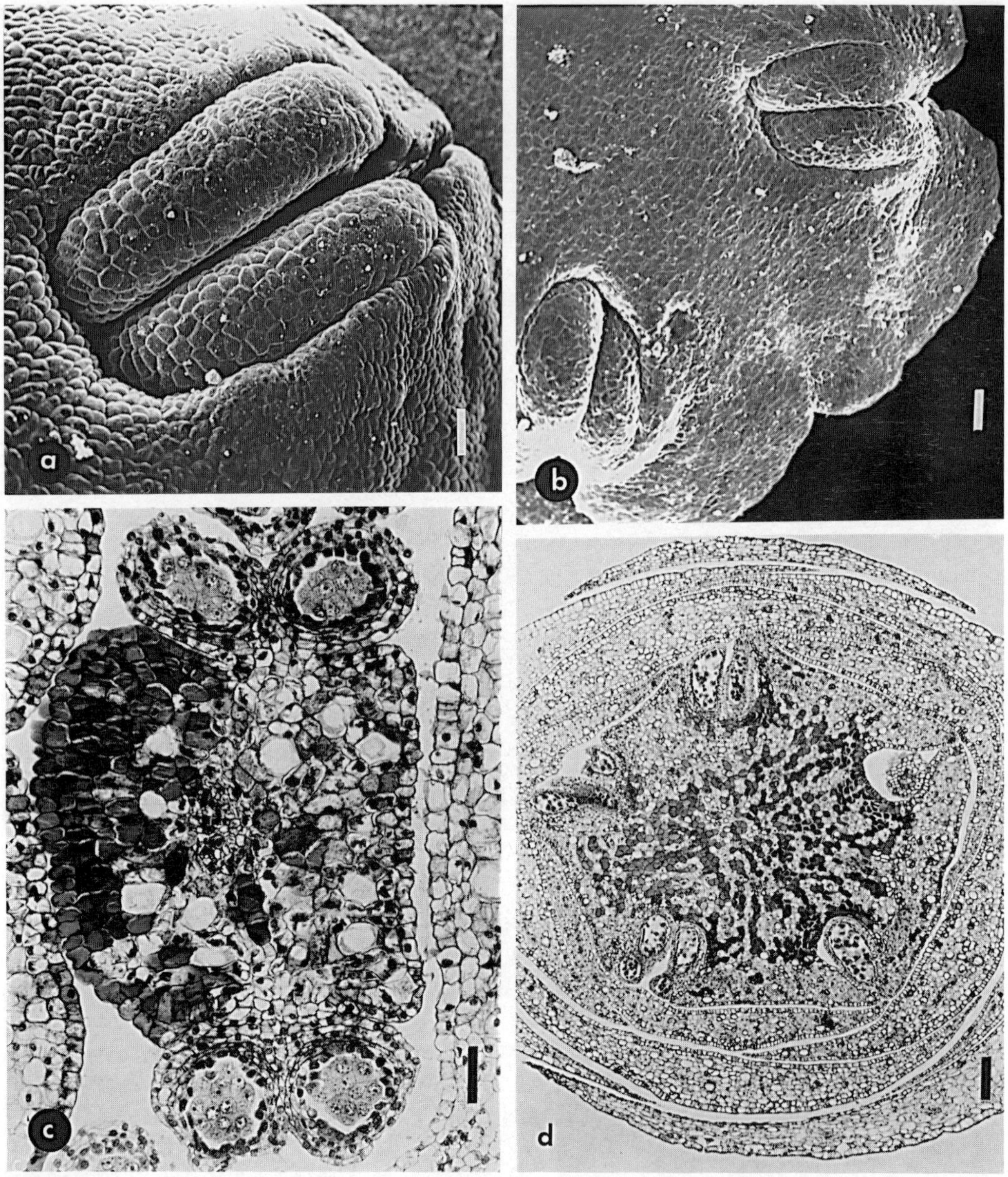

Fig. 6. *Schisandra glabra*, differentiation in stamens and staminal shield. *a, b* SEM micrographs, *c–d* photomicrographs of sections. *a* Staminal shield (lower left) and microsporangia sunken below edges of connectives; *b* anther sacs of two adjacent anthers in pit between two connectives; *c* differentiated tissue in stamen (transverse section) with lateral anthers; *d* flower (transverse section) cut at level of staminal shield and pits containing anthers between stamen connectives. – Bars: *a, c* 50 μm, *b, d* 100 μm

surface layer remain relatively small, divide anticlinally (Fig. 5b) and eventually become filled with tannins (Fig. 5e, f). The subsurface cells divide repeatedly in the periclinal plane to form files of cells perpendicular to the surface (Fig. 5f), and many of these contain crystals or oil. The apical residuum (Fig. 6a, b) becomes continuous with the broad stamen connectives, and shares with these the same type of cellular structure, comprising the stamen shield, about 1000–1400 µm diameter (Fig. 1b).

**Carpel organography.** The carpellate flower has 12–15 tepals, plus 25–30 free conduplicate carpels. The phyllotaxy changes between tepal and carpel initiation from 2/5 to 3/8 spiral phyllotaxy (Fig. 7a–c). After tepal initiation is completed, the floral apex is high-convex (Fig. 7a, b), 180–190 µm in diameter, and 56–60 µm high above the uppermost tepals. The apex increases in size during carpel initiation to about 200 µm diameter, while the height is unchanged. The cell configuration is mantle and core (Fig. 7b). The mantle diminishes in thickness as carpel initiation proceeds (Fig. 7d).

**Carpel initiation.** The carpel primordia are initiated singly and in acropetal succession on the conical receptacle (Fig. 7a, c). The initiatory divisions are periclinal to oblique in the T2 and the adjacent mantle, low on the flanks of the floral apex (at arrows, Fig. 7b). Surface cells of the mantle undergo anticlinal divisions nearby. Further division and cell enlargement around the initiatory site result in the formation of a protuberance (Fig. 7b). Procambium is present below the base of the young primordium (Fig. 7d). In transverse section the young carpel primordia are spherical (Fig. 8a) and their helical order of succession is evident. The older carpel primordia (Fig. 7a, c) show apical and marginal growth.

After carpel initiation is complete, the remaining apical residuum (Fig. 7e) is low-convex, with only two or three meristematic layers. The apical residuum is about 130 µm in diameter and about 40 µm high. Later the apical residuum increases (Fig. 7e) as the meristematic cells continue to divide anticlinally in the outermost layer, and anticlinally and periclinally in the subsurface layers. The cells toward the base of the apical residuum begin to differentiate through vacuolation, cell enlargement, and tannin accumulation (Fig. 7f).

**Carpel development.** Primordia of different sizes and ages surround the axis (Fig. 8a). The young carpel primordium (Fig. 8b, upper primordium) has become a protuberance by anticlinal division of the surface layer, and by anticlinal and periclinal division in the subsurface layer. The primordium grows by general cell division throughout until it is 50–60 µm high, when apical and subapical initials become active and increase its height (Fig. 8b). Vacuolation and cell enlargement are evident early in the abaxial protoderm (Fig. 8b, lower primordium), indicating that "new" protoderm on the heightening primordium is laid down by the apical initials. The time at which apical and subapical initial activity cease is difficult to determine, but apical activity is still present at a height of 250–260 µm.

Marginal growth begins in the carpel primordium at a height of 80–112 µm (Fig. 8c). The carpel primordium becomes incurved as a consequence. Marginal initials (Fig. 8d) divide anticlinally to perpetuate the protoderm. The submarginal initials (at arrow, Fig. 8d) divide anticlinally and periclinally to produce the internal tissue. The cleft deepens as the margins extend in depth.

The carpel margins become appressed and conduplicate, and a locule forms by the time it is about 520 µm high (Fig. 9a, c). There is a small cross zone (Fig. 9c) that contributes little to the organ. Two ovule primordia are initiated on either side of the carpel midrib by periclinal divisions in the adaxial hypodermal layer of the carpel walls (Fig. 8e), when the carpel is 345–360 µm high. Other adjacent cells divide and contribute to the ovule primordium. The placental position of the

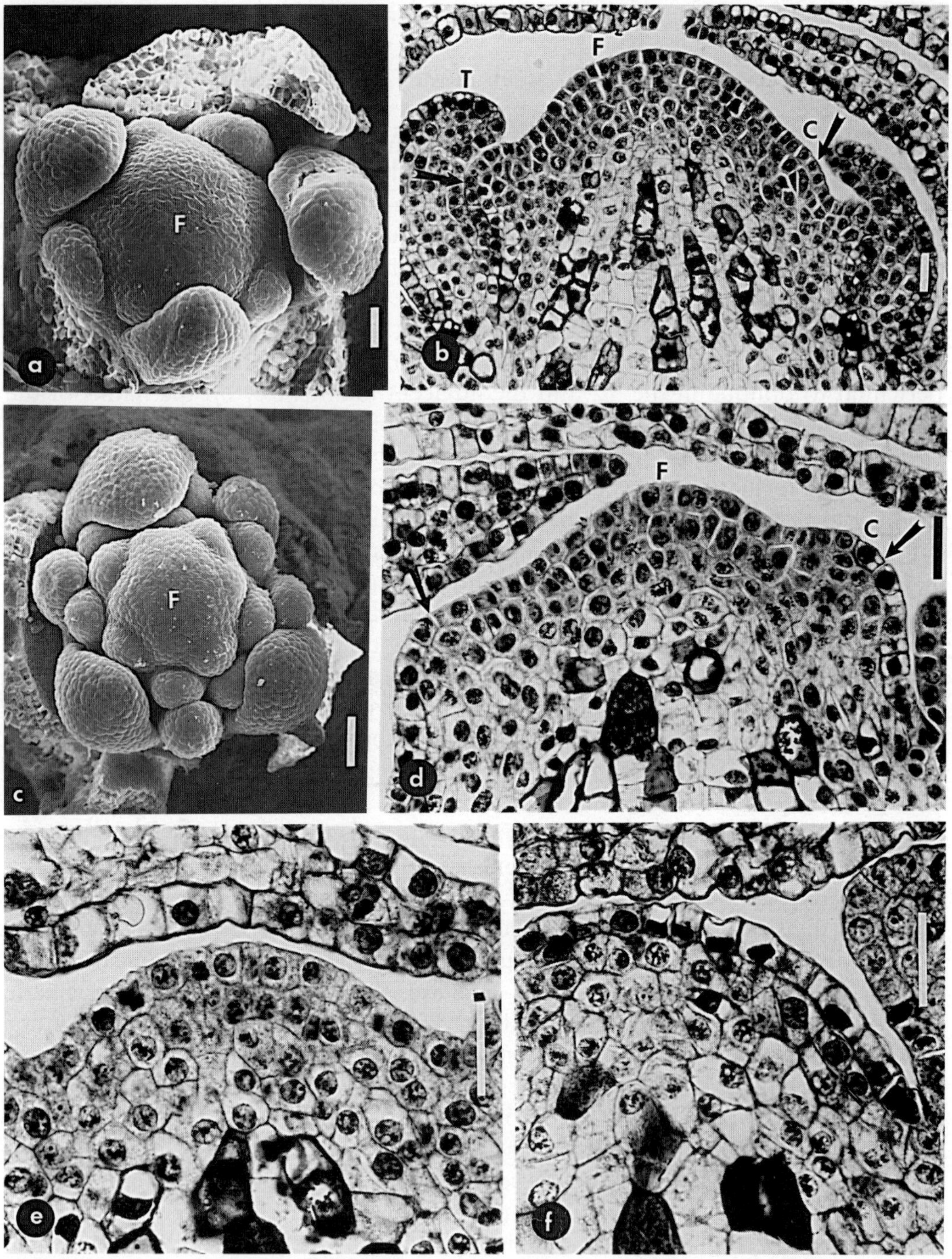


Fig. 7. *Schisandra glabra*, carpellate flower development. *a*, *c*, SEM micrographs, *b*, *d–f* photomicrographs of sections. *a* Flower at beginning of carpel initiation (near-polar view); *b* floral apex at carpel initiation (at arrows, longitudinal section), tepal primordia are present lower on flanks; *c* floral apex that has produced at least 16 carpel primordia, tepals around periphery are broad and incurved; *d* floral apex of carpellate flower after carpels have all initiated, two carpels are present high on flanks (at arrows); *e*, *f* successive stages in cellular differentiation of the apical residuum. – *C* carpel; *F* floral apex; *T* tepal. – Bars: *a*, *b*, *d–f* 50 μm, *c* 100 μm

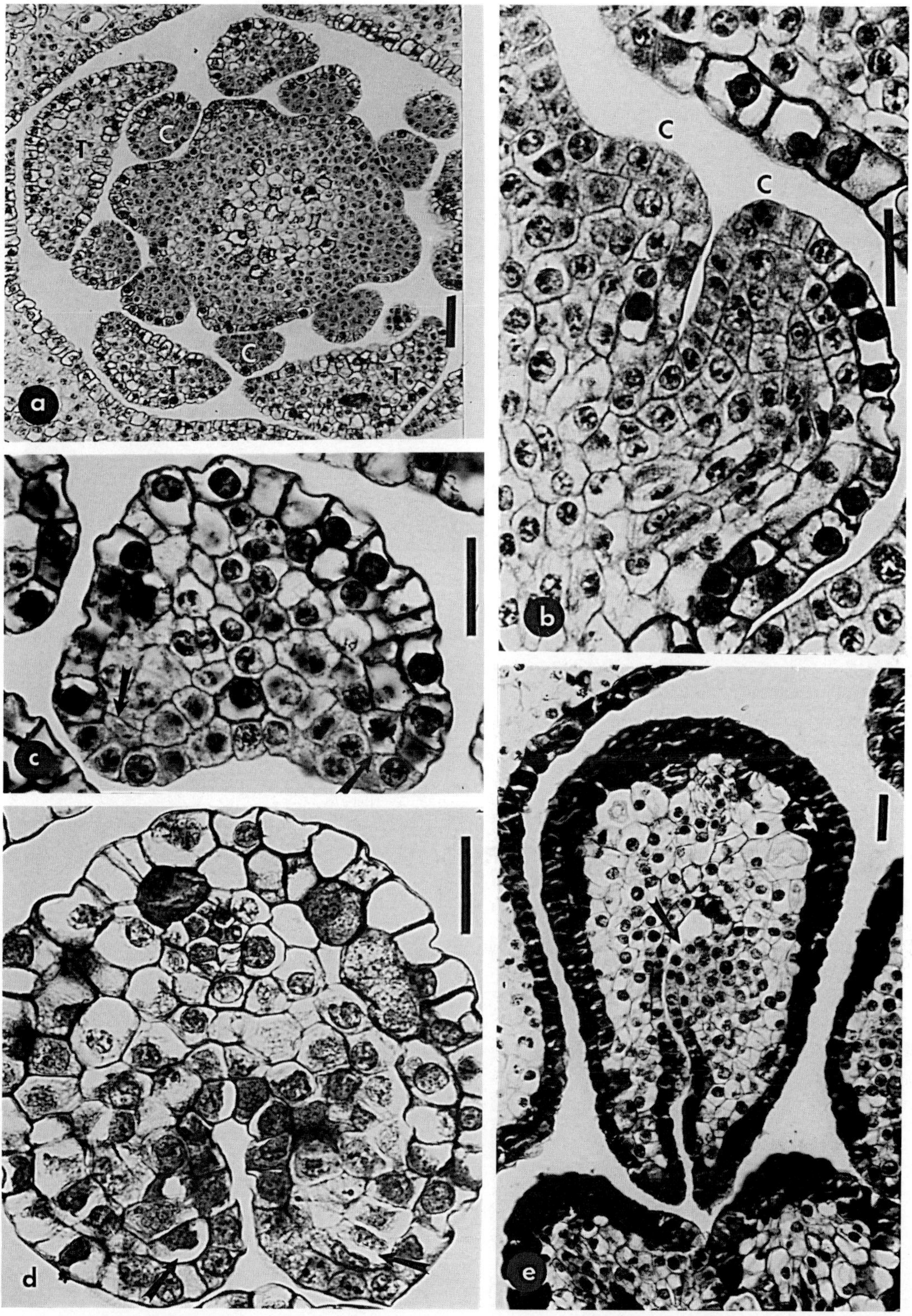


Fig. 8. *Schisandra glabra*, carpel development in carpellate flower. All figures are photomicrographs of sections. *a* Floral apex with numerous carpel primordia of various sizes and tepal primordia (transverse section); *b* two carpel primordia (longitudinal section), the younger above the older; *c*, *d* carpel primordia (transverse sections) of two stages, in *c* submarginal initials are just beginning their activity, while in *d* active marginal and submarginal initials (at arrows) are present, as well as a procambial strand; *e* carpel (transverse section) showing conduplicate laminae and ovule primordium (at arrow). – *C* carpel; *T* tepal. – Bars: *a* 100 μm, *b*–*e* 50 μm

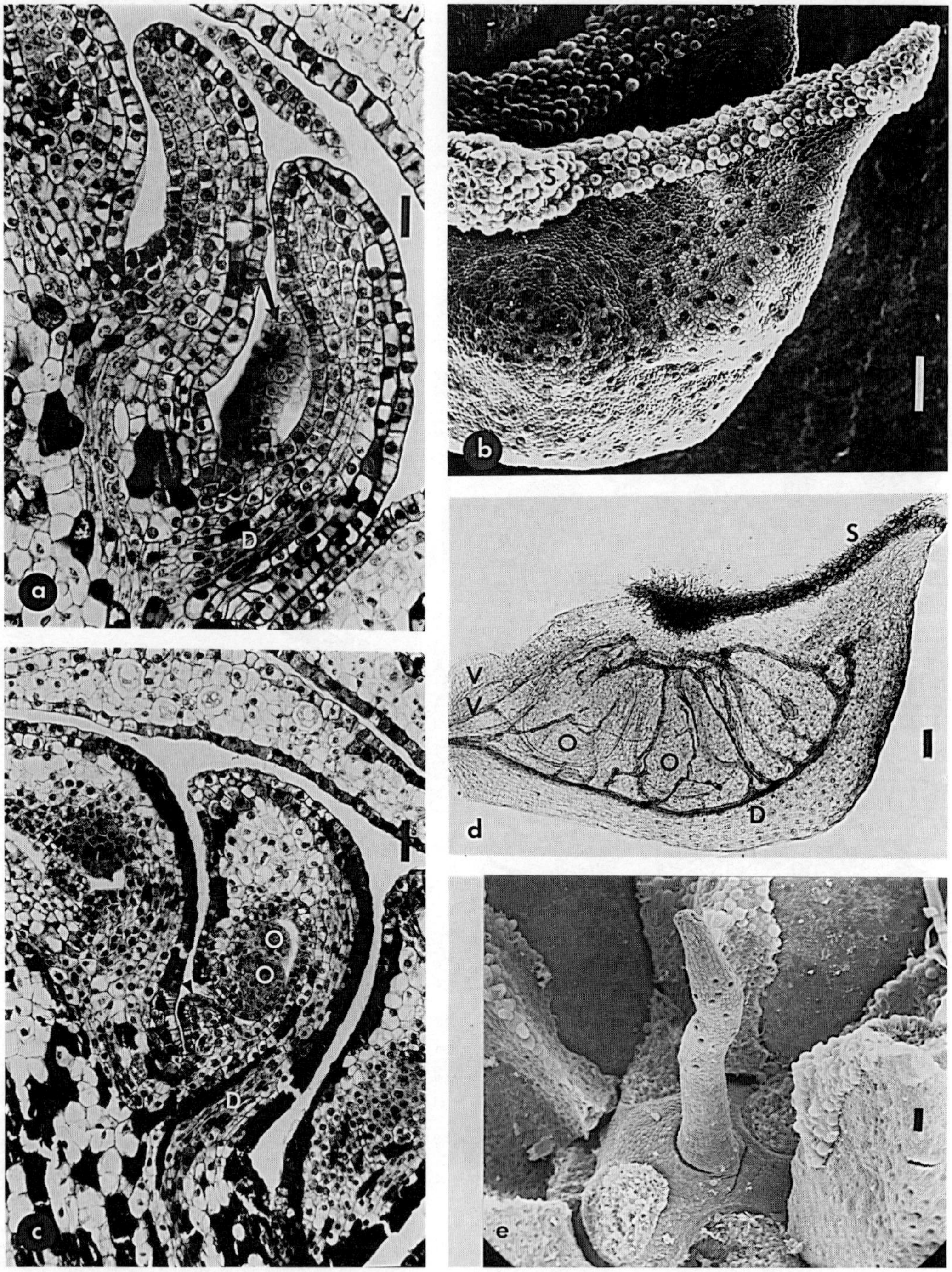


Fig. 9. *Schisandra glabra*, carpel and apical residuum development in carpellate flower; *a*, *c* are photomicrographs, *d* is a clearing, and *b*, *e* are SEM micrographs. *a* Carpel primordia at locule formation (longitudinal section), one lateral wall is shown (at arrow); *b* carpel at anthesis with stigmatic crest; *c* carpel with cross zone (at arrowhead) and two ovules, the carpellary wall is undergoing differentiation; *d* carpel at anthesis, with vascular system and stigmatic tissue stained, dorsal and ventral bundles and their branches are visible; *e* terminal spikelike structure that has resulted from growth of the apical residuum. – *D* Dorsal bundle; *O* ovule; *S* stigmatic tissue; *V* ventral bundles. – Bars: *a*, *c* 50 μm, *b*, *d*, *e* 100 μm

ovules, well back from the margins (Fig. 8e), has been called "lateral-laminar" (TAKHTAJAN 1969, STEBBINS 1974). As the ovules enlarge, they become tilted obliquely downward into the enlarging locule. At anthesis, the ovules are anatropous, bitegmic and crassinucellate.

Carpel enlargement is accompanied by cell differentiation; vacuolation and tannin accumulation are evident earliest in the abaxial part of the carpel primordium (Fig. 8d, e). At anthesis, the decurrent stigmatic crest (Fig. 9b, d) formed by the appressed carpel margins is prolonged distally into an acute, nonvascularized pseudostyle (BAILEY & SWAMY 1951) and proximally into a pendent appendage (shown in part in Fig. 9b). After fertilization, carpel enlargement continues, together with elongation of the receptacle or gynophore (Fig. 1e).

**Vascularization in the carpel.** Procambium is present below the base of the initiating carpel when it is about 40 µm high (Fig. 7b). It differentiates acropetally and continuously into the dorsal bundle of the primordium at a height of 50–60 µm. The procambial dorsal strand is seen in a carpel about 1250 µm high (Fig. 8d). The two ventral bundles form slightly after the median dorsal bundle, and they develop acropetally and continuously. All three bundles branch repeatedly (Fig. 9d). The ovules are vascularized by branches of both the ventral and dorsal bundles (Fig. 9d). The ovular trace extends into the funiculus.

Not all of the carpels of a flower persist. Many individual carpels or even entire fruiting pedicels abscise from the plant during June and July. In late August and early September, mature fruits are pendent on slightly thickened pedicels (Fig. 1e). The aggregate fruit may include one to 20 sessile maturing carpels. At the summit of the flower is a spinelike structure (Fig. 9e), the last product of the floral apex. The receptacle axis bearing the fruits varies in length from 0.6–6.2 cm. Each berry-like carpel is red, subglobose, and 6.5–10 mm in diameter. Each contains one or two reniform-ellipsoid reddish-brown superposed seeds, each with a rugulose seed-coat and a lateral hilum.

## Discussion

**Systematics, morphology and anatomy.** The landmark systematic work on *Schisandra* is the monograph on *Schisandraceae* by A. C. SMITH (1947). He and I. W. BAILEY and their colleagues at Harvard produced a remarkable series of investigations of the diverse vesselless Ranalean taxa that had been included in *Magnoliales.* SMITH (1947) designated sections, defined on staminal characters, particularly whether the stamens were free or fused in some way. BAILEY & NAST (1948) described the vegetative anatomy of leaf and stem of 22 species each of *Schisandra* and *Kadsura* (*Schisandraceae*), and 34 of *Illicium* (*Illiciaceae*). Two groups of Ranalean plants were divided on pollen type (monocolpate or tricolpate). These three genera belong to the tricolpate group, but have oil cells like the monocolpate group. As such, these genera were considered important for study. These authors concluded that *Schisandra* and *Kadsura* cannot be separated on vegetative characteristics, and that neither belongs in *Magnoliaceae.* WOOD (1958) described *S. glabra*, as the only member of the family in the southeastern United States. OZENDA (1949) described aspects of vegetative anatomy of two species of *Schisandra.*

The nature of the carpel was discussed by BAILEY & SWAMY (1951) and by SWAMY & PERIASAMY (1964). A major concern was the submarginal position of ovules, contrary to the presumed marginal position according to the classical theory of the flower (WILSON & JUST 1939). Another revised concept was the shape of the carpel in transection: with conduplicate or folded laminae, rather than arcuate ones. The shape of the *Schisandra* carpel was noted by BAILEY & SWAMY (1951) to be a modification of the *Drimys piperita* type in that the former has an arched dorsal surface, with the ventral surface consequently reduced in extent.

BAILEY & SMITH (1942) and BAILEY & NAST (1943) described conduplicate carpels in *Degeneriaceae* and *Winteraceae*. SWAMY & PERIASAMY (1964) were concerned with the details of closure of the conduplicate carpel: edge to edge, or involuted first and then the abaxial surfaces fused. They also discussed the problem of marginal placentation, as an integral part of the classical view of the carpel. We now know that ovules are not truly marginal, but submarginal, but this fact does not refute the classical view.

Embryology was described in *Schisandra chinensis* (YOSHIDA 1962, SWAMY 1964), *S. grandiflora* (KAPIL & JALAN 1964), and *S. repanda* (HAYASHI 1963). HAYASHI (1960) reviewed embryology in *Schisandraceae* as well as other magnoliaceous taxa. BHANDARI (1971) compared the different interpretations of embryo sac development in the genus. The embryo sac is of the *Polygonum* type in common with most *Magnoliaceae*. BHANDARI's data contradict any close relationship between *Schisandra* and *Illicium*. Pollen of *Schisandra* was described by HAYASHI (1960), JALAN & KAPIL (1964), and PRAGLOWSKI (1976). Cytological studies were reported on *Schisandra* ($n = 14$, while *Illicium* has $n = 13$) (OKADA 1975, STONE 1968, WHITAKER 1933). EHRENDORFER & al. (1968) reported that the chromosome number may be either 13 or 14 in *Schisandra*.

**Floral apex and its eventual specialization.** Conversion from a tunica-corpus configuration to one of mantle and core, as in *Schisandra glabra*, has been reported frequently in other taxa such as *Aquilegia* by GRÉGOIRE (1938), who originated the terms mantle ("manchon") and core. A great size increase of the apical meristem accompanies the change in configuration in most cases.

Other primitive plants show modifications of the apical residuum after organogeny. In floral development of *Illicium floridanum* (*Illiciaceae*; ROBERTSON & TUCKER 1979) the apical residuum forms a conical structure, the internal structure of which resembles that of the staminal shield of *Schisandra*. Subsurface cells of the apical meristem undergo repeated periclinal divisions to form files of cells. In *Myristica fragrans* (*Myristicaceae*; ARMSTRONG & TUCKER 1986) the apical residuum of the floral apex continues to grow after stamen initiation in the staminate flower. It actually may be zonal growth below the apical residuum. The apex is carried upward as the stamen primordia elongate, producing a synandrous condition. The anthers are attached to the receptacle along their length. ENDRESS (1990) pointed out the frequency of synandry in male flowers and in primitive flowers. It is found throughout the *Myristicaceae* and *Canellaceae*, and some *Annonaceae*. ENDRESS (1990) also gives examples in *Monimiaceae*, *Lauraceae*, *Chloranthaceae*, *Rafflesiaceae*, *Lardizabalaceae*, *Menispermaceae*, and *Eupomatiaceae*.

Fleshy structures in other Ranalean flowers such as *Magnolia* have been interpreted by GOTTSBERGER (1974) as specialized beetle lures. Nothing is known about whether the staminal shield in *Schisandra* might have a similar role.

**The specialized staminal shield.** ENDRESS (1990) has pointed out that the types of specializations differ between unspecialized and specialized flowers. Synandry, one specialization, results from zonal growth of the receptacle below the level of attachment of the stamens. Synandry is common in primitive flowers of *Magnoliidae* (ENDRESS 1990), and more common in male unisexual flowers than hermaphroditic ones. ENDRESS (1990) pointed out that the frequency of synandry in male flowers may be due to the lack of constraint imposed by the gynoecium in a hermaphroditic flower.

SCHAEPPI (1976) showed the synandrial androecium of *Dioscoreophyllum cumminsii* (*Menispermaceae*) as a barrel-shaped receptacle to which are attached the 12 pairs of pollen sacs. In the male flower of *Myristica fragrans* (*Myristicaceae*; ARMSTRONG & TUCKER (1986) there is a similar elongate receptacle to which the microsporangia are attached along their length. Developmentally, the stamen primordia are initiated laterally around the floral meristem, which then proceeds to elongate together with the stamens.

The connective in *Schisandra* is very broad. Other primitive plants similarly showing broad connectives are members of *Chloranthaceae* (ENDRESS 1987), *Degeneria* (*Degeneriaceae*, BAILEY & SMITH 1942), *Akebia* (*Lardizabalaceae*, PAYNE & SEAGO 1968), and several taxa of *Magnoliaceae* (CANRIGHT 1952).

BAILLON (1867–1869: 133–192) illustrated and described the flower of *S. glabra* (as *S. coccinea*) and several other species including some with free stamens. OZENDA (1949) also showed free staments of *S. henryi* in which the single vascular bundle branches repeatedly in a broad, distally flared connective with lateral anthers.

**Floral vasculature.** The carpellary vasculature of *Schisandra* is unusual (OZENDA 1949, EAMES 1961, LEINFELLNER 1966, MELVILLE 1969) in that ovular traces are contributed by either the dorsal or ventral strands, rather than strictly ventral as in most carpels, e.g., those of *Magnoliaceae* (CANRIGHT 1960) and *Drimys* (TUCKER & GIFFORD 1975).

**Apocarpy.** Other primitive taxa for which the complete floral ontogeny has been described include *Magnolia* (ERBAR & LEINS 1981, 1983), *Michelia* (TUCKER 1960, 1961), *Myristica* (ARMSTRONG & TUCKER 1986), *Austrobaileya* (ENDRESS 1983), *Drimys* and *Pseudowintera* in *Winteraceae* (SAMPSON & KAPLAN 1970, TUCKER 1959, TUCKER and GIFFORD 1966a, b, 1975), *Akebia* in *Lardizabalaceae* (PAYNE & SEAGO 1968), and *Nigella* in *Ranunculaceae* (KAUSSMANN & NEITZEL 1972).

**Unisexual flowers: primitive character state?** Unisexuality has generally been considered a derived character state (EAMES 1961). However, it is a specialization that occurs in some of the families considered primitive: *Ceratophyllaceae* (*Ceratophyllum*), *Cercidiphyllaceae* (*Cercidiphyllum*), *Chloranthaceae*, *Myristicaceae*, *Magnoliaceae* (*Kmeria*, DANDY 1927), *Piperaceae* (*Piper* pro parte), *Ranunculaceae* (*Clematis*, *Thalictrum*), *Schisandraceae*, and *Winteraceae* (*Tasmannia*). Unisexuality also is common in many of the most derived families: *Asteraceae*, *Betulaceae*, *Euphorbiaceae*, *Fagaceae*, *Gramineae*, and *Orchidaceae* among others. The condition occurs in about 168 angiosperm families including 26 monocotyledonous families. Sexual condition of angiosperm taxa seems to be a labile character that can shift in response to adaptive pressures, regardless of the general level of specialization of other characteristics.

This work was part of a dissertation project by the second author. Its preparation is supported in part by National Science Foundation grant DEB-9207671 to S. C. T. The authors thank Ben Martin for his help and encouragement, and Andrew Douglas for his technical assistance.

## References


Armstrong, J. E., Tucker, S. C., 1986: Floral development in *Myristica* (*Myristicaceae*). – Amer. J. Bot. **73**: 1131–1143.

Bailey, I. W., Nast, C. G., 1943: The comparative morphology of the *Winteraceae*. II. Carpels. – J. Arnold Arbor. **24**: 472–481.

– – 1948: Morphology and relationships of *Illicium*, *Schisandra*, and *Kadsura*, 1. Stem and leaf. – J. Arnold Arbor. **29**: 77–89.

– Smith, A. C., 1942: *Degeneriaceae*: a new family of flowering plants from Fiji. – J. Arnold Arbor. **23**: 356–365.

– Swamy, B. G. L., 1951: The conduplicate carpel of dicotyledons and its initial trends of specialization. – Amer. J. Bot. **38**: 373–379.

Baillon, H., 1867–1869: Histoire des plantes. 1. – Paris: Librairie de L. Hachette et Cie.

Bhandari, N. N., 1971: Embryology of the *Magnoliales* and comments on their relationships. – J. Arnold Arbor. **52**: 1–39, 285–304.

Canright, J. E., 1952: The comparative morphology and relationships of the *Magnoliaceae*. I. Trends of specializations in the stamens. – Amer. J. Bot. **39**: 484–497.

– 1990: The comparative morphology and relationships of the *Magnoliaceae*. III. Carpels. – Amer. J. Bot. **47**: 145–155.

Cronquist, A., 1981: An integrated system of classification of flowering plants. – New York: Columbia University Press.

Dandy, J. E., 1927: The genera of the *Magnolieae*. – Kew Bull. **1927**: 257–264.

Eames, A. J., 1961: Morphology of the angiosperms. – New York: McGraw-Hill.

Ehrendorfer, F., Krendl, F., Habeler, E., Sauer, W., 1968: Chromosome numbers and evolution in primitive angiosperms. – Taxon **17**: 337–353.

Endress, P. K., 1983: The early floral development of *Austrobaileya*. – Bot. Jahrb. Syst. **103**: 481–497.

– 1987: The *Chloranthaceae*: reproductive structures and phylogenetic position. – Bot. Jahrb. Syst. **109**: 153–226.

– 1990: Evolution of reproductive structures and functions in primitive angiosperms (*Magnoliidae*). – Mem. New York Bot. Gard. **55**: 5–34.

Erbar, C., Leins, P., 1981: Zur Spirale in Magnolien-Blüten. – Beitr. Biol. Pfl. **56**: 225–241.

– – 1983: Zur Sequenz von Blütenorganen bei einigen Magnoliiden. – Bot. Jahrb. Syst. **103**: 433–449.

Gottsberger, G., 1974: The structure and function of the primitive angiosperm flower – a discussion. – Acta Bot. Neerl. **23**: 461–471.

Grégoire, V., 1938: La morphogénèse et l'autonomie morphologique de l'appareil floral. 1. Le carpelle. – La Cellule **47**: 287–452.

Hayashi, Y., 1960: On the microsporogenesis and pollen morphology in the family *Magnoliaceae*. – Sci. Rep. Tóhoku Univ., Ser. IV, Biol., **26**, 45–52.

– 1963: The embryology of the family *Magnoliaceae* sens. lat. II. Megasporogenesis, female gametophyte, and embryogeny of *Schisandra repanda* and *Kadsura japonica*. – Sci. Rep. Tôhoku Univ., Ser. IV, Biol., **29**: 403–411.

Jalan, S., Kapil, R. N., 1964: Pollen grains of *Schisandra* Michaux. – Grana Palynol. **5**: 216–221.

JOHANSEN, D. A., 1940: Plant microtechnique. – New York: McGraw-Hill.
KAPIL, R. N., JALAN, S., 1964: *Schisandra* MICHAUX – its embryology and systematic position. – Bot. Not. **117**: 285–306.
KAUSSMANN, B., NEITZEL, H., 1972: Ein Beitrag zur Morphologie des Gynoeceums von *Nigella damascena* aus histogenetischer Sicht. – Flora **161**: 30–45.
LEINFELLNER, W., 1966: Über die Karpelle verschiedener *Magnoliales* III. *Schisandra* (*Schisandraceae*). – Österr. Bot. Zeit. **113**: 563–569.
MELVILLE, R., 1969: Studies in floral structure and evolution. I. The *Magnoliales*. – Kew Bull. **23**: 133–180.
OKADA, H., 1975: Karyomorphological studies of woody *Polycarpicae*. – J. Sci. Hiroshima Univ., Ser. B, Div. 2, **15**: 115–200.
OZENDA, P., 1949: Recherches sur les dicotylédones apocarpiques. Contribution à l'étude de angiospermes dites primitives. – Lab. Ecole Normale Supérieure, Publ. Sér. Biol., **2**: 1–183.
PAYNE, W. W., SEAGO, J. L., 1968: The open conduplicate carpel of *Akebia quinata* (*Berberidales*: *Lardizabalaceae*). – Amer. J. Bot. **55**: 575–581.
PRAGLOWSKI, J., 1976: *Schisandraceae*. – In NILSSON, S., (Ed.): World Pollen and Spore Flora. **5**, pp. 1–32. – Stockholm: Almkvist & Wiksell.
ROBERTSON, R. E., TUCKER, S. C., 1979: Floral ontogeny of *Illicium floridanum*, with emphasis on stamen and carpel development. – Amer. J. Bot. **66**: 605–617.
RODIN, R. J., DAVIS, R. E., 1967: The use of papain in clearing plant tissues for whole mounts. – Stain Technol. **42**: 203–206.
SAMPSON, F. B., KAPLAN, D. R., 1970: Origin and development of the terminal carpel in *Pseudowintera traversii*. – Amer. J. Bot. **57**: 1185–1196.
SASS, J. E., 1951: Botanical microtechnique. 3rd edn. – Ames, Iowa: Iowa State College Press.
SCHAEPPI, H., 1976: Über die männlichen Blüten einiger Menispermaceen. – Beitr. Biol. Pfl. **52**: 207–215.
SMITH, A. C., 1947: The families *Illiciaceae* and *Schisandraceae*. – Sargentia **7**: 1–224.
STEBBINS, G. L., 1974: Flowering plants. Evolution above the species level. – Cambridge, Mass.: Belknap Press, Harvard University.
STONE, D. E., 1968: Cytological and morphological notes on the southeastern endemic, *Schisandra glabra* (*Schisandraceae*). – J. Elisha Mitchell Sci. Soc. **84**: 351–356.
SWAMY, B. G. L., 1964: Macrogametophytic ontogeny in *Schisandra chinensis*. – J. Indian Bot. Soc. **43**: 391–396.
– PERIASAMY, K., 1964: The concept of the conduplicate carpel. – Phytomorphology **14**: 319–327.
TAKHTAJAN, A., 1969: Flowering plants, origin, and dispersal. – Edinburgh: Oliver & Boyd.
TUCKER, S. C., 1959: Ontogeny of the inflorescence and the flower in *Drimys winteri* v. *chilensis*. – Univ. Calif. Publ. Bot. **30**: 257–336.
– 1960: Ontogeny of the floral apex in *Michelia fuscata*. – Amer. J. Bot. **47**: 266–276.
– 1961: Phyllotaxis and vascular organization of the carpels of *Michelia fuscata*. – Amer. J. Bot. **48**: 60–71.
– GIFFORD, E. M., Jr., 1966a: Organogenesis in the carpellate flower of *Drimys lanceolata*. – Amer. J. Bot. **53**: 433–442.
– – 1966b: Carpel development in *Drimys lanceolata*. – Amer. J. Bot. **53**, 671–678.
– – 1975: Carpellary vasculature and the ovular supply in *Drimys*. – Amer. J. Bot. **62**: 191–197.
WHITAKER, T. W., 1933: Chromosome number and relationship in the *Magnoliales*. – J. Arnold Arbor. **14**: 376–385.
WILSON, C. L., JUST, T., 1939: The morphology of the flower. – Bot. Rev. **5**: 97–131.

Wood, C. E., 1958: Genera of woody *Ranales*. – J. Arnold Arbor. **39**: 296–346.
Yoshida, O., 1962: Embryologische Studien über *Schisandra chinensis*. – J. College Arts Sci., Chiba University, **3**: 459–462.

Addresses of the authors: Shirley C. Tucker, Department of Botany, Louisiana State University, Baton Rouge, Louisiana, USA J. Allen Bourland, Department of Biology, G. Wallace State University, Dothan, Alabama, USA.

Accepted January 11, 1994 by P. K. Endress

Pl. Syst. Evol. [Suppl.] 8: 159–173 (1994)

# Floral aspects of *Barclaya* (*Nymphaeaceae*): pollination, ontogeny and structure[1]

PAULA S. WILLIAMSON and EDWARD L. SCHNEIDER

Received November 23, 1993

**Key words:** *Nymphaeaceae*, *Barclaya*. – Floral morphology, vascular anatomy, pollen.

**Abstract:** *Barclaya*, endemic to Southeast Asia, includes four species: *B. longifolia* WALLICH. *B. motleyi* HOOKER f., *B. kunstleri* (KING) RIDLEY, and *B. rotundifolia* HOTTA. The genus has been traditionally assigned monotypic status in the family *Barclayaceae* or tribe *Barclayeae* primarily based on the occurrence of a hypogynous calyx, originally interpreted as an involucre in an otherwise epigynous flower, orthotropous ovules, lack of an aril, and inaperturate pollen. Cladistic and molecular studies support the association of *Barclaya* with *Euryale*, *Victoria*, *Nymphaea*, *Ondinea*, and *Nuphar* in the family *Nymphaeaceae*. Floral structure reveals (1) vasculature and developmental support for the hypothesis that the hypogynous appendages are sepals, (2) that pollen is zonasulculate, (3) that floral ontogeny is similar to that previously described for epigynous members of the *Nymphaeaceae* s. str., and (4) anatomical and morphological similarities with other *Nymphaeaceae* s. str. The cleistogamous and chasmogamous flowers are self-pollinating, the latter perhaps facilitated by flies in emergent flowers.

The *Nymphaeales*, a paleoherb group, is assigned a key, basal, pivotal evolutionary position in nearly every old and modern classification system. Historically, the *Nymphaeaceae* sensu lato include nine genera: *Barclaya* WALLICH, *Brasenia* SCHREB., *Cabomba* AUBL., *Euryale* SALISB., *Nelumbo* (TOURN.) ADANS., *Nuphar* SM., *Nymphaea* (TOURN.) L., *Ondinea* den HARTOG, and *Victoria* LINDL. Although the nine genera have been lumped into a single family (BENTHAM & HOOKER 1862, CASPARY 1891, WETTSTEIN 1935, GUNDERSEN 1950, BUCHHEIM, 1964), other systematists recognize three families: the *Cabombaceae* including *Brasenia* and *Cabomba*, the *Nelumbonaceae* including *Nelumbo*, and the *Nymphaeaceae* including the remaining six genera (BESSEY 1915, TAKHATAJAN 1959, 1980, GOLENIEWSKA-FURMANOWA 1970, HUTCHINSON 1973, THORNE 1976, CRONQUIST 1981, LES & al. 1991). This assemblage of six genera is referred to as the *Nymphaeaceae* sensu stricto. ITO (1987) also recognized three families, but they are the *Ceratophyllaceae* (*Cabomba*, *Brasenia*, and *Ceratophyllum*), *Nelumbonaceae* (*Nelumbo*) and *Nymphaeaceae* (sensu stricto). While many classification schemes place *Barclaya* within the *Nymphaeaceae*, the genus

[1] Morphological studies of the *Nymphaeaceae* XX.

has also been segregated into a separate monotypic family, the *Barclayaceae*, due to the occurrence of a hypogynous calyx, also interpreted as an involucre subtending an epigynous flower, orthotropous ovules, the absence of a seed aril, and inaperturate pollen (Li 1955, Walker 1974). Numerous studies have used floral characters to examine phylogenetic relationships among these genera (Moseley 1961, 1965, 1971, Richardson 1969, Heinsbroek & van Heel 1969, Schneider 1976, 1983, Weidlich 1976a, b, 1980, van Heel 1977, Ito 1983, 1984, 1986a, b, 1987, Moseley & al. 1984, Moseley & Uhl 1985, Williamson & Moseley 1989, Moseley & al. 1993). This paper specifically examines aspects of pollination, floral ontogeny, and structure of the flower to assess the systematic placement of *Barclaya*.

## Material and methods

Field observations and collections were made in July 1984, July 1988, and March 1990. *Barclaya kunstleri* and *B. rotundifolia* were observed in Johore, Malaysia in roadside streams midway between Kluang and Mersing. *Barclaya motleyi* was observed in shallow rainforest streams between Mersing and Jemaluang, Johore. *Barclaya longifolia* was observed 1.6 km east of Bukit Wang, near Jitra, Kedah. All species with the exception of *B. longifolia* were monitored on a monthly basis by Robert Blankinship, a graduate student at Southwest Texas State University, from September 1991 to June 1992. Voucher specimens are housed at the Santa Barbara Botanic Garden Herbarium (SBBG) and the Southwest Texas State University Herbarium (SWT).

Specimens of *Barclaya* were collected from small streams in the tropical rainforest of Malaysia and preserved in 50% ethanol. Following fixation, flowers at various developmental stages and fruits were prepared for light microscopy following standard techniques (Johansen 1940). Tissue was dehydrated in a tertiary-butyl alcohol series, then infiltrated and embedded with paraplast-plus TM. The material was sectioned at 10–15 µm using a rotary microtome and stained with Harris hematoxylin, safranin, and fast green. Photomicrographs were taken with a Zeiss Ultraphot and Hasselblad 550C, using T-MAX 100 film.

Material was also prepared for examination with scanning electron microscopy. Following fixation in 50% ethanol, samples were dehydrated in a graded ethanol series to amyl acetate or absolute ethanol. Following dehydration samples were critical point dried with $CO_2$, mounted on SEM stubs using copper adhesive tape or Pella #16084 carbon conductive tabs and coated with a layer of gold in a Denton or Hummer X vacuum sputter coating unit. Samples were observed with Cambridge 90-B and Bausch & Lomb NanoLab LE 2100 scanning electron microscopes at 15–25 kV. Micrographs were taken using Polaroid 52 and 55 positive/negative film.

Cladograms were generated using the PAUP computer program version 3.0s (Swofford 1991). Twelve floral characters were selected for the cladistic analysis. All characters were unordered because no prior basis for ordering the states could be identified. The branch and bound option was used to find the most parsimonious cladogram. A majority-rule consensus tree was prepared. *Nelumbo* was selected as the outgroup for rooting trees since phyletic studies by Moseley & Uhl (1985), Ito (1987) and Les & al. (1991) indicate that this taxon is taxonomically distinct from the other eight genera.

## Results

**Taxonomic history of *Barclaya*.** In 1826, Nathaniel Wallich, Director of the East India Company's Botanic Garden at Calcutta, discovered a new aquatic herb near Rangoon, Burma, growing abundantly in pools and along the margins of small, slow flowing rivers. WALLICH considered the genus to be allied to members of the *Nymphaeaceae*, noting that the taxon differed from *Euryale* in the hypogynous calyx, and from *Nymphaea* by its tubular, corolla-like receptacle with the stamens attached to the inner surface. WALLICH's description of the genus contained in a letter sent to HENRY COLEBROOKE, a director of the East India Company, was read at a meeting of the Linnaean Society of London on May 1, 1827 and published in the June 1827 issue of the Philosophical Magazine (TAYLOR & PHILLIPS 1827, RAPHAEL 1970). WALLICH designated the new taxon *Hydrostemma linguiforme*, the name appearing on sheets housed in the Wallich Herbarium at Kew (KEW) and at the British Museum of Natural History (BM). A more complete description, contained in the letter, appeared in the December 1827 Transactions of the Linnaean Society, where the previously published name *Hydrostemma linguiforme* had been changed to *Barclaya longifolia* to commemorate ROBERT BARCLAY, a highly respected friend and benefactor of WALLICH (WALLICH 1827). MABBERLEY (1982), noting that the validly published name *Hydrostemma* antedated *Barclaya*, proposed two new combinations: *Hydrostemma longifolium* (WALL.) MABB. and *H. motleyi* (HOOK. f.) MABB. STONE (1982) made the combination *H. kunstleri* (KING) B. C. STONE. CRUSIO & BOGNER (1984) proposed to conserve *Barclaya* against *Hydrostemma* based on the long-standing usage and better known generic name *Barclaya*. Their proposal was approved by a vote of ten to one by the International Botanical Congress Committee on Nomenclature (BRUMMITT 1987).

**Habit and habitat.** *Barclaya* is endemic to SE Asia. It occurs in Burma, Thailand, Malaysia, Singapore, and Indonesia. It grows in or along the edges of streams in the tropical rain forest. Because rapid deforestation of southeast Asia is resulting in increased turbidity of habitats, *Barclaya* is becoming threatened (STONE 1978). At present four species are recognized.

*Barclaya longifolia* was described by WALLICH (1827). This species occurs submersed in small- to medium-size streams. Leaves are linear-lanceolate, cordate at the base, with an obtuse apex. Blades are 12–30 cm long and 2–5 cm wide, with slightly wavy margins (Fig. 1A). Flowers are solitary, typically cleistogamous, and usually remain at or below the water surface. This species is well known in the aquatic plant trade industry and has been frequently discussed in aquarium books and journals in addition to the scientific literature on aquatic plants (HEINE 1958, van BRUGEN 1961, PAFFRATH 1972, de WIT 1983).

*Barclaya motleyi* was described by J. D. HOOKER (1862). The species was observed by MOTLEY growing in sand-bottom rivers near Banjoreang in the Southern Borneo peninsula and named *B. rotundifolia*. Before MOTLEY's specimens and letter of description reached Kew, he and his family were killed by Muslim settlers in Southern Borneo. HOOKER then renamed and published the species *B. motleyi* in memory of the plant's discoverer. The species possesses broad, orbicular blades, 6–12 cm in diameter, which occur just beneath the water surface (Fig. 1B). During the seasonal dry period, lasting from February to March, the streams may be void of water, but the plant persists with cuticularized leaves exposed on the damp litter. In addition to leaf shape, this species differs from *B. longifolia* in producing emergent chasmogamous flowers with longer, narrower sepals, petals which are light yellow to pink, and fewer stamens.

*Barclaya kunstleri* was first described by KING (1889) as a variety of *B. motleyi*. The varietial name was in honor of KUNSTLER who was employed by KING as a

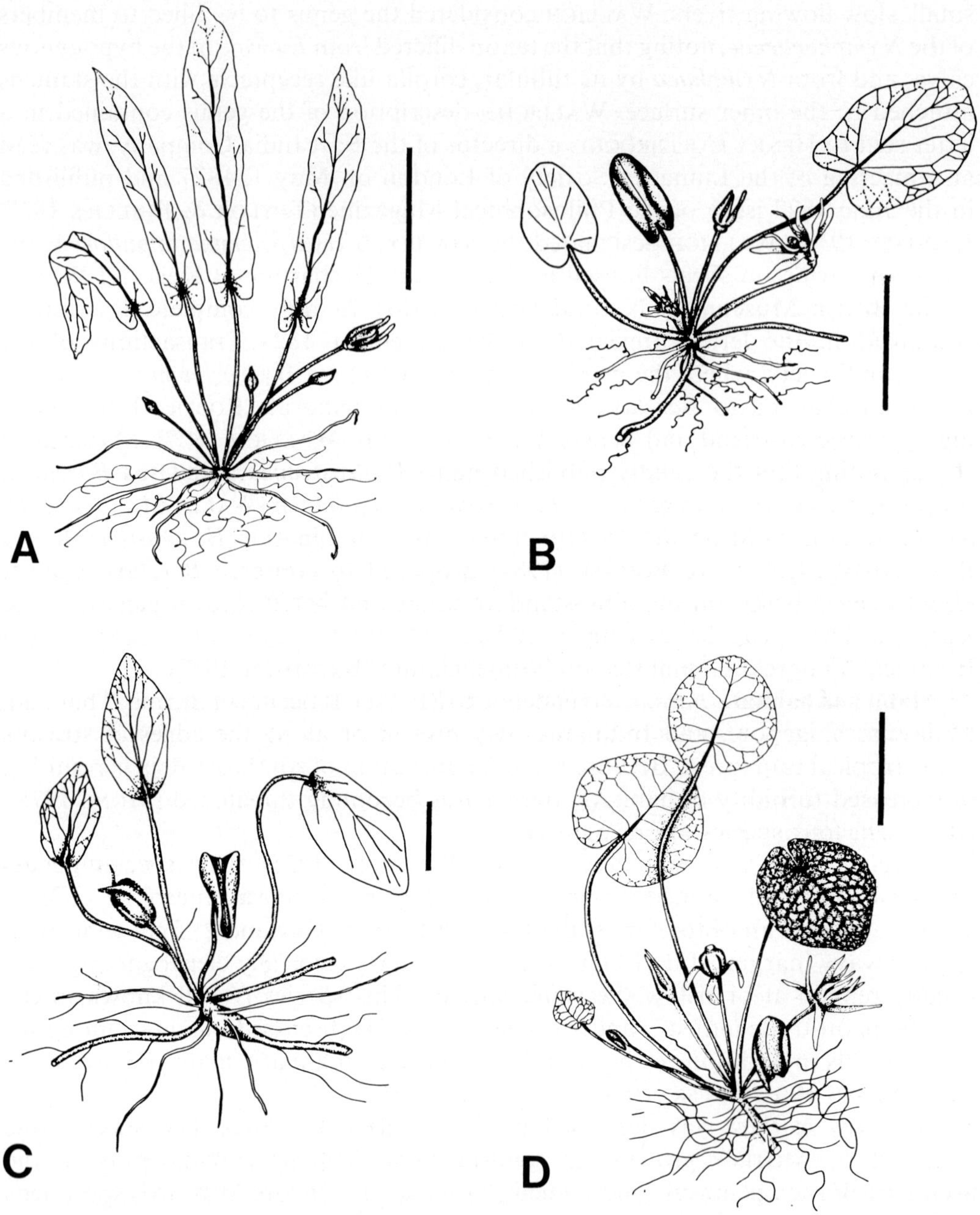


Fig. 1. *Barclaya* spp. *A B. longifolia*, bar: 14 cm. *B B. motleyi*, bar: 7 cm. *C B. kunstleri*, bar: 3.5 cm. *D. B. rotundifolia*, bar: 14 cm

plant collector. Later, Ridley (1922) elevated the taxon to the species level based on the thin, glabrous leaves which resembled those of *B. longifolia*. Ridley further noted, that the flowers were similar to those of *B. motleyi*. Plants occur submersed in shallow streams typically with depth of 1–3 dm and in areas with minimal stream flow. Leaves resemble those of *Cryptocoryne* which commonly occur in this habitat.

The blades are ovate to elliptic, typically 4–7 cm long and 2–5 cm wide, with a broad base and rounded to obtuse apex (Fig. 1C).

*Barclaya rotundifolia* was described by HOTTA (1966) who discovered the species during an expedition to Sarawak and Brunei. This study provides an extension of the species range into West Malaysia. The specific epithet refers to the round shaped leaves of this species. Unlike the other three species which are submersed, *B. rotundifolia* grows as an emergent marsh-like species in muddy, swampy areas along stream banks. Blades are round, 8–16 cm in diameter, deeply cordate at the base, with a somewhat puckered surface (Fig. 1D).

Basic floral morphology is similar in all four species. The sepals are attached at the base of the ovary (Fig. 2A). Petals form a corolla tube fused around the ovary. The floral tube extends above the ovary and distally, the petals are separate. Staminodia and numerous fertile stamens with short filaments and pendulent anthers are attached to the petals. Carpellary appendages arch over a funnel shaped stigmatic cup in a tent like fashion leaving a small central opening.

**Nature of the calyx.** Four to five keeled appendages are positioned at the base of the flower. Some observers have interpreted these appendages to be an involucre subtending an epigynous flower (HOOKER 1852, HU 1968) while others have considered them to represent members of a calyx (WALLICH 1827, HOOKER 1862, RIDLEY 1922, TAMURA 1982). Evidence from floral development and vasculature indicates that these hypogynous structures are sepals. Our observations have revealed that the mode and sequence of initiation of these appendages is the same as described for the sepals of other *Nymphaeaceae* sensu stricto genera (MOSELEY 1961, 1965; SCHNEIDER 1976, WILLIAMSON & MOSELEY 1989) with the anterior sepal initiated first, followed by simultaneous initiation of the two lateral sepals, followed lastly by initiation of the posterior sepal. In a slightly later developmental stage, it can be seen that vascular supplies to the sepals arise from an anastomosing mass of vascular tissue that also supplies traces to the remaining floral organs (Fig. 2B).

**Pollen and pollination.** Inaperturate pollen has been reported in some species of *Barclaya* (WALKER 1974, SCHNEIDER 1978). Reexamination of pollen using scanning electron microscopy, however, has revealed the occurrence of zonasulculate pollen in *B. kunstleri* (Fig. 2C), *B. longifolia*, *B. motleyi* and *B. rotundifolia.* During the study period flowers of *B. longifolia* were observed to be submersed, cleistogamous, and self-pollinating. Both submersed and emergent flowers were observed in *B. kunstleri.* Open flowers were not observed, suggesting that this species is also cleistogamous. The cleistogamous flowers are presumably self-pollinated based on the occurrence of numerous fruits in the population. A detailed study of pollination, however, has not been completed. *Barclaya motleyi* produces chasmogamous flowers that are elevated above the water surface. Field observations to study pollinator visitation were made in July 1988. No floral visitors were documented during the three week observation period. Exclusion experiments indicate that this species is also capable of selfing. The flowers of *B. rotundifolia* are also aerial, and chasmogamous. Sepals are greenish to white with patches of maroon at the base. Petals and stamens are purplish. Individual flowers bloom for three days, opening each morning and closing at dusk. The flowers have a pungent, fermenting odor. Unidentified small and medium sized flies were collected around the flowers of this species in July 1988, March 1990, and June 1992. Occasionally flies were also

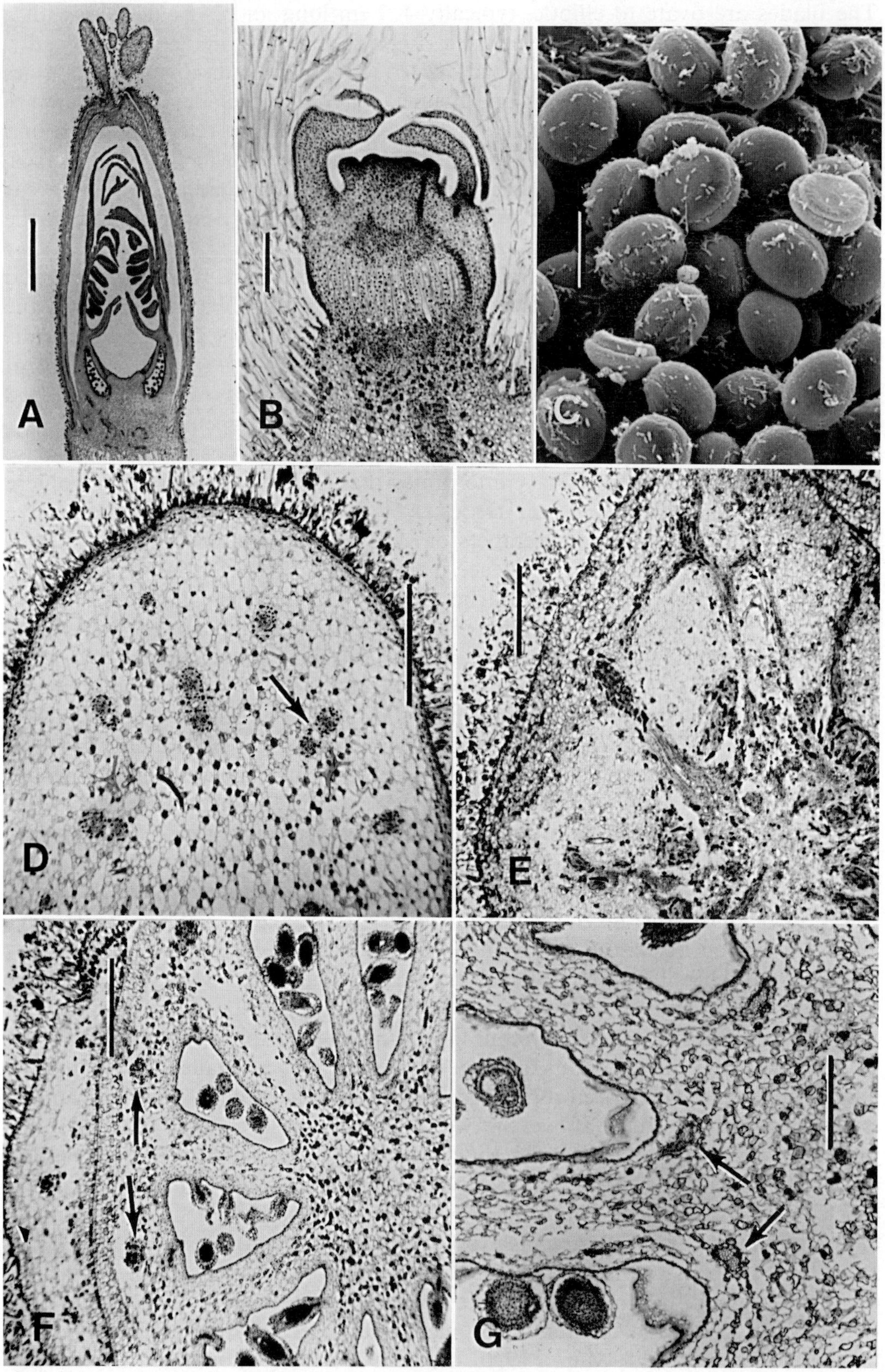
A
B
C
D
E
F
G

found suffocated in the mucilage that covers the surface of the stigmatic cup, lines the appressed carpellary surfaces, and fills the locules. The occurrence of flies in the flowers, the floral architecture, color, and odor suggest that flies may facilitate pollination.

**Floral anatomy.** This study focuses on 12 floral anatomical characters used in a cladistic analysis of the *Nymphaeaceae* sensu lato (MOSELEY & al. 1993). Criteria for selecting characters were the ease of comprehension, the possibility of clear illustration, and the presence of a character and its modification in all taxa or its absence in one taxon. The characters examined are: 1. Vascular supply to the base of the peduncle. 2. Arrangement of stelar structures in the peduncle. 3. Structure of major vascular bundles in the peduncle. 4. Structure of the receptacular vascular plexus. 5. Origin of traces to sepals. 6. Origin of supplementary sepal traces. 7. Nature of vascular structures originating from the plexus. 8. Source of petal traces. 9. Structure of a single petal trace. 10. Source of supplementary veins in petals. 11. Occurrence of a staminal pseudostele. 12. Occurrence of supplementary ventral carpellary veins. Character states are shown in Table 1.

The following description of the floral anatomy of *Barclaya* will emphasize the characters utilized in the cladistic analysis. The description is based principally on an investigation of flowers of *B. rotundifolia*. Some aspects of floral structure of *B. motleyi* have been described by TAMURA (1982). The peduncle is vascularized by small, single vascular bundles and larger, major vascular structures (Fig. 2D). Each major vascular structure consists of two radially aligned, collateral bundles with the inner bundle inverted in relation to the outer one. A region of phloem and a region of xylem occurs in each of the two paired bundles. A large, well defined, central protoxylary lacuna is positioned between the two bundles. Remains of secondary wall of stretched xylary elements are apparent within the lacuna. The central protoxylary lacunae maintain their identity and remain well defined through the peduncle up to the receptacle. Similar major vascular structures occur in other nymphaeaceous taxa; the central protoxylary lacuna may represent a reduced vascular bundle or a component of the centripetal bundle (MOSELEY & al. 1993). Typically, ten of these major vascular structures are present and are arranged in a ring. Five major vascular structures with well developed inner bundles alternate with five major vascular structures in which the inner bundle is smaller. The paired collateral bundles composing major vascular structures are comparable to those termed satellite bundles present in *Nymphaea tetragona* (ITO 1983, 1984). Variable numbers (typically eight to ten) of smaller, single vascular bundles occur along the

◄

Fig. 2. *A* Longitudinal section of *B. longifolia* flower illustrating stigmatic cup, carpellary appendages, pendent stamens, and the hypogynous position of the calyx, bar: 0.2 cm. *B.* Floral bud of *B. rotundifolia*, note density of trichomes and procambial principal vascular supply to petals originating from receptacular plexus, bar: 0.26 mm. *C* Zonasulculate pollen from *B. kunstleri*, bar: 0.01 mm. *D* Peduncle cross section of *B. rotundifolia*, arrow denotes major vascular supply, bar: 1.3 mm. *E* Vascular traces to a sepal, originating from the receptacular plexus in *B. rotundifolia*, bar: 0.7 mm. *F* Cross section through syncarpous gynoecium of *B. rotundifolia* showing laminar placentation and principal vascular supplies (at arrows), bar: 0.65 mm. *G* Ventral carpellary supplies (at arrows) in *B. rotundifolia*, bar: 0.27 mm

Table 1. Selected floral characters and their states used in the cladistic analysis of the *Nymphaeaceae* sensu lato

(1) Vascular supply to base of peduncle:
0, stelar origin; 1, stelar and cortical origin
(2) Arrangement of stelar structures in peduncle:
0, reduced axial bundle complex; 1, major branch of stele
(3) Structure of major vascular bundles in peduncle:
0, single collateral; 1, two radially aligned; 2, two tangentially aligned
(4) Structure of receptacular vascular plexus:
0, cylindrical stele; 1, anastomosing stele;
2, stele with partial vascular center; 3, stele with vascular center
(5) Origin of traces to sepals:
0, distal to plexus; 1, from plexus and peduncle bundles or interior bundles
(6) Origin of supplementary sepal traces:
0, none; 1, stelar origin; 2, cortical origin
(7) Nature of vascular supplies originating from plexus;
0, single bundle to single organ; 1, single bundle to multiple organs (PVB);
2, double bundle to multiple organs (GVS)
(8) Source of petal traces:
0, cylindrical stele; 1, anastomosing stele; 2, PVB; 3, GVS
(9) Structure of single petal trace:
0, single bundle; 1, two radially aligned bundles
(10) Source of supplementary veins in petals:
0, none; 1, stelar origin; 2, cortical origin
(11) Occurrence of staminal pseudostele:
0, absent; 1, present
(12) Occurrence of supplementary ventral carpellary veins:
0, absent; 1, rare and poorly developed; 2, well developed

same or different radii as the major vascular structures and at slightly more peripheral positions. These smaller bundles are collateral and normally oriented. A protoxylary lacuna is present in each of these bundles. The smaller bundles typically merge with major vascular structures at a level proximal to the receptacle.

The epidermis of the peduncle in *B. rotundifolia* is densely covered with uniseriate, multicellular trichomes. Four layers of collenchyma tissue occur beneath the epidermis. The remainder of the ground tissue is aerenchymatous, similar to peduncles of *Nuphar* (MOSELEY 1971). Large, symmetrically arranged air canals, characteristic of the peduncles of *Euryale, Ondinea, Nymphaea* and *Victoria* (MOSELEY 1961, SCHNEIDER 1976, WILLIAMSON & MOSELEY 1989), are lacking in *Barclaya.* Many parenchyma cells within the vascular bundles and ground tissue contain a dark-staining material. Astrosclereids occur scattered throughout the peduncle.

The major vascular structures are the primary supplies to the receptacular vascular tissue. The tissues of the major vascular structures expand and anastomose to form a cylindrical receptacular vascular plexus (Fig. 2E). Strands of xylem and phloem occur within the central portion of the receptacle and are surrounded by the cylindrical mass of the plexus. Vascular tissue is not continuous across the center of the receptacle as in flowers of *Nymphaea*, *Nuphar*, and *Euryale* (MOSELEY 1961, 1971; MOSELEY & al. 1993).

Four to five sepals occur attached at the base of the flower. Each sepal receives three traces, a median and two laterals, from the plexus (Fig. 2F). Each trace consists of a single, collateral bundle. The three traces to a given sepal originate as discrete traces, however each lateral trace anastomoses forming a bridge with the lateral trace of an adjacent sepal near the periphery of the receptacle. The laterals then regain their separate identities and continue into the base of the sepal. Sepals of *Victoria* (SCHNEIDER 1976) and larger flowered species of *Nymphaea* (MOSELEY 1961) receive supplementary stelar traces from the major peduncle bundles below the plexus or from peripheral bundles that do not join the plexus. Supplementary sepal traces have not been observed in flowers of *Barclaya*.

Several large, collateral bundles termed principal vascular bundles originate from the plexus (Fig. 2B). These bundles extend through the floral tube (Fig. 2F) and supply all remaining floral organs. Ventral carpellary bundles separate from the principal vascular bundles below the locular region, and slope inwardly to positions centripetal to the locules (Fig. 2G). The principal vascular bundles also give rise to numerous small supernumerary dorsal bundles positioned near the locules. Branches from the ventral carpellary bundles and principal vascular bundles form a system of septal carpellary veins that supply the numerous orthotropous ovules (Fig. 2F, G). No supplementary ventral carpellary veins, such as described for *Nymphaea* (Moseley 1961), were observed in *Barclaya*. Placentation is laminar. The locules are filled with a clear mucilage that becomes thick and white when exposed to ethanol. Each ovule receives one trace which may be supplied by a ventral bundle, septal carpellary vein, principal vascular bundle or supernumerary dorsal bundle. The ventral bundles and septal carpellary veins continue into the carpellary appendages where they terminate. At levels distal to the ovary, principal vascular bundles in the floral tube supply traces to both petals and stamens (Fig. 3A). Traces are single, collateral bundles. Each petal is supplied with three separate traces. Each stamen receives one trace. Each staminodium also receives a single trace. Adaxial and/or abaxial tier(s) of supplementary veins noted in petals of *Victoria* (SCHNEIDER 1976) and large-flowered taxa of *Nymphaea* (HEINSBROEK & van HEEL 1969, van HEEL 1977) were not observed in petals of *Barclaya*.

During our study we were particularly intrigued by the process of floral organography (i.e., the spatial and temporal initiation of the appendicular organs). Following sepal formation and the onset of petal initiation (Fig. 3B) the floral apex remains momentarily flattened, similar to the process described for *Nymphaea* (MOSELEY 1961) and *Victoria* (SCHNEIDER 1976). The center of the floral primordium, however, becomes quiescent with further upward growth being peripheral, thus forming the basic epigynous architecture through the formation of petals (i.e., the floral tube; Fig. 3C) and epipetalous stamens (Fig. 3D). The pronounced upward expansion above the position of the sepals undoubtedly accounts for the interpretation by some investigators of the calyx being an involucre. No prominent projecting floral apex or its remnant as described in *Nymphaea* (MOSELEY 1961), *Ondinea* (SCHNEIDER 1983, WILLIAMSON & MOSELEY 1989) or *Victoria* (SCHNEIDER 1976) was observed.

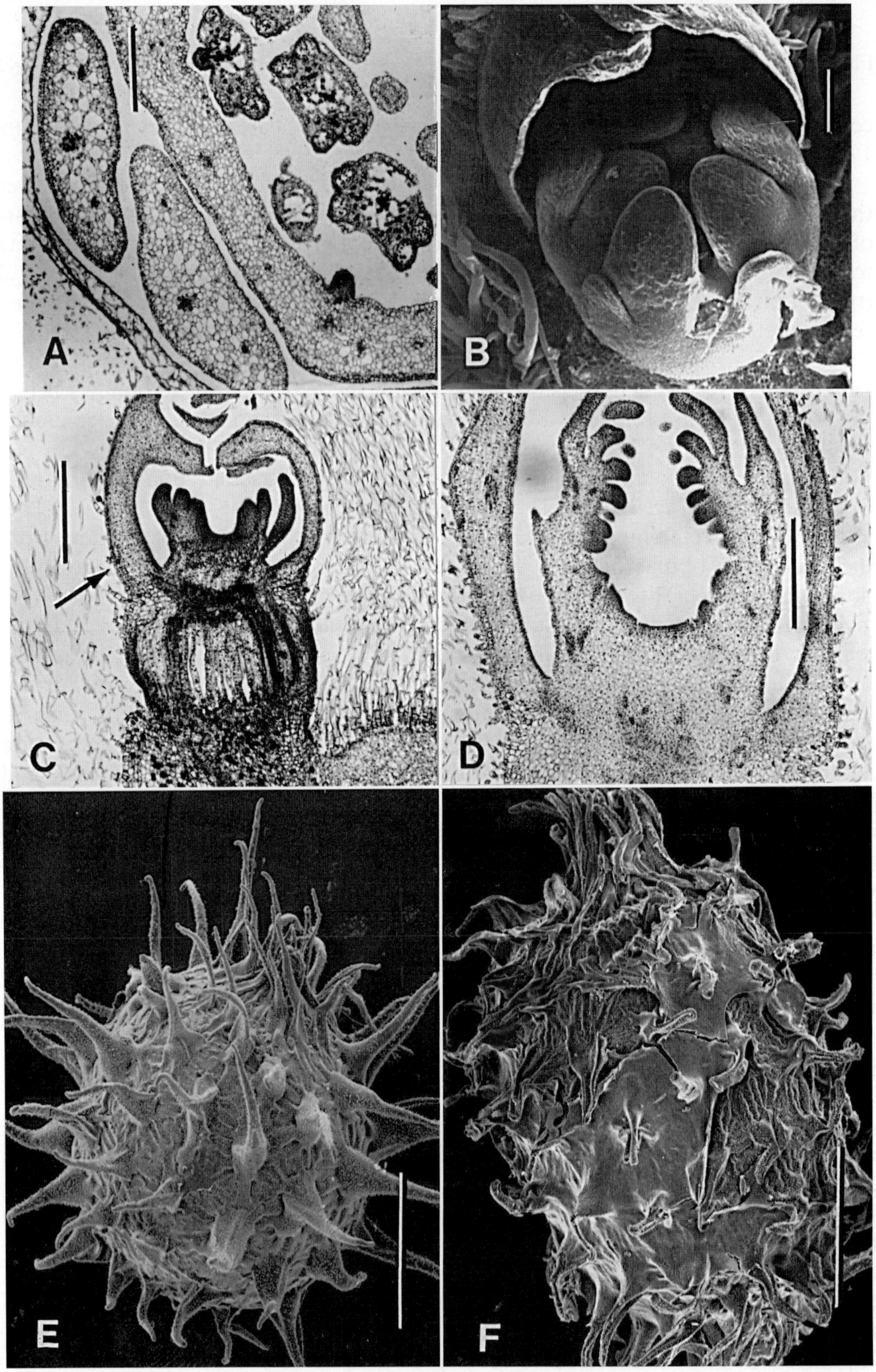


Fig. 3. *A* Cross section through flower of *B. rotundifolia* depicting three-veined petals, floral tube with primary vascular supplies, and stamens with single vein, bar: 0.5 mm. *B* Floral bud of *B. motleyi* showing basal, connate nature of the floral tube, bar: 151 μm. *C B. rotundifolia* illustrating zonal growth to the periphery of the original floral apex and the initiation of petals, sepal indicated by arrow, bar: 0.5 mm. *D B. rotundifolia* stage in floral development showing staminal and carpel initiation, bar: 0.7 mm. *E* Seed of *B. longifolia*, bar: 0.4 mm. *F* Seed of *B. rotundifolia*, bar: 0.3 mm

Table 2. Genus matrix of twelve floral character states used in the cladistic analysis of the *Nymphaeales* sensu lato. Numerical designations for each character state are shown in Table 1

| Taxon | Character state | | | | | | | | | | | |
|---|---|---|---|---|---|---|---|---|---|---|---|---|
| | 1 | 2 | 3 | 4 | 5 | 6 | 7 | 8 | 9 | 10 | 11 | 12 |
| *Barclaya* | 0 | 0 | 1 | 2 | 1 | 0 | 1 | 2 | 0 | 0 | 0 | 0 |
| *Brasenia* | 0 | 0 | 2 | 1 | 1 | 0 | 0 | 1 | 0 | 0 | 0 | 1 |
| *Cabomba* | 0 | 0 | 2 | 1 | 1 | 0 | 0 | 1 | 0 | 0 | 0 | 1 |
| *Euryale* | 0 | 0 | 1 | 3 | 1 | 0 | 2 | 3 | 1 | 0 | 0 | 0 |
| *Nelumbo*[1] | 1 | 1 | 0 | 0 | 0 | 2 | 0 | 0 | 0 | 2 | 1 | 0 |
| *Nuphar* | 0 | 0 | 1 | 3 | 1 | 0 | 1 | 2 | 0 | 0 | 0 | 2 |
| *Nymphaea* | 0 | 0 | 1 | 3 | 1 | 1 | 2 | 3 | 1 | 1 | 0 | 2 |
| *Ondinea* | 0 | 0 | 1 | 2 | 1 | 0 | 1 | 2 | 0 | 0 | 0 | 0 |
| *Victoria* | 0 | 0 | 1 | 2 | 1 | 1 | 2 | 3 | 1 | 1 | 0 | 0 |

The seeds of *B. kunstleri* and *B. longifolia* are globose, c. 1 mm in diameter, non-arillate and spiny (Fig. 3E). The spines probably aid in dispersal and anchorage of the seeds in the substrate. Seeds are perispermous, with a small amount of endosperm, and contain a minute, dicotyledonous embryo (Schneider 1978). In *B. motleyi* and *B. rotundifolia* (Fig. 3F) the seeds are more oblong.

**Cladistic analysis.** Table 2 indicates the 12 character states for each of the nine taxa used in our cladistic analysis. In generating cladograms all characters were unordered. The branch and bound option was used to find the most parsimonious cladogram. *Nelumbo* was selected as the outgroup since phyletic studies indicate that this taxon is distinct from the *Nymphaeaceae* sensu stricto and *Cabombaceae* (Ito 1987, Les & al. 1991).

The cladistic analysis of the 12 floral characters and nine taxa resulted in six equally parsimonious trees, each with 23 steps and a consistency index of 0.913. A majority-rule consensus tree was prepared (Fig. 4C). In all six trees, a *Euryale, Nymphaea, Victoria* clade was identifiable based on the occurrence of two radially aligned bundles composing a gynoecial vascular strand as in *Nymphaea* and *Victoria*, or two associated bundles, not necessarily in contact, as in *Euryale* (Moseley pers. comm.). These strands originate from the plexus and supply petals and other appendicular organs. In five trees a *Barclaya–Ondinea* sister clade was identifiable and in four of the six trees a *Brasenia–Cabomba* clade was basal to the other six genera. *Nuphar* was equally positioned as either a sister clade to the *Euryale, Nymphaea, Victoria* clade or basal to the *Barclaya–Ondinea* clade. The consensus tree lends support to the recognition of a single monophyletic group of *Brasenia, Cabomba, Barclaya, Euryale, Nuphar, Nymphaea, Ondinea,* and *Victoria* composing two families, the *Cabombaceae* (*Brasenia* and *Cabomba*) and *Nymphaeaceae*. At present no evidence is available to support the placement of *Barclaya, Euryale,* or *Nuphar* in monotypic families. These data are the first to support a close phyletic relationship between the Australian genus *Ondinea* and the Indo Malaysian *Barclaya*.

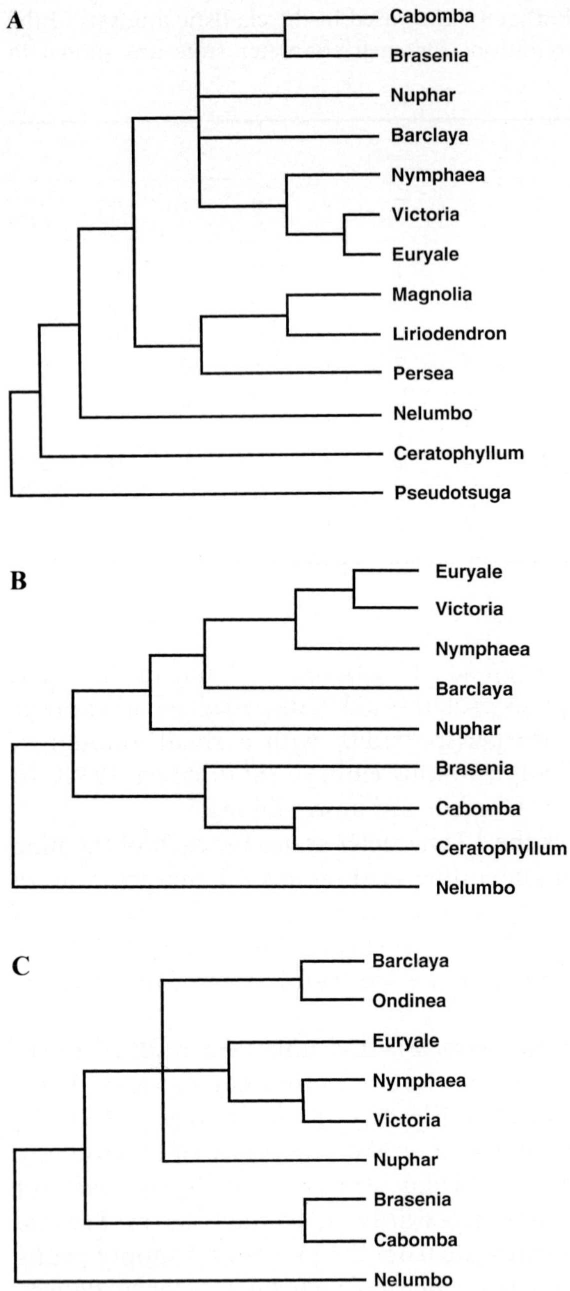


Fig. 4. Cladograms of the *Nymphaeales*. *A* Strict consensus tree found by parsimony analysis of *rbc*L data in water lily genera and woody *Magnoliidae*, rooted using *Pseudotsuga* as an outgroup (after LES & al. 1991). *B* Cladogram constructed by ITO (1987) using 17 characters of gross morphology, anatomy, and palynology. *C* A majority rule consensus tree found by analysis of twelve floral vasculature characters rooted using *Nelumbo* as the outgroup

## Discussion

The identification of zonasulculate (rather than anasulcate) pollen in *Barclaya* and the interpretation of the hypogynous appendages as sepals, based on comparative organographic and vasculature observations, lend support to inclusion of *Barclaya* in the *Nymphaeaceae* rather than segregation into a monotypic family.

Figure 4A shows a cladogram generated by LES & al. (1991) using *rbc*L sequence data, and Fig. 4B one by ITO (1987) using data from anatomy, morphology, and palynology. The analyses by LES & al. (1991) and ITO (1987) do not include *Ondinea*. The consensus tree (Fig. 4C) generated in our current investigation using floral anatomical data is consistent with the topology of the cladograms from the studies by these authors. In summary, both of these studies, as well as our study, all using different data sets, support the monophyletic nature of the *Nymphaeales* and the recognition of two families, the *Nymphaeaceae* sensu stricto and *Cabombaceae*.

The authors extend thanks to Ms AMY MAHLOCH for preparing the line drawings of *Barclaya*, to Dr BEN STONE for logistic assistance in making collections and in herbarium sheet preparation, to Dr HARRINGTON WELLS (Tulsa University), JOE VAUGHAN (Central Oklahoma State University), Ms LaTITIA TAYLOR ODOMS, and Mr BOB BLANKINSHIP for field assistance, to Ms GEORGIA RICH for preparation of light microscope slides, to Dr JENNIFER THORSCH and Dr SHERWIN CARLQUIST for assistance in obtaining photomicrographs, and to Dr. MOTOMI ITO and Dr. PETER ENDRESS for helpful suggestions. – This research was supported by grants from the National Science Foundation (DEB-8102041) to Dr. ED SCHNEIDER, Southwest Texas State University, and the Southern Ute Tribal Education Committee.

## References


BENTHAM, G., HOOKER, J. D., 1862: Genera plantarum, I. – London: Reeve.

BESSEY, C. F., 1915: The phylogenetic taxonomy of the flowering plants. – Ann. Missouri Bot. Gard. **2**: 109–164.

BRUMMITT, R. K., 1987: Report of the committee for *Spermatophyta*: 33. – Taxon **36**: 734–762.

BUCHHEIM, G., 1964: *Nymphaeineae*. – In MELCHIOR, H. (Ed.): A. ENGLER's Syllabus der Pflanzenfamilien, 12th edn. – Berlin: Borntraeger.

CASPARY, R., 1891: *Nymphaeaceae*. – In ENGLER, A., PRANTL, K., (Eds): Die natürlichen Pflanzenfamilien **3**, pp. 1–10. – Leipzig: Engelmann.

CRONQUIST, A., 1981: An integrated system of classification of flowering plants. – New York: Columbia University Press.

CRUSIO, W. E., BOGNER, J., 1984: Proposal 750. 2515 *Barclaya* WALLICH vs. *Hydrostemma* WALLICH (*Nymphaeaceae*/*Barclayaceae*). – Taxon **33**: 517–519.

DEWIT, H. C. D., 1983: Aquarium planten II. – Het handboek voor de aquariumliefhebber, **6**: 83. – Baarn: Hollandia.

GOLENIEWSKA-FURMANOWA, M., 1970: Comparative leaf anatomy and alkaloid content in the *Nymphaeaceae*. – Monogr. Bot. **31**: 1–55.

GUNDERSEN, A., 1950: Families of dicotyledons. – Waltham, Mass: Chronica Botanica.

HEINE, H., 1958: *Barclaya longifolia* WALLICH, eine neueingeführte wertvolle Aquarienpflanze. – Aquarien Terrarien-Z. **11**: 314–317; 345–346.

HEINSBROEK, P. G., van HEEL, W. A., 1969: Note on the bearing of the pattern of vascular bundles on the morphology of the stamens of *Victoria amazonica* (POEPP.) SOWERBY. – Proc. K. Nederl. Akad. Wet., Amsterdam, Ser. C, **72**: 431–444.

HOOKER, W. J., 1852: *Barclaya longifolia*, WALL. – Icones Plantarum **9**: Pl. 809–10.

HOOKER, J. D., 1862: *Nymphaeaceae*, XIV. Illustrations of the floras of the Malayan Archipelago and of tropical Africa. – Trans. Linn. Soc. **23**: 155–172.

HOTTA, M., 1966: Notes on Bornean plants, I. – Acta Phytotax. Geobot. **22**: 9–10.

HU, S. Y., 1968: The genus *Barclaya* (*Nymphaeaceae*). – In: Studies in the flora of Thailand 48. – Dansk Bot. Ark. **23**: 534–541.

HUTCHINSON, J., 1973: The families of flowering plants, 3rd edn. – Oxford: Clarendon.

Ito, M., 1983: Studies in the floral morphology and anatomy of *Nymphaeales*. I. The morphology of vascular bundles in the flower of *Nymphaea tetragona* Georgi. – Acta Phytotax. Geobot. **34**: 18–26.
– 1984: Studies in the floral morphology and anatomy of *Nymphaeales*. II. Floral anatomy of *Nymphaea tetragona* Georgi. – Acta Phytotax. Geobot. **35**: 94–102.
– 1986a: Studies in the floral morphology and anatomy of *Nymphaeales*. III. Floral anatomy of *Brasenia schreberi* Gmel. and *Cabomba caroliniana* A. Gray. – Bot. Mag. (Tokyo) **99**: 169–184.
– 1986b: Studies in the floral morphology and anatomy of *Nymphaeales*. IV. Floral anatomy of *Nelumbo nucifera*. – Acta Phytotax. Geobot. **37**: 82–96.
– 1987: Phylogenetic systematics of the *Nymphaeales*. – Bot. Mag. (Tokyo) **100**: 17–35.
Johansen, D. A., 1940: Plant microtechnique. – Minneapolis: McGraw-Hill.
King, J., 1889: *Barclaya motleyi* var. *kunstleri*. – As. Soc. Bengal **58**: 390.
Les, D. H., Garvin, D. K., Wimpee, C. F., 1991: Molecular evolutionary history of ancient aquatic angiosperms. – Proc. Natl. Acad. Sci. USA **88**: 10119–10123.
Li, H. L., 1955: Classification and phylogeny of the *Nymphaeaceae* and allied families. – Amer. Midl. Naturalist **54**: 33–41.
Mabberley, D. J., 1982: William Roxburgh's 'Botanical description of a new species of *Swietenia* (Mahogany)' and other overlooked binomials in 36 vascular plant families. – Taxon **31**: 65–73.
Moseley, M. F., 1961: Morphological studies of the *Nymphaeaceae* II. The flower of *Nymphaea*. – Bot. Gaz. **122**: 233–259.
– 1965: Morphological studies of the *Nymphaeaceae* III. The floral anatomy of *Nuphar*. – Phytomorphology **15**: 54–84.
– 1971: Morphological studies of the *Nymphaeaceae* VI. Development of the flower of *Nuphar*. – Phytomorphology **21**: 253–283.
– Uhl, N. W., 1985: Morphological studies of the *Nymphaeaceae* sensu lato XV. The anatomy of the flower of *Nelumbo*. – Bot. Jahrb. Syst. **106**: 61–98.
– Mehta, I. J., Williamson, P. S., Kosakai, H., 1984: Morphological studies of the *Nymphaeaceae* (sensu lato) XIII. Contributions to the vegetative and floral structure of *Cabomba*. – Amer. J. Bot. **71**: 902–924.
– Schneider, E. L., Williamson, P. S., 1993: Phylogenetic interpretations from selected floral vasculature characters in the *Nymphaeaceae* sensu lato. – Aquatic Bot. **44**: 325–342.
Paffrath, K., 1972: Nochmals *Barclaya longifolia*. – Aquarien Terrarien – Z. **25**: 372–374.
Raphael, S., 1970: The publication dates of the Transactions of The Linnean Society of London, Series I, 1791–1875. – Biol. J. Linn. Soc. **2**: 61–76.
Richardson, F. C., 1969: Morphological studies of the *Nymphaeaceae* IV. Structure and development of *Brasenia schreberi* Gmel. – Univ. Calif. Publ. Bot. **47**: 1–101.
Ridley, H. N., 1922: The flora of the Malay Peninsula. *Polypetalae*. – London: Reeve and Co.
Schneider, E. L., 1976: The floral anatomy of *Victoria* Schomb. (*Nymphaeaceae*). – Bot. J. Linn. Soc. **72**: 115–148.
– 1978: Morphological studies of the *Nymphaeaceae*. IX. The seed of *Barclaya longifolia* Wall. – Bot. Gaz. **139**: 223–230.
– 1983: Morphological studies of the *Nymphaeaceae*. XIV. The floral biology of *Ondinea* den Hartog. – Austral. J. Bot. **31**: 371–382.
Stone, B. C., 1978: *Barclaya*, the Malaysian "waterlily". – Malayan Nat., Sept.: 20–22.
– 1982: A new combination for *Barclaya kunstleri* (King) Ridley of the *Nymphaeaceae*. – Gard. Bull. **35**: 69–71.

SWOFFORD, D. L., 1991: PAUP: Phylogenetic analysis using parsimony, version 3.0s. – Computer program distributed by the Illinois Nat. Hist. Survey, Champaign, Illinois.
TAKHTAJAN, A. L., 1959: Die Evolution der Angiospermen. – Stuttgart, Jena: Fischer.
– 1980: Outline of the classification of flowering plants. – Bot. Rev. **46**: 226–359.
TAMURA, M., 1982: Relationship of *Barclaya* and classification of *Nymphaeales*. – Acta Phytotax. Geobot. **23**: 336–345.
TAYLOR, R., PHILIPS, R., 1827: *Hydrostemma* WALL. – Phil. Mag., n.s., **1**: 454.
THORNE, R. F., 1976: A phylogenetic classification of the angiosperms. – Evol. Biol. **9**: 35–106.
VAN BRUGGEN, H. W. E., 1961: *Barclaya motleyi* HOOKER. f. – Het. Aq. **33**: 6–8.
VAN HEEL, W. A., 1977: The pattern of vascular bundles in the stamens of *Nymphaea lotus* L. and its bearing on stamen morphology. – Blumea **23**: 345–348.
WALKER, J. W., 1974: Aperture evolution in the pollen of primitive angiosperms. – Amer. J. Bot. **61**: 1112–1136.
WALLICH, N., 1827: Description of a new genus of plants belonging to the order *Nymphaeaceae*: in a letter to H. T. COLEBROOKE. – Trans. Linn. Soc. **15**: 442–449.
WEIDLICH, W.D., 1976a: The organization of the vascular system in the stems of the *Nymphaeaceae* I. *Nymphaea* subgenera *Castalia* and *Hydrocallis*. – Amer. J. Bot. **63**: 499–509.
– 1976b: The organization of the vascular system in the stems of the *Nymphaeaceae* II. *Nymphaea* subgenera *Anecypha*, *Lotos* and *Brachyceras*. – Amer. J. Bot. **63**: 1365–1379.
– 1980: The organization of the vascular system in the stems of the *Nymphaeaceae* III. *Victoria* and *Euryale*. – Amer. J. Bot. **76**: 1779–1794.
WETTSTEIN, R., 1935: Handbuch der Systematischen Botanik. – Leipzig: Deuticke.
WILLIAMSON, P. S., MOSELEY, M. F., 1989: Morphological studies of the *Nymphaeaceae* sensu lato XVII. Floral anatomy of *Ondinea purpurea* subspecies *purpurea* (*Nymphaeaceae*). – Amer. J. Bot. **76**: 1779–1794.

Addresses of the authors: PAULA S. WILLIAMSON, Department of Biology, Southwest Texas State University, San Marcos, Texas 78666 USA – EDWARD L. SCHNEIDER, Santa Barbara Botanic Garden, 1212 Mission Canyon Road, Santa Barbara, California 93105, USA.

Accepted February 21, 1994 by P. K. ENDRESS

Pl. Syst. Evol. [Suppl.] 8: 175–183 (1984)

# Evolutionary aspects of the floral structure in *Ceratophyllum*

PETER K. ENDRESS

Received November 12, 1993

**Key words:** Angiosperms, *Magnoliidae*, *paleoherbs*, *Nymphaeales*, *Ceratophyllales*, *Ceratophyllum*. – Floral evolution.

**Abstract:** A number of floral features of *Ceratophyllum* conform with the present view on primitive angiosperm flowers and may therefore support the hypothesis of molecular systematists and palaeobotanists that the genus represents the basal clade of extant angiosperms: the flowers are minute; the organ number and phyllotaxis are highly variable – spiral (Fibonacci or Lucas pattern) or whorled (3-4-merous whorls) – but there is a single carpel; the stamens are not differentiated into anther and filament; the organs surrounding the stamens or the carpel are not closely integrated into the floral architecture and they may be interpreted either as bracts or tepals. However, other simple features may still be seen as apomorphies in the context of underwater pollination, especially the lack of lignified tissues in floral organs (except for two horns on each anther).

Recently, *Ceratophyllum* (*Ceratophyllaceae*, *Ceratophyllales*) has been proposed to represent the basal clade of extant angiosperms. This view is apparently supported by both palaeobotanical (DILCHER 1989, 1992; CREPET & al. 1993) and molecular biological aspects (LES & al. 1991, 1992; CHASE & al. 1993a, b; QIU & al. 1993). Therefore, a re-examination of the floral structure of extant *Ceratophyllum* from new material and from the literature seems timely. The genus comprises six extant species in three sections (LES 1989, 1993).

## Material and methods

Flowers of *Ceratophyllum demersum* L., cultivated in the Botanical Garden of the University of Zurich and fixed in FAA, were studied in different developmental stages with microtome serial sections (stained with safranin and astra blue) and with the SEM (critical point dried and Au/Pd sputter coated).

## Results

The miniature flowers (ca. 1 mm) of *Ceratophyllum* are unisexual. The male flowers have 3–46 stamens (SEHGAL & MOHAN RAM 1981, WILMOT-DEAR 1985, LES 1993) (Figs. 1–3), the female flowers have a single carpel (Fig. 4). Carpel or stamens are

A
B
C
D
E
F
G
H

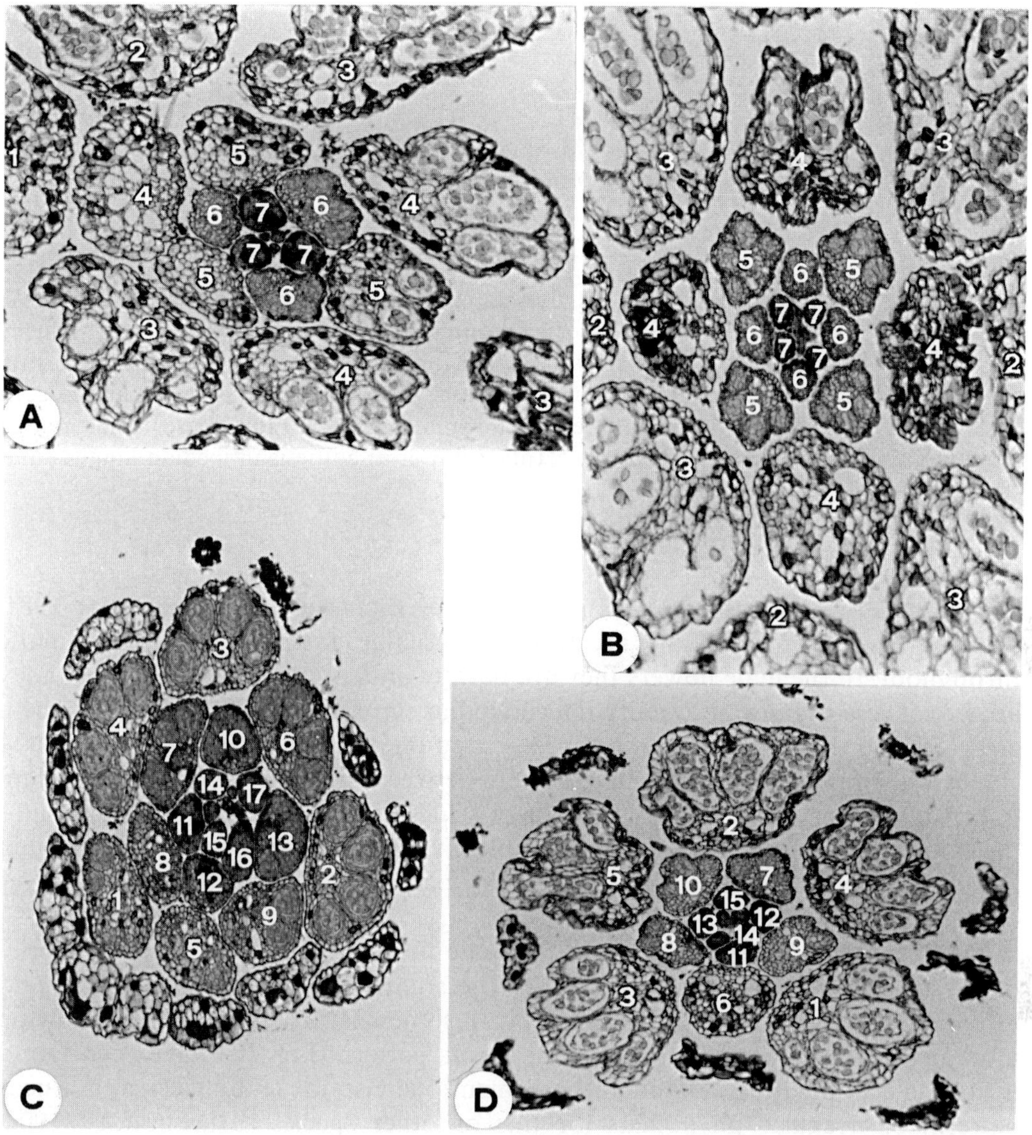


Fig. 2. *Ceratophyllum demersum.* Male flowers in transverse section. *A–B* Androecium with whorled phyllotaxis, successive whorls numbered. *A* Whorls trimerous, ×80. *B* Whorls tetramerous, ×80. *C–D* Androecium with approximately spiral phyllotaxis, successive stamens numbered. *C* Lucas pattern, ×80. *D* Fibonacci pattern, ×60

◀

Fig. 1. *Ceratophyllum demersum.* Male flower development. *A–B* Floral buds from above, with the inner stamen primordia still lacking; the bracts showing large-celled secretory tips. *A* ×300, *B* ×250. *C* Older floral bud, with the bracts and stamens conniving; the secretary tips of the stamens beginning to break off, remains of secretion visible, ×80. *D* Mature flower from below, showing the basally united tepals and the still closed anthers, ×18. *E* Mature flower from above, showing the developmental gradient of the stamens, ×18. *F* Mature flowers from the side, with the anthers opened by irregular slits, ×20. *G* Mature flower from above, with the outer stamens abscised, ×25. *H*. Same stage from the side, ×25

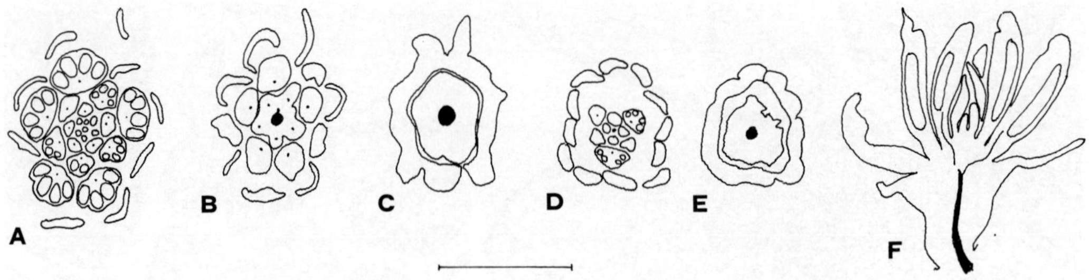


Fig. 3. *Ceratophyllum demersum.* Male flowers; vascular tissue black. Bar: 1 mm. *A–C* Flower shortly before anthesis in transverse section series. *A* Level with tepals and stamens free. *B* Floral base, innermost stamens united. *C* Floral base, tepals united, forming a ring around the floral centre. *D–E* Flower at anthesis with the outer stamens abscised in transverse section series. *D* Level with tepals and stamens free. *E* Floral base, tepals united, forming a ring around the floral centre. *F* Intact flower at anthesis in median longitudinal section

surrounded by a series of 6–13 narrow scales (SEHGAL & MOHAN RAM 1981, WILMOT-DEAR 1985). The scales are devoid of vascular tissue and are parenchymatic throughout. In the male flowers they are basally united; in the female flowers they are free. These organs are usually interpreted as tepals. LES (1993) interprets them as bracts and the flowers as perianthless, based on an unpublished thesis by ABOY (1936) who found additional female flowers between the scales and the gynoecium (see also CARUEL 1876). At any rate, these scales are only loosely integrated into the floral architecture. They do not completely cover stamens or carpel in young stages, but their secretory tip may have a protective function for the earliest stages of stamens or carpel (Figs. 1A–C, 4A).

**Male flowers.** The phyllotaxis of the male flowers is variable. In my material of *Ceratophyllum demersum* flowers with the stamens in trimerous (Fig. 2A) and tetramerous whorls (Fig. 2B) and others with somewhat irregular spirals following more or less the Fibonacci (Fig. 2D) or Lucas pattern (Fig. 2C) have been found in a single population. Other authors mentioned spiral (Fibonacci and Lucas pattern) or irregular phyllotaxis for this or other species (STRASBURGER 1902, RUTISHAUSER & SATTLER 1987). Although in the literature stamen whorls are mentioned, I am not aware of previous developmental studies that demonstrated whorled phyllotaxis. The outer stamens do not alternate with the tepals nor are they superposed to them (Figs. 1A–B, 3B). It is difficult to assess a rule in the positional relationships of tepals and stamens.

Each stamen is served by a vascular bundle with (ontogenetically) undifferentiated xylem but with differentiated phloem. Each stamen has a secretory connective protrusion that breaks off during development, with only an inconspicuous scar remaining at anthesis (Fig. 1C–F). The connective protrusion is papillose, the epidermal cells are thin-walled, and the cytoplasm stains darker with safranin than the anther tissue below the protrusion. At anthesis two or more unicellular horns with lignified cell walls are atop each stamen. The secretory connective protrusion and horns resemble the structure of the leaf and tepal tips. The secretion of the glandular part may have a protective function for young developmental stages of anthers. The horns may function to hold the stamens upright below the water surface when they release pollen grains, but this has not been studied in detail.

*Ceratophyllum* shows floral adaptations to underwater-pollination. The endothecium is reduced and the thecae open by irregular slits (SHAMROV 1981, 1983a, b; ENDRESS & HUFFORD 1989) (Fig. 1F, H). The stamens abscise during anthesis and ascend to the water surface while releasing pollen grains (ROZE 1892, JONES 1931) or grains that have produced a pollen tube (SEHGAL & MOHAN RAM 1981, LES 1988b). The stamens have large intercellular spaces in the ventral part of the connective (see also STRASBURGER 1902, SEHGAL & MOHAN RAM 1981, SHAMROV 1983a).

Within a flower the stamens show a strong developmental gradient from the periphery to the centre: When the outermost stamens are mature, the inner ones are still very juvenile (see also STRASBURGER 1902, SHAMROV 1981) (Figs. 2, 3A, F). As a consequence, mature stamens may be available for some time in a single flower. The inner stamens remain retarded and pollen is not produced. I found up to 14 sterile inner stamens, whereby the innermost ones remained in a minute primordial stage, as seen in microtome sections. After abscission of the oldest stamens in a flower, the tepals, which spread at the beginning of anthesis (Figs. 1F, 4A–B, F), bend again into a more upright position, as they were in bud (Figs. 1H, 4D).

This behaviour of successive stamen maturation is unusual for an angiosperm flower and would be more characteristic of an inflorescence. However, there is no additional evidence in favour of the interpretation of the male flowers of *Ceratophyllum* as inflorescences consisting of bractless monandrous flowers (as SCHLEIDEN suggested in 1837). The arguments by SCHLEIDEN in favour of an inflorescence do not hold. He mentioned the superposition of outermost stamens and surrounding organs (which is not really the case) and the convex basis of stamen insertion (which occurs in many normal flowers, especially in magnoliids).

Although there is considerable variability in stamen number per flower, the lowermost numbers that are given in the literature (three or five stamens, see above) may be an artifact in that old flowers with only the innermost stamens left may have been counted. It is difficult to determine the number of fertile stamens of a flower, because at the stage when the outermost stamens are still present, it is not evident how many of the inner, younger stamens will reach maturity.

**Female flowers**. The female flowers have a single ascidiate carpel with a single pendent, orthotropous, unitegmic, crassinucellar ovule in a median position. The carpel has a more or less abaxial position in the flower. However, strangely enough, the ventral part of the carpel bearing the ovule, becomes more extended than the dorsal part and forms a tapering thread at anthesis (SCHLEIDEN 1837, SHAMROV 1983a) (Fig. 4B). This brought DE KLERCKER (1885) to the interpretation of the gynoecium as consisting of two carpels. However, in young stages the ventral part is not longer or even shorter than the dorsal part (STRASBURGER 1902, SHAMROV 1983a), thus, at this early stage, shows the normal condition as expected for a single carpel (Fig. 4A). The entrance into the closed stylar canal of the mature carpel appears as a transverse slit in the lower half of the style (including the ventral appendage) (Fig. 4B–C). The surface of style and stigma is smooth. SHAMROV (1983a) defined the stigmatic region as extending from immediately above the transverse slit to about halfway up the ventral extension of the carpel (for *C.*

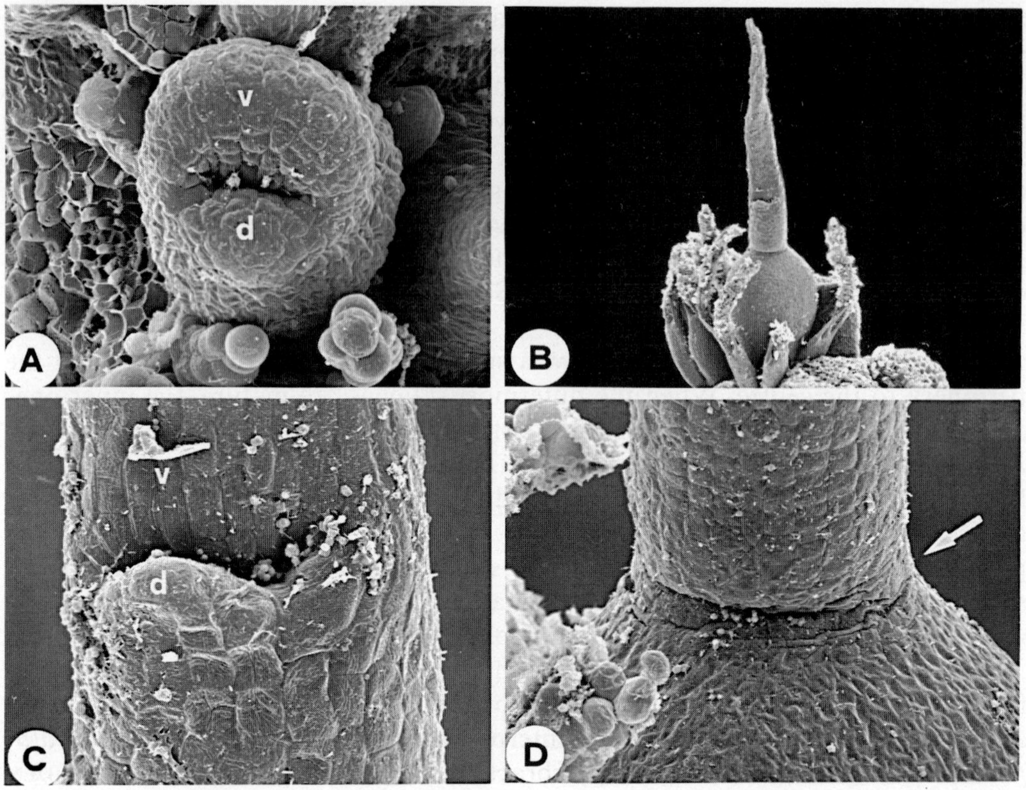


Fig. 4. *Ceratophyllum demersum*. Female flowers; *d* dorsal side, *v* ventral side. *A* Floral bud from above, showing tepals with large-celled tips and carpel with dorsal and ventral side of still equal length, ×350. *B–D* Mature flower from the side. *B* Entire flower, showing tepals with secretary tips and carpel from dorsal side, ×35. *C* Same flower, entrance into stylar canal, magnified, ×350. *D* Same flower, abscission region of style (arrow), magnified, ×250

*pentacanthum*). The carpel is served by a dorsal and a ventral vascular bundle, which, however, do not extend into the style (see also ABOY 1936). The ovule is served by the ventral bundle. At the time when the embryo sac is mature, the two vascular bundles lack lignified xylem elements and contain only one or two phloem elements. After anthesis the style falls off at a preformed abscission zone (Fig. 4D).

## Discussion

As indicated in the introduction, it has been argued by several authors that *Ceratophyllum* represents the last remnant of the most basal extant clade of the angiosperms. However, it has to be taken into account that the extant representatives of the most basal clade of the angiosperms do not necessarily show the structures that are closest to the basal structures. They may be much more modified than representatives of other clades and may not reveal much about the earliest structures. If herbaceous plants of moist habitats were the first angiosperm clades of which there are still living representatives, as has been suggested by, e.g., DOYLE & HICKEY

(1976), it is most probable that extant representatives of the paleoherbs with the same habitat have retained, in general, more primitive features than the submerged water plant *Ceratophyllum*. However, Lower and mid-Cretaceous fossil leaves and fruits similar to those of extant *Ceratophyllum*, as reported by DILCHER (1989, 1992), make it probable that the clade had submerged aquatic representatives as early as in the Lower Cretaceous. It has also been argued that *Ceratophyllum* may have survived, because in the stable aquatic habitat evolution is slow and the extinction rate relatively low (DILCHER 1989, LES & al. 1991). Slow evolution in underwater plants is also explained by their predominant asexual reproduction (LES 1988b, 1991).

What are the relationships with (other) paleoherbs? Earlier assumptions of a close relationship with *Cabomba* have been questioned on various grounds (LES 1988a). The vegetative similarities of *Ceratophyllum* and *Cabomba* are also present in other submerged water plants of very different affinities (see, e.g., COOK 1978). What is the contribution of floral structure? Besides sharing some general plesiomorphies, the simplified underwater flowers of *Ceratophyllum* are different from the above-water flowers of *Cabomba* in structural features that are related to the different biology (see also LES 1988a). The flowers do not have any means of pollinator attraction. The floral vascular system is simple as opposed to *Nymphaeales* with emerged and much larger flowers (e.g., MOSELEY & al. 1993). Simplification is also obvious in histology. The floral vascular tissue and the endothecium are devoid of lignified cell walls. Only the two horns of each anther have lignified cell walls. The vascular tissue needs closer investigation. Although a parallel may be found in some alismatids where representatives with underwater pollination have equally simplified flowers, in contrast to emerged relatives, the actual differences do not allow a definite conclusion of a closer relationship of *Ceratophyllum* and *Cabomba*. Also embryologically, *Ceratophyllum* does not stand out but shares a number of plesiomorphies with *Nymphaeales* and other paleoherbs (DE KLERCKER 1885; STRASBURGER 1902; BATYGINA & al. 1980, 1982; JOHRI & al. 1992).

Although floral features do not seem to contribute much to the question of affinities with other extant clades, there are a number of features in *Ceratophyllum* that do conform well with present views on primitive angiosperm flowers (see, e.g., ENDRESS 1990, FRIIS & ENDRESS 1990): the floral organs and the entire flowers are minute; the floral organ number and the phyllotaxis of the floral organs are variable but there is a single carpel with a single ovule; the stamens are not differentiated into anther and filament; the organs surrounding the stamens or the carpel are not closely integrated into the floral architecture and they may be seen either as bracts or tepals.

I thank Prof. E. M. FRIIS and Dr E. L. SCHNEIDER for helpful suggestions. R. SIEGRIST prepared the microtome sections, U. JAUCH (Institute of Plant Biology) helped with the SEM, and A. ZUPPIGER did the photographic work. I thank all of them.

## References

ABOY, H. E., 1936: A study of the anatomy and morphology of *Ceratophyllum demersum*. – M.Sc. Thesis, Cornell University, Ithaca.

BATYGINA, T. B., KRAVTSOVA, T. I., SHAMROV, I. I., 1980: The comparative embryology of some representatives of the orders *Nymphaeales* and *Nelumbonales*. – Bot. Ž. (Moscow, Leningrad) **65**: 1071–1087.

– SHAMROV, I. I., KOLESOVA, G. E., 1982: Embryology of the *Nymphaeales* and *Nelumbonales*. II. The development of the female embryonic structures. – Bot. Ž. (Moscow, Leningrad) **67**: 1179–1195.

CARUEL, T., 1876: Sui fiori di *Ceratophyllum*. – Nuovo Giorn. Bot. Ital. **8**: 28–32.

CHASE, M. W., QIU, Y.-L., ALBERT, V. A., 1993a: Molecular phylogenetics of angiosperms: cladistic analyses of nucleotide sequence variation in a plastid locus. – Amer. J. Bot. **80** [6, Suppl.]: 122.

– SOLTIS, D. E., OLMSTEAD, R. G., MORGAN, D., LES, D. H., MISHLER, B. D., DUVALL, M. R., PRICE, R. A., HILLS, H. G., QIU, Y.-L., KRON, K. A., RETTIG, J. H., CONTI, E., PALMER, J. H., MANHART, J. R., SYTSMA, K. J., MICHAELS, H. J., KRESS, W. J., KAROL, K. G., CLARK, W. D., HEDRÉN, M., GAUT, B. S., JANSEN, R. K., KIM, K.-J., WIMPEE, C. F., SMITH, J. F., FURNIER, G. R., STRAUSS, S. H., XIANG, Q.-Y., PLUNKETT, G. M., SOLTIS, P. S., SWENSEN, S. M., WILLIAMS, S. E., GADEK, P. A., QUINN, C. J., EGUIARTE, L. E., GOLENBERG, E., LEARN, G. H., Jr., GRAHAM, S. W., BARRETT, S. C. H., DAYANANDAN, S., ALBERT, V. A., 1993b: Phylogenetics of seed plants: an analysis of nucleotide sequences from the plastid gene *rbc*L. – Ann. Missouri Bot. Gard. **80**: 528–580.

COOK, C. D. K., 1978: The *Hippuris* syndrome. – In STREET, H. E., (Ed.): Essays in plant taxonomy, pp. 163–176. – London: Academic Press.

CREPET, W. L., NIXON, K. C., STEVENSON, D. W., FRIIS, E. M., 1993: The relationships of seed plants in reference to angiosperm outgroups. – Amer. J. Bot. **80**: [6, Suppl.]: 123.

DILCHER, D. L., 1989: The occurrence of fruits with affinities to *Ceratophyllaceae* in Lower and mid-Cretaceous sediments. – Amer. J. Bot. **76** [6, Suppl.]: 162.

– 1992: The concept of the flower and its most primitive expression. – The KATHERINE ESAU International Symposium – Plant structure: concepts, connections & challenges. – DAVIS: University of California, Abstr.

DOYLE, J. A., HICKEY, L. J., 1976: Pollen and leaves from the mid-Cretaceous Potomac group and their bearing on early angiosperm evolution. – In BECK, C. B., (Ed.): Origin and early evolution of angiosperms, pp. 139–206. – New York: Columbia University Press.

ENDRESS, P. K., 1990: Evolution of reproductive structures and functions in primitive angiosperms (*Magnoliidae*). – Mem. New York Bot. Gard. **55**: 5–34.

– HUFFORD, L. D., 1989: The diversity of stamen structures and dehiscence patterns among *Magnoliidae*. – Bot. J. Linn. Soc. **100**: 45–85.

FRIIS, E. M., ENDRESS, P. K., 1990: Origin and evolution of angiosperm flowers. – Adv. Bot. Res. **17**: 99–162.

JOHRI, B. M., AMBEGAOKAR, K. B., SRIVASTAVA, P. S. (Eds.), 1992: Comparative embryology of angiosperms, **1**. – Berlin: Springer.

JONES, E. N., 1931: The morphology and biology of *Ceratophyllum demersum*. – Univ. Iowa Studies Nat. Hist. **13**: 11–55.

DE KLERCKER, J.-E.-F., 1885: Sur l'anatomie et le développement de *Ceratophyllum*. – K. Svenska Vet.-Akad. Handl. Bihang **9**(10): 1–23.

LES, D. H., 1988a: The origin and affinities of the *Ceratophyllaceae*. – Taxon **37**: 326–345.

– 1988b: Breeding systems, population structure, and evolution in hydrophilous angiosperm. – Ann. Missouri Bot. Gard. **75**: 819–835.

– 1989: The evolution of achene morphology in *Ceratophyllum* (*Ceratophyllaceae*), IV. Summary of proposed relationships and evolutionary trends. – Syst. Bot. **14**: 254–262.

– 1991: Genetic diversity in the monoecious hydrophile *Ceratophyllum* (*Ceratophyllaceae*). – Amer. J. Bot. **78**: 1070–1082.

– 1993: *Ceratophyllaceae*. – In KUBITZKI, K., ROHWER, J. G., BITTRICH, V., (Eds): The families and genera of vascular plants, **II**, pp. 246–250. – Berlin: Springer.
– GARVIN, D. K., WIMPEE, C. F., 1991: Molecular evolutionary history of ancient aquatic angiosperms. – Proc. Natl. Acad. Sci. USA **88**: 10119–10123.
– – – 1992: A phylogeny of the ancient genus *Ceratophyllum* (*Ceratophyllaceae*) derived from DNA sequence data. – Amer. J. Bot. **79** [6, Suppl.]: 151.
MOSELEY, M. F., SCHNEIDER, E. L., WILLIAMSON, P. S., 1993: Phylogenetic interpretations from selected floral vasculature characters in the *Nymphaeaceae* sensu lato. – Aquat. Bot. **44**: 325–342.
QIU, Y.-L., CHASE, M. W., LES, D. H., PARKS, C. R., 1993: Molecular phylogenetics of the *Magnoliidae*: cladistic analyses of nucleotide sequences of the plastid gene *rbc*L. – Ann. Missouri Bot. Gard. **80**: 587–606.
ROZE, E., 1892: Sur le mode fécondation du *Najas major* ROTH. et du *Ceratophyllum demersum* L. – Bull. Soc. Bot. France **39**: 361–364.
RUTISHAUSER, R., SATTLER, R., 1987: Complementarity and heuristic value of contrasing models in structural botany. II. Case study on leaf whorls: *Equisetum* and *Ceratophyllum*. – Bot. Jahrb. Syst. **109**: 227–255.
SCHLEIDEN, M. J., 1837: Beiträge zur Kenntniss der Ceratophylleen. – Linnaea **11**: 513–542.
SEHGAL, A., MOHAN RAM, H. Y., 1981: Comparative developmental morphology of two populations of *Ceratophyllum* L. (*Ceratophyllaceae*) and their taxonomy. – Bot. J. Linn. Soc. **82**: 383–356.
SHAMROV, I. I., 1981: Some peculiar features of the development of the anther in *Ceratophyllum demersum* and *C. pentacanthum* (*Ceratophyllaceae*). – Bot. Ž. (Moscow, Leningrad) **66**: 1464–1472.
– 1983a: Anthecological investigation of three species of the genus *Ceratophyllum* (*Ceratophyllaceae*). – Bot. Ž. (Moscow, Leningrad) **68**: 1357–1366.
– 1983b: The structure of the anther and some peculiar features of the microsporogenesis and pollen grain development in the representatives of the genus *Ceratophyllum* (*Ceratophyllaceae*). – Bot. Ž. (Moscow, Leningrad) **68**: 1662–1667.
STRASBURGER, E., 1902: Ein Beitrag zur Kenntniss von *Ceratophyllum submersum* und phylogenetische Erörterungen. – Jahrb. Wiss. Bot. **37**: 477–526.
WILMOT-DEAR, M., 1985: *Ceratophyllum* revisited – a study in fruit and leaf variation. – Kew Bull. **40**: 243–271.

Address of the author: PETER K. ENDRESS, Institut für Systematische Botanik der Universität, Zollikerstrasse 107, CH-8008 Zürich, Switzerland.

Accepted February 2, 1994 by E. M. FRIIS

Pl. Syst. Evol. [Suppl.] 8: 185–191 (1994)

# Petal evolution in *Ranunculaceae*

KEIKO KOSUGE

Received November 15, 1993

**Key words:** Angiosperms, *Ranunculaceae*, *Magnoliaceae*. – Petal development, petal evolution.

**Abstract:** The ontogeny of flowers in *Ranunculaceae* indicates a close similarity between petal and stamen primordia. The various shapes of petals are derived from stamen-like petals consisting of a long stalk and a fleshy limb by the partial or total reduction of the stalk and the adaxial wall. In this family the petals originally offer nectar as an insect attractant. Various types of nectary cup evolved independently in each phyletic group adapting to different kinds of pollinators. In some genera, the petals develop long spurs. In others they replace sepals as the showy organ with the reduction of the nectary part, and in the extreme they become completely flat structures by the loss of nectar secretion.

Flowers show an extreme variability in the structure and in their topological differentiation in angiosperms. The fundamental morphology of flower and floral elements has been disputed for more than a century. The morphological variability of petals can be explained by their late appearance in the phylogenetic history. They are formed secondarily from other floral elements already in existence. Different suggestions have been made concerning the morphological interpretation and evolutionary development of the petals.

In the *Ranunculaceae*, the phylogenetically lowermost group among herbaceous angiosperms, various shapes of petals are already differentiated. As the petal in *Ranunculaceae*, *Berberidaceae* and allied families usually secretes nectar, PRANTL (1887) considered it as a modified stamen and called it "Honigblatt" to distinguish it from organs of bract origin. HIEPKO (1965) made an excellent study of the morphology and ontogeny of the perianth in *Polycarpicae*.

In this paper, the separate studies (KOSUGE & TAMURA 1988, 1989) on the ontogeny of petals in *Ranunculaceae* by scanning electron microscopy are summarized and the evolution of petals in this family is discussed with regard to ontogenetic and phylogenetic aspects.

## Primordia of petals

In primitive angiosperms, petals are considered to have a dual origin: in *Magnoliales* and *Illiciales* they are of bract origin, while in *Nymphaeales* and *Ranunculales* they are modified stamens (TAKHTAJAN 1980).

In the *Magnoliaceae*, the flower usually consists of three or more trimerous whorls of perianth, and numerous spirally arranged stamens and carpels. Although *Liriodendron tulipifera* has nectariferous petals, which were differentiated from sepals, the early development of the perianth is nearly the same as that in the homochlamydeous flower of *Magnolia*. Each sepal of the trimerous whorl appears as a crescent-shaped primordium at the base of a dome-shaped floral apex, then the floral apex becomes triangular (Fig. 1). The outer petal primordia are formed at the angles alternating with the sepal primordia and the inner whorl of petals quickly follows opposite the sepals (Fig. 2). The growth of sepals is very fast and they soon overarch the other floral appendages. The hemispherical stamen and carpel primordia are initiated in a rapid spiral succession (Figs. 3, 4). At the time of carpel initiation, the petals cover the floral apex like sepals (Fig. 5).

In the *Ranunculaceae*, numerous patterns of flower types and petal shapes occur. *Aquilegia* shows a particular specialization in a eucyclic flower and long spurred petals.

The ontogeny of the flower in *Aquilegia buergeriana* is nearly the same as that described by PAYER (1857) for *A. vulgaris* and TEPFER (1953) for *A. formosa*.

Five crescent-shaped sepal primordia initiate in a spiral sequence at the base of a high dome-shaped apex (Fig. 6). Shortly after the inception of the last sepal, petal and stamen primordia arise in a cyclic pattern (Figs. 7, 8). At the time of initiation, the petal primordia are nearly hemispherical and resemble the stamen primordia. Later, the primordia become more flat like a thick disk (Fig. 11). The centre of the disk becomes depressed, indicating the beginning of spur formation, and two swellings appear near the lateral margin at the base (Figs. 10, 12). After the inception of stamens, two whorls of inner staminodium primordia (a peculiar character of *Aquilegia*, *Semiaquilegia* and *Urophysa*) and five carpel primordia are initiated (Fig. 9). The floral apex is covered by the sepals at the time of the carpel initiation. Although the floral elements initiate centripetally, the growth of carpel, staminodia and filaments, the development of microsporangia, and anther dehiscence occur in a centrifugal order (Fig. 10), and the petals elongate shortly before anthesis.

Thus, there is a clear ontogenetic difference between *Liriodendron* and *Aquilegia* in the shape and size, the arrangement and the growth of petal primordia. This difference may suggest that the petals are of distinct origins. In *Ranunculaceae*, the petal primordia have the same shape and size as the stamen primordia (PAYER 1857, TEPFER 1953, HIEPKO 1965, SATTLER 1973, KOSUGE & TAMURA 1988, 1989), and they are arranged on the same parastichies. The retarded development of the petals is a relatively common phenomenon in other families (PAYER 1857, SATTLER 1973), and has been explained as the arrested condition of stamens (e.g., DE CANDOLLE 1817, TROLL 1928, ARBER 1937). The growth of sessile large petals, such as in *Anemonopsis*, *Consolida* and *Delphinium*, is less delayed than that in *Aquilegia*. In these genera the inception and development of stamens are in the same centripetal order as observed in *Nymphaea alba* (PAYER 1857). As suggested by many authors, the ontogenetic and anatomical evidence indicates that petals of *Ranunculaceae* are similar to stamens (PAYER 1857, SMITH 1928, HIEPKO 1965, TAMURA 1965, 1984) and they represent the outermost members of the androecium.

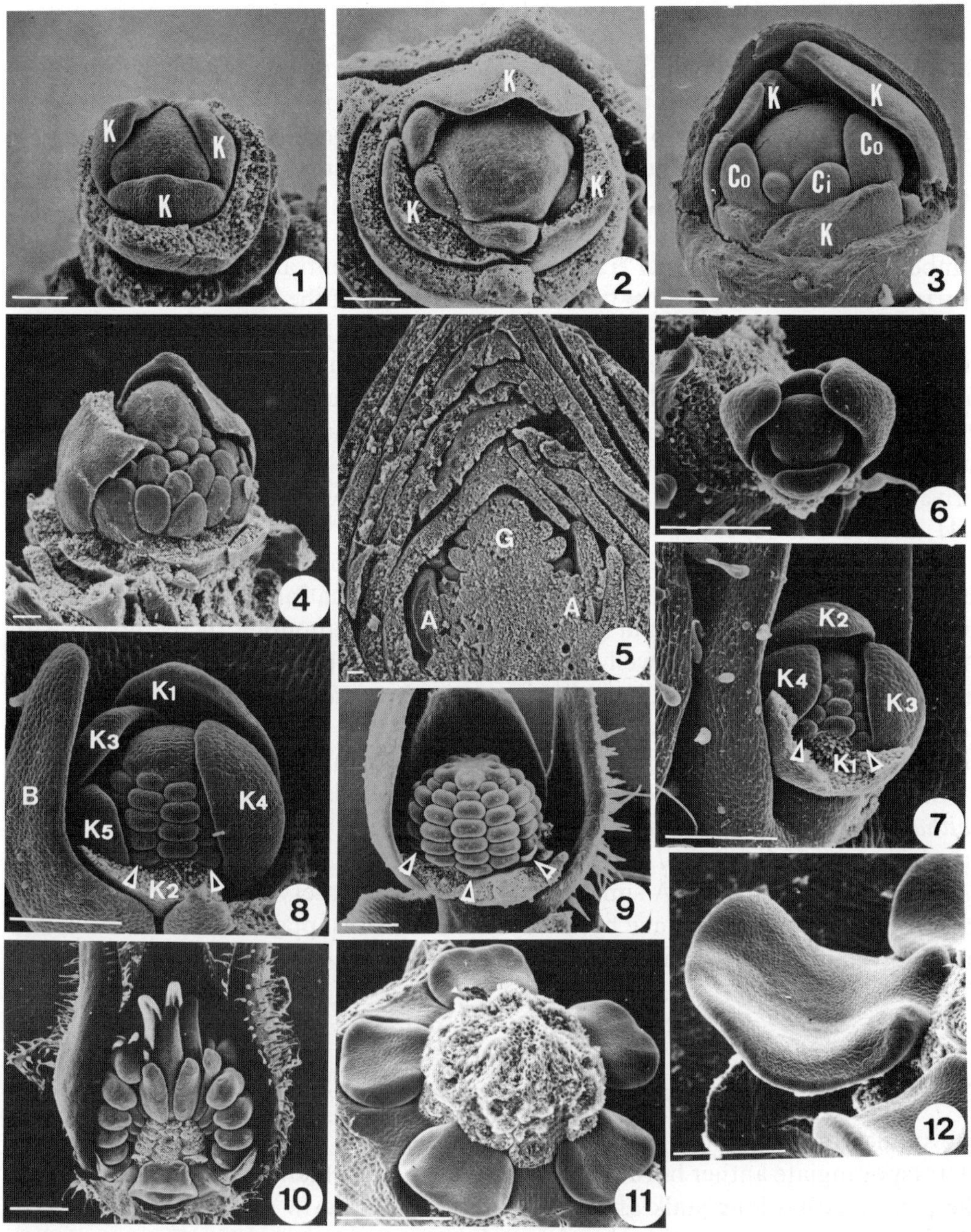


Figs. 1–5. Initiation of floral organs in *Liriodendron tulipifera*. – Fig. 1. Top view of a floral bud bearing a whorl of 3 sepals (K). – Fig. 2. Floral bud with a whorl of sepals and two whorls of petals. – Fig. 3. Side view of a floral bud after the inception of outer and inner whole of petals (Co, Ci). – Fig. 4. Initiation of carpels. – Fig. 5. Side view of longitudinal section of a floral bud. *A* Stamen, *G* carpels. – Figs. 6–12. Initiation of floral organs in *Aquilegia buergeriana*. – Fig. 6. Top view of a floral bud at the time of initiation of sepals. – Fig. 7. Side view of a floral bud showing the formation of stamen primordia. – Fig. 8. Initiation of staminodium primordia, arrows indicate petal primordia. – Fig. 9. Initiation of carpel primordia, petal primordia (arrows) become thick disks. – Fig. 10. Floral organs grow in a centrifugal order. – Fig. 11. Petals appear as thick disks. – Fig. 12. Depressed petal with two swells at the base. – Bars: 0.05 mm

## Phylogeny of *Ranunculaceae*

The chromosome types and the carpel morphology are important characters in phylogenetic considerations of the *Ranunculaceae*. Until early in this century, this family had been divided into two groups, subfam. *Ranunculoideae* with achenes or uniovular carpels, and subfam. *Helleboroideae* with follicles or multiovular carpels (Hutchinson 1923, Janchen 1949). However, anatomical studies have revealed that the uniovular carpel is derived from the multiovular one by reduction of ovule numbers and fusion of carpel traces, i.e. dorsal and ventral bundles (Smith 1928, Chute 1930, Eames 1931, Kumazawa 1938). Instead of the classification by carpel types, Langlet (1932) proposed the subdivision of the family based on the chromosome type: subfam. *Ranunculoideae* with R(anunculus)-type chromosomes and subfam. *Thalictroideae* with T(halictrum)-type chromosomes. Thereafter, the karyological types and the basic number of chromosomes have been referred to as the most important character, and the uniovular carpel has been considered to be derived independently from multiovular ones within each group (e.g., Gregory 1941, Tamura 1966, 1990).

Phylogenetic analyses of sequence data obtained from the nuclear gene alcohol dehydrogenase (adh) statistically supported the classification of the *Ranunculaceae* based on the karyological data (Sawada & Kosuge, unpubl.). In the group with T-type chromosomes, two well-defined clades correspond to Langlet's (1932) subdivision: tribe *Thalictreae*, including *Enemion*, *Isopyrum*, *Semiaquilegia*, *Aquilegia*, *Thalictrum*; and tribe *Coptideae*, including *Coptis* and *Xanthorhiza*. In the group with R-type chromosomes, genera with multiovular carpels are also present.

In the *Ranunculaceae*, flowers are devoid of petals in some genera, such as *Caltha*, *Anemone*, *Enemion*, and *Thalictrum*. In others, petals show various forms: cup-shaped in *Helleborus*, *Isopyrum*, and *Coptis*, and long-spurred in *Aconitum*, *Delphinium*, and *Aquilegia*. A similar state of petals is observed in different phyletic groups. Accordingly, they have been formed or have disappeared sporadically, and they evolved in parallel after the phylogenetic splitting of ancestral *Ranunculaceae* (Tamura 1965, 1984).

## Course of the evolution of petals in *Ranunculaceae*

The stamen of this family is the most common type in angiosperms, consisting of a tetrasporangiate anther borne on a thin single-veined filament. The most stamen-like petals, with a long stalk and fleshy limb, are found in *Kingdonia* and *Coptis*, which have peculiar primitive characters (Foster & Arnott 1960, Tamura 1981). In *Kingdonia uniflora*, the young petal çlosely resembles the stamen, and the adaxial groove of the knob-like limb where nectar is secreted seems to correspond to the connective (Kosuge & al. 1989). In *Coptis quinquefolia* the limb of the petal is shaped like a thick cup. At first the petal primordia become more or less flat and differentiate into the stalk and limb. The limb becomes depressed by marginal growth and delay of growth in the centre (Fig. 13). The growth of the adaxial mound of limb is retarded at the middle part (Fig. 14). The limb becomes a shallow cup-like structure by the growth along the margin (Fig. 15). In *Xanthorhiza simplicissima*, which is closely related to *Coptis*, the limb of petal primordia expands in both lateral directions (Figs. 16, 17) and the adaxial side turns upward forming the nectary cup (Fig. 18). In the stamens of *Kingdonia* and *Xanthorhiza* both anther halves are attached to each other at the top, and the anther wall forms a cup by the latrose-extrose anther dehiscence.

Compared with the stamen, the limb and the stalk of the petal are homologous to the anther and the filament of the stamen respectively. In the cup-shaped or tubular petals in *Eranthis*, *Helleborus*, and *Dichocarpum*, the adaxial wall of the limb arises as two lateral swells (KOSUGE & TAMURA 1989). These two swells seem to correspond to the adaxial locules. The retarded growth between the two swells may be caused by the late activation of the growth and by pressure by the adjacent stamen on the same radius.

In *Clematis* sect. *Atragene*, stamens are transformed into petals gradually and they seem to be of a filament origin (TAMURA 1965).

The petals in *Ranunculaceae* have been considered to be peltate structures (SCHRÖDINGER 1909, LEINFELLNER 1958, HIEPKO 1965, TAMURA 1965). It is difficult to decide whether petals are really peltate, because the adaxial part arises later but grows faster than the abaxial part (KOSUGE & TAMURA 1989). In the early development, however, petal primordia are more or less flat and depressed by the marginal growth and the retarded growth at the bottom, which resembles the development of the ascidiform carpel primordia (VAN HEEL 1983).

The various shapes of petals are derived from the differential development of the adaxial wall (KOSUGE & TAMURA 1988, 1989). A tubular or cup-shaped limb is formed by the continuous growth of the adaxial and abaxial margin (e.g., Type I,

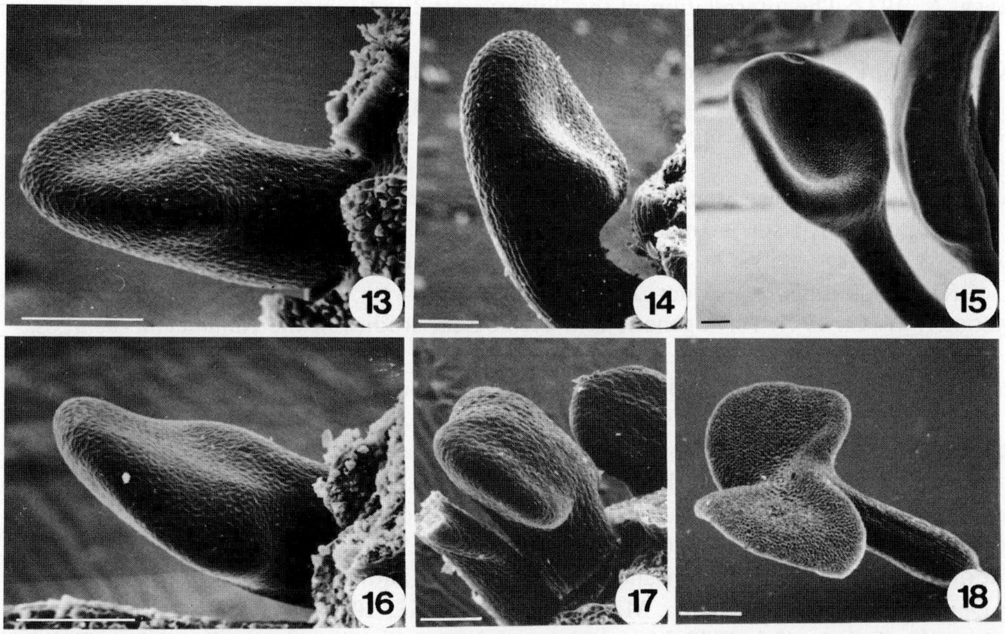


Figs. 13–15. Development of petal in *Coptis quinquefolia*. – Figs. 13, 14. The limb depressed in the centre. – Fig. 15. Cup-shaped limb. – Figs. 16–18. Development of petal in *Xanthorhiza simplicissima*. – Figs. 16, 17. The limb expands in lateral directions. – Fig. 18. Adaxial wall curves up. – Bars: 0.1 mm

II, and III petals in *Aconitum*, *Eranthis*, *Helleborus*, *Isopyrum*, *Coptis*) (Kosuge & Tamura 1988). If the middle part of the adaxial meristem is not activated or if it is completely suppressed, a spoon-like limb (*Trollius*) or petals with a lateral appendage (Type V petal in *Aconitum*, *Aquilegia*) are formed. If the adaxial meristem does not grow continuously with the abaxial one, the adaxial wall develops into a central flap, ridge or pit (Type IV petals in *Aconitum*, *Cimicifuga*, *Ranunculus*). The flat limb without the adaxial appendages results by the total reduction of the adaxial meristem (Type VI in *Aconitum*, *Anemonopsis*, *Adonis*).

In *Ranunculaceae*, sepals play an important role as optically attractive organs and petals originally offer nectar as an insect attractant. The adaxial appendage of the petal evolved to form a nectary cup or sometimes to conceal the nectar. However, in some genera such as *Aquilegia*, *Consolida*, *Delphinium*, and *Ranunculus*, the limb of the petal increased in size and replaced the sepal as an optically attractive organ involving the reduction of the stalk and the adaxial wall. In *Isopyroideae*, the bottom of the limb is situated at the top of the stalk in *Dichocarpum*. The stalk is reduced and the bottom of limb moves backward in *Isopyrum*, then the bottom develops the spur in unguiculate tubular petals of *Semiaquilegia*. Furthermore, the spur elongates forming a deep nectary cup in the sessile petal of *Aquilegia*. In *Ranunculus* and allied genera, petals enlarge and replace the sepals. The nectary cup is reduced as a small pit or shallow pocket at the base in *Ranunculus*, and the petals are nectarless in *Adonis*, as mentioned by Hiepko (1965).

**References**


Arber, A., 1937: The interrelation of the flower: a study of some aspects of morphological thought. – Biol. Rev. **12**: 157–184.

Candolle, A. P., de, 1817: Considérations générales sur les fleurs doubles, et en particulier sur celles de la famille des *Renonculacées*. – Mém. Phys. Chim. Soc. Arcueil **3**: 385–404.

Chute, H. M., 1930: The morphology and anatomy of the achene. – Amer. J. Bot. **17**: 703–723.

Eames, A. J., 1931: The vascular anatomy of the flower with refutation of the theory of carpel polymorphism. – Amer. J. Bot. **18**: 147–188.

Foster, A. S., Arnott, H. J., 1960: Morphology and dichotomous vasculature of the leaf of *Kingdonia uniflora*. – Amer. J. Bot. **47**: 684–698.

Gregory, W. C., 1941: Phylogenetic and cytological studies in the *Ranunculaceae* Juss. – Trans. Amer. Philos. Soc., n.s., **31**: 443–521.

Heel, W. A., van, 1983: The ascidiform early development of free carpels, a S.E.M.-investigation. – Blumea **28**: 231–270.

Hiepko, P., 1965: Vergleichend-morphologische und entwicklungs-geschichtliche Untersuchungen über das Perianth bei den *Polycarpicae*. – Bot. Jahrb. Syst. **84**: 359–508.

Hutchinson, J., 1923: Contributions towards a phylogenetic classification of flowering plants. 1. – Kew Bull. **1923**: 65–89.

Janchen, E., 1949: Die systematische Gliederung der Ranunculaceen und Berberidaceen. – Denkschr. Akad. Wiss. Wien, Math. – naturwiss. Kl., **108**(4): 1–82.

Kosuge, K., Tamura, M., 1988: Morphology of the petal in *Aconitum*. – Bot. Mag. (Tokyo) **101**: 223–237.

– 1989: Ontogenetic studies on petals of the *Ranunculaceae*. – J. Jap. Bot. **64**: 65–74.

– Pu, F.-D., Tamura, M., 1989: Floral morphology and relationships of *Kingdonia*. – Acta Phytotax. Geobot. **40**: 61–67.

KUMAZAWA, M., 1938: On the ovular structure in the *Ranunculaceae* and *Berberidaceae*. – J. Jap. Bot. **14**: 10–25.

LANGLET, O., 1932: Über Chromosomenverhältnisse und Systematik der *Ranunculaceae*. – Svensk Bot. Tidskr. **26**: 381–400.

LEINFELLNER, W., 1958: Beiträge zur Kronblattmorphologie VIII. Der peltate Bau der Nektarblätter von *Ranunculus*, dargelegt an Hand jener von *Ranunculus pallasii* SCHLECHT. – Österr. Bot. Z. **105**: 184–192.

PAYER, J. B., 1857: Traité d'organogénie comparée de la fleur. – Paris: Masson.

PRANTL, K., 1887: Beiträge zur Morphologie und Systematik der Ranunculaceen. – Bot. Jahrb. Syst. **9**: 225–273.

SATTLER, R., 1973: Organogenesis of flowers. – Toronto: University of Toronto Press.

SCHRÖDINGER, R., 1909: Der Blütenbau der zygomorphen Ranunculaceen und seine Bedeutung für die Stammesgeschichte der Helleboreen. – Abh. K. K. Zool.-Bot. Ges. Wien **4**(5): 1–63.

SMITH, G. H., 1928: Vascular anatomy of Ranalian flowers II. *Ranunculaceae* (continued), *Menispermaceae*, *Calycanthaceae*, *Annonaceae*. – Bot. Gaz. **85**: 152–177.

TAKHTAJAN, A. L., 1980: Outline of the classification of flowering plants (*Magnoliophyta*). – Bot. Rev. **46**: 225–359.

TAMURA, M., 1965: Morphology, ecology and phylogeny of the *Ranunculaceae* IV. – Sci. Rep. Osaka Univ. **14**: 53–71.

– 1966: Morphology, ecology and phylogeny of the *Ranunculaceae* VI. – Sci. Rep. Osaka Univ. **15**: 13–35.

– 1981: Morphology of *Coptis japonica* and its meaning in phylogeny. – Bot. Mag. (Tokyo) **94**: 165–176.

– 1984: Phylogenetical consideration on the *Ranunculaceae*. – Korean J. Pl. Taxon. **14**: 33–42.

– 1990: A new classification of the family *Ranunculaceae* 1. – Acta Phytotax. Geobot. **41**: 93–101.

TEPFER, S. S., 1953: Floral anatomy and ontogeny in *Aquilegia formosa* var. *truncata* and *Ranunculus repens*. – Univ. Calif. Publ. Bot. **25**: 513–648.

TROLL, W., 1928: Organisation und Gestalt im Bereich der Blüte. – Berlin: Springer.

Address of the author: Dr KEIKO KOSUGE, Department of Biology, Faculty of Science, Kobe University, Kobe 657, Japan.

Accepted January 31, 1994 by P. K. ENDRESS

Pl. Syst. Evol. [Suppl.] 8: 193–208 (1994)

# Flowers in *Magnoliidae* and the origin of flowers in other subclasses of the angiosperms. I. The relationships between flowers of *Magnoliidae* and *Alismatidae*

CLAUDIA ERBAR and PETER LEINS

Received November 2, 1993

**Key words:** *Magnoliidae*, *Monocotyledoneae*. – Floral development, androecial patterns, origin of the monocotyledons.

**Abstract:** In the flowers of many *Magnoliidae*, the androecium consists of numerous, spirally arranged stamens (e.g., *Illicium*). The androecia in the *Aristolochiaceae* and the monocotyledonous *Alismatales*, with paired arrangement of the first six stamens (additional stamens arising collateral-centrifugally, or collaterally, or centrifugally, or centripetally), seem to be quite different from spiral androecia. Comparative ontogenetic studies, however, reveal that the paired arrangement of stamens can be derived from the early stages of a spiral androecium following a perianth consisting of trimerous whorls. In *Magnolia denudata*, with whorled perianth, this pattern of six stamens is caused by a break in the basic ontogenetic spiral (precocious inception of the eighth stamen), whereas in *Annonaceae* (e.g., *Artabotrys hexapetalus*, *Annona montana*) the first six stamen primordia are formed simultaneously. – Based on the androecial pattern (and other features), we may perhaps assume that *Annonaceae*-like dicotyledons were the ancestors of the *Aristolochiaceae* as well as of the monocotyledonous *Alismatales*.

The classical taxonomic division of the angiosperms into the two classes, monocotyledons and dicotyledons, dates from the late seventeenth century (TAKHTAJAN 1969). Fossil record indicates that the origin of the monocotyledons from the dicotyledons was a very early dichotomy in the evolution of the angiosperms (DOYLE 1973). The fact that so-called monocotyledonous features occur outside the *Monocotyledoneae*, particularly within the *Magnoliidae*, reflects the close relationship of the two groups. Very recently, molecular data (analyses of DNA sequences of the chloroplast gene *rbc*L; CHASE & al. 1993, QIU & al. 1993) support the view that the major division is not between dicotyledons and monocotyledons, but rather the division appears to be correlated with the pollen type: triaperturate versus uniaperturate (see also HUBER 1977, 1982). Molecular data further do not exclude that the monocotyledons are derived from woody *Magnoliidae* with monosulcate pollen grains. The immediate sister group of the monocotyledons are the "paleoherbs" (*Aristolochiales*, *Piperales*, and *Nymphaeales*; Fig. 4A in CHASE & al. 1993).

In spite of this correspondence, floral ontogenetic studies revealed a basic difference: the organs are spirally arranged in polymerous flowers of many *Magnoliidae*, whereas there is no evidence for a spiral origin of stamens (and carpels) in monocotyledonous flowers, and certainly not in *Alismatales* (SINGH & SATTLER 1972, 1973, 1974, 1977; SATTLER & SINGH 1973, 1978; LEINS & STADLER 1973). Trimerous flowers are one of the unifying traits of the *Monocotyledoneae*. Trimerous flowers are also found in many members of the *Magnoliidae*, e.g., in the *Aristolochiaceae*. However, in other *Magnoliidae*, in members of *Magnoliaceae* and *Annonaceae*, trimerous whorls are restricted to the perianth, whereas the members of androecium and gynoecium are initiated in a spiral sequence. This observation leads to two questions: **What is the order of early stamen initiation in a multistaminate spiral androecium, after initiation of trimerous, alternating perianth whorls? Does this pattern provide information about the formation of non-spiral androecia in some *Magnoliidae* and in the *Monocotyledoneae*?**

During the last decade we compared the early androecial pattern of members of the *Magnoliidae* and the *Monocotyledoneae* (ERBAR & LEINS 1981, 1983; ERBAR 1988; LEINS & ERBAR 1985, 1991a, b, 1994). With this contribution we want to present an extended survey.

## On the androecial pattern in *Magnoliales*

In the flowers of *Illicium*, as an example, a regular spiral arrangement with the "golden" divergence angle prevails throughout the flower (Figs. 1, 2; ERBAR & LEINS 1983). In a spiral pattern, subsequent organs are initiated at more or less equal spatial and temporal (?) intervals (equal "plastochrons"). If the spiral pattern follows the limiting divergence of the Fibonacci series, the angle between subsequent organs is about 137, 5°[1]. A transition from spiral to trimerous whorls can be shown examplarily within the *Magnoliaceae* (ERBAR & LEINS 1981, 1983; ERBAR 1988). *Magnolia stellata* exhibits a spiral pattern comparable to *Illicium*: all organ primordia follow more or less the "golden" divergence angle (Figs. 3–5). Rhythmisation of the spiral sequence leads to a whorled perianth in *Magnolia denudata* (Figs. 9–11) and *Liriodendron tulipifera* (Figs. 6–8). In other words, after three short plastochrons

[1]The "golden" angle of 137, 5° is the average value of the whole spiral sequence in one flower. Individual divergence angles may differ considerably from this average (about 125° to 150°).

▶

Figs. 1, 2. *Illicium anisatum* L. (*Illiciaceae*). Spiral initiation of perianth members, stamen primordia and carpel primordia (36–43 in Fig. 2); numbering according to the sequence (±limiting divergence). – Figs. 3–5. *Magnolia stellata* (SIEB. & ZUCC.) MAXIM. (*Magnoliaceae*). Spiral pattern in the formation of perianth and stamen primordia. – Figs. 6–8. *Liriodendron tulipifera* L. (*Magnoliaceae*). Three trimerous perianth whorls (the outer one removed) with still clear indication of a spiral sequence (black numerals) are followed by stamen primordia in a spiral sequence more or less according to the limiting divergence (white numerals)

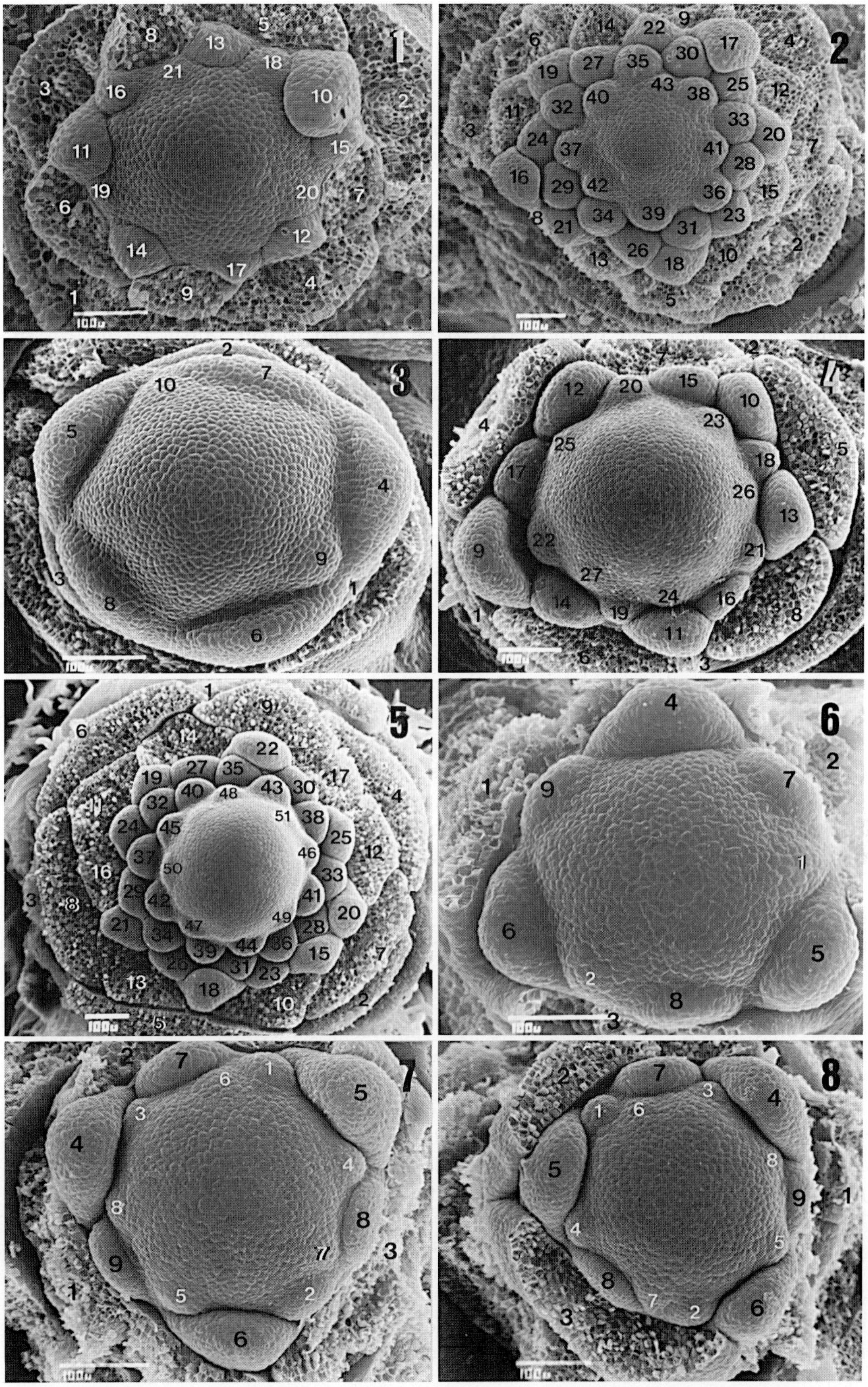
1
2
3
4
5
6
7
8

follows one relatively longer plastochron, which is again followed by three short plastochrons and a longer one, and again by three short plastochrons. Each longer plastochron seems to be responsible for the alternation of three organs (Erbar & Leins 1983, Erbar 1988).[2] A spiral perianth has changed to a perianth consisting of three trimerous whorls. In *Liriodendron tulipifera* the trimerous perianth shows in each whorl still clear indications of a spiral sequence (Figs. 6–7). In addition, the last three perianth members (7–9) do not alternate exactly with the preceding whorl and thereby mediate between the whorled perianth and the spiral androecium: stamen 1 arises in limiting divergence to perianth member 9 (Fig. 6). All further primordia are initiated in keeping this limiting divergence (Figs. 7, 8). Also in *Magnolia denudata* the whorled perianth shows a more or less spiral sequence within each whorl, but there seems to be an even longer interval between the initiation of subsequent whorls (Figs. 9–11). Judged from the presence of parastichies, the members of androecium and gynoecium arise spirally (Fig. 13). **Young** developmental stages of the androecium in *Magnolia denudata*, however, show breaks in this spiral (Erbar & Leins 1981, Erbar 1988). A frequent case is shown in Figs. 12a–d[3]: The stamen primordia are numbered according to their **position** on the basic spiral, which follows more or less the "golden" divergence angle, irrespective of their **temporal sequence**. The primordia 1, 2, 3, 4, 5 correspond to the temporal sequence in the basic spiral. After this there is a break: primordium 8 of the basic spiral arises before primordia 6 and 7. Primordium 8 is somewhat larger, and as the older primordium it has arisen further down at the floral apex than the primordia 6 and 7. Possible causes for this break in the sequence include the small size of the stamen primordia in relation to the perianth members and the entire apex, and the large gaps between the three perianth members last formed.

A comparison of *Magnolia denudata* with *Illicium* and *Liriodendron* with regard to the sequence and position of those six primordia that follow the first nine primordia makes it easier to understand this early pattern (Fig. 22; cf. Erbar & Leins 1983). These six primordia (marked black in the diagrams, Fig. 22) belong to the perianth in *Illicium* but are the first six stamens in *Liriodendron* and *Magnolia denudata*. The pattern in *Liriodendron* differs from the regular spiral in the flowers of *Illicium* in that the stamen primordia 6, 7 and 8 (dotted in the diagram) arise nearly simultaneously (see Fig. 8). Sometimes primordium 8 seems to be slightly

---

[2]According to the "field theory" each just initiated primordium is surrounded by an inhibitory field which decreases as the growing primordium increases in age. Thus the latest developed primordia exercise a regulating influence on the place where the next one will be initiated. There is, however, another inhibitory gradient extending from the top of the floral apex to its base. A longer interval after the inception of, e.g., three primordia results in a clear decrease of their inhibitory fields. Therefore the influence of the inhibitory gradient of the floral apex becomes dominant. Consequently, the minimum distance from the top of the apex, necessary for organ formation, will first become available in the gaps between the latest three developed primordia.

[3]SEM-photos allow the detailed investigation of spiral sequences. The exact arrangement of all organs is best ascertained if the flower buds are photographed in top view and in several side views. By this method each primordium is registered at least twice as regards its position at the floral apex as well as its position to neighbouring primordia.

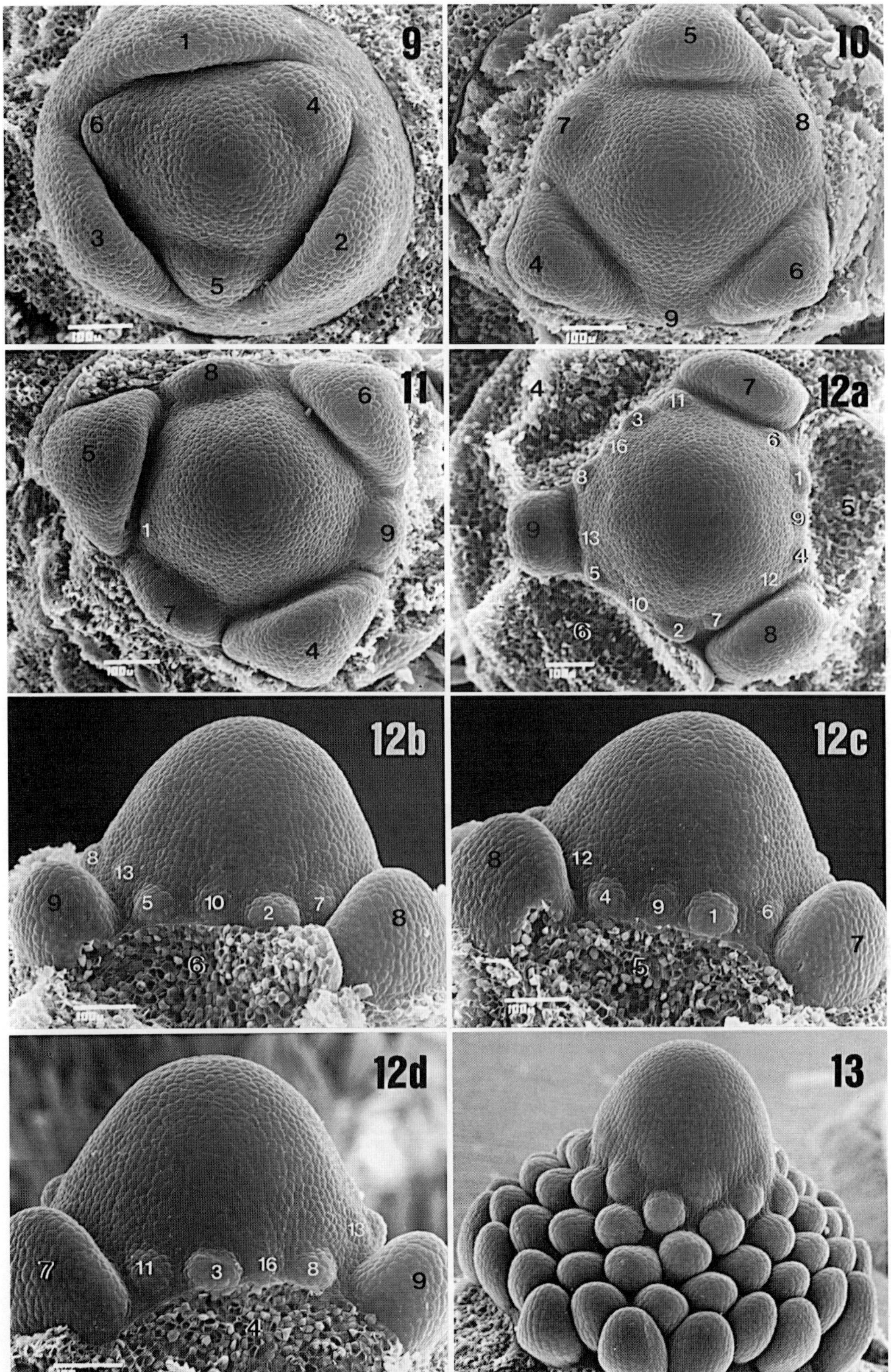


Figs. 9–13. *Magnolia denudata* DESR. (*Magnoliaceae*). – Figs. 9–11. Formation of the three perianth whorls. – Fig. 12. Top view (*a*) and three side views (*b–d*) of the same flower bud, perianth removed except for the three inner members (7, 8, 9; black numerals); stamen primordia numbered (white) according to their approximate position on the basic spiral (limiting divergence), irrespective of their temporal sequence (cf. Fig. 22). – Fig. 13. Parastichies in the androecium (and gynoecium)

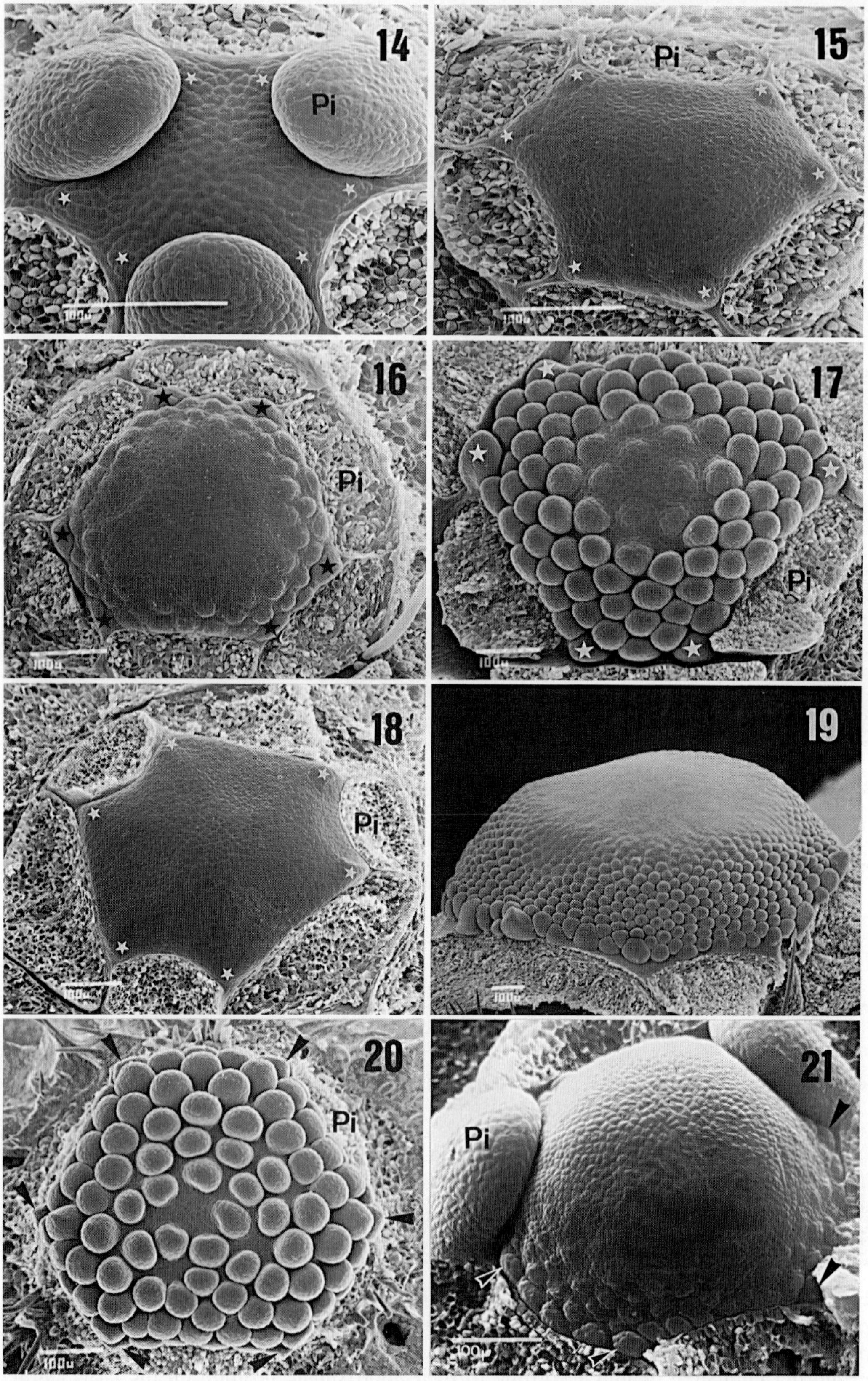
14
15
16
17
18
19
20
21
Pi
Pi
Pi
Pi
Pi
Pi
Pi
100μ
100μ
100μ
100μ
100μ
100μ
100μ
100μ

earlier (see Fig. 7). In addition, stamen primordium 8 is more distant from stamen primordium 3 than in *Illicium*. Thus a trend becomes apparent, which leads to the conditions in *Magnolia denudata*, in which one can see large gaps between the perianth members last formed and a considerable size difference between the primordia of perianth and androecium. In *Magnolia denudata*, stamen primordium 8 of the basic spiral now arises before stamen primordia 6 and 7, and more or less at the same level as stamen primordium 3. We want to emphasize once more that the series *Illicium-Liriodendron-Magnolia denudata* can only serve as a model, which shows the kind of pattern that may arise if a spiral follows a trimerous arrangement. A break in the basic spiral, namely the precocious inception of the eighth stamen, causes an arrangement (in *Magnolia denudata*) transitional to the early androecial pattern in the more advanced *Annonaceae*, e.g., in *Artabotrys hexapetalus* or *Annona montana*. There the first six stamen primordia – initiated after a trimerous perianth – are formed simultaneously. In *Artabotrys hexapetalus* (Figs. 14, 15; see also ENDRESS 1987a, who used the synonym *Artabotrys uncinatus*) and *Annona montana* (Fig. 18) these six primordia are arranged exactly in the corners of the hexagonal floral apex. This is presumably also the case in *Polyalthia suberosa* (Fig. 20) and *Monodora crispata* (Fig. 21; LEINS & ERBAR 1994), and the same has been reported for *Popowia whitei* (RONSE DECRAENE & SMETS 1990). For the last three species, however, the initial stage of androecial development, with only the first six stamen primordia, has not been documented. In *Annonaceae*, further stamen primordia originate on the floral apex centripetally in a relatively rapid succession (Figs. 16, 17, 19). We could never observe a regularly spiral sequence of stamen primordia, but in some areas there are more or less distinct parastichies. An acyclic, acropetal sequence of stamen primordia and the partial arrangement of the primordia in parastichies have been found in all species investigated. We can assume that there are morphogenetical correlations between the irregularities in the stamen arrangement and the small size of stamen primordia relative to the entire floral apex. The irregular stamen arrangement may also be correlated with the short plastochrons as well as with the large gaps between the perianth members (see ERBAR & LEINS 1981, 1983; LEINS & ERBAR 1994; ENDRESS 1987a, 1990).

## On the androecial pattern of *Aristolochiales* and *Alismatales*

The early pattern of the androecium in *Annonaceae* ultimately can be derived from the early stages of still spiral magnolioid androecia formed after a perianth consisting

◄

Figs. 14–17. *Artabotrys hexapetalus* (L.f.) BHANDARI (*Annonaceae*). – Figs. 14, 15. Simultaneous formation of the first six stamens in the corners of a hexagonal floral apex. – Figs. 16, 17. Non-spiral, acropetal inception of further stamens. In Figs. 14–17 the first six stamens are marked by asterisks. – Figs. 18, 19. *Annona montana* MACF. (*Annonaceae*). – Fig. 18. Inception of the first six stamens (asterisks). – Fig. 19. Acropetal, but somewhat "chaotic" initiation of further stamens. – Fig. 20. *Polyalthia suberosa* (ROXB.) BENTH. & HOOK.f. (*Annonaceae*). Inception of stamen and carpel primordia (the arrows indicate the first six stamen primordia). – Fig. 21. *Monodora crispata* ENGL. & DIELS (*Annonaceae*). Acropetal, but somewhat "chaotic" inception of the stamens (the arrows point to "corner primordia"). – *Pi* inner perianth member

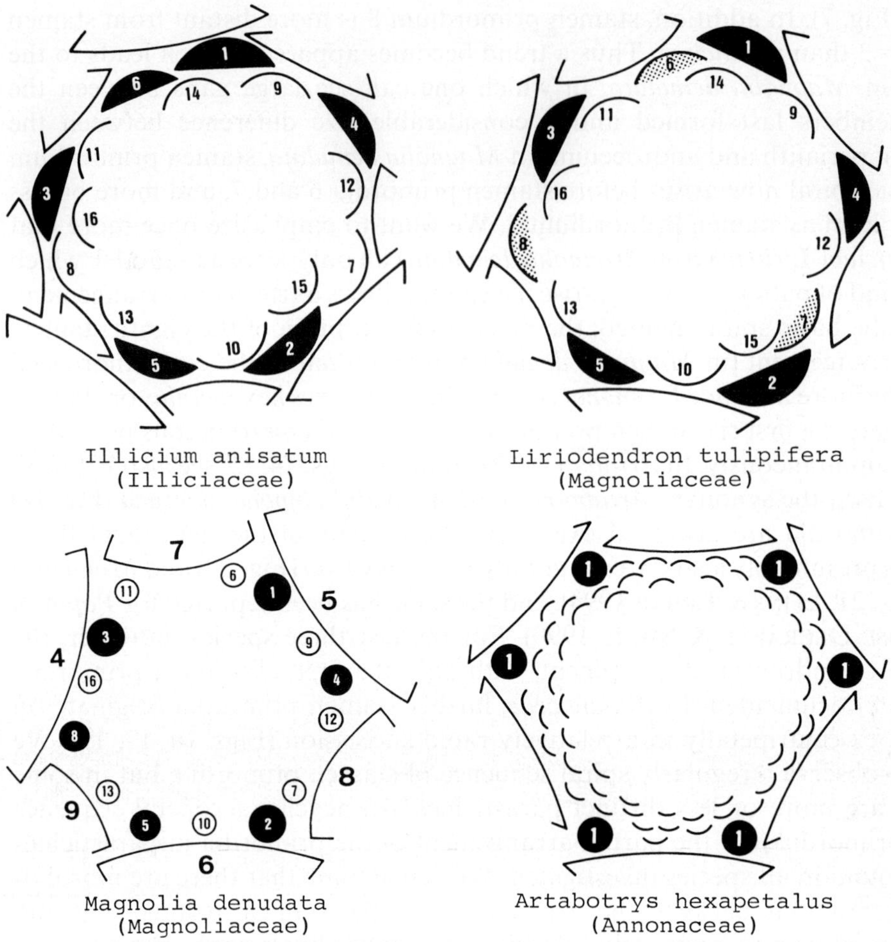


Fig. 22. Floral developmental diagrams showing the transition from a spiral to a trimerous arrangement of the outer floral parts; adjustment of a "three pairs"-pattern at the beginning of the androecial development (series *Illicium anisatum-Liriodendron tulipifera-Magnolia denudata-Artabotrys hexapetalus* chosen as a model). Numbering according to the position on the basic spiral; for further explanations see text

of trimerous whorls (Fig. 22). It is now easy to connect the quite different androecia in the *Aristolochiaceae* and the monocotyledonous *Alismatales* with the early androecial pattern of *Annonaceae* (LEINS & ERBAR 1985, ERBAR 1988). In most *Aristolochiaceae* and *Alismatales*, the androecial development starts with the simultaneous inception of six stamen primordia (each marked by an asterisk in Figs. 23–30)

Figs. 23–30. Early androecial pattern (left side) and further stamen inception (right side) of members of the *Aristolochiaceae* and *Alismatales*. The androecial development begins with the simultaneous inception of six stamen primordia (each marked by an asterisk). – Figs. 23, 24. *Asarum caudatum* LINDLEY (*Aristolochiaceae*). – Figs. 25, 26. *Alisma plantago-aquatica* L. (*Alismatales*). – Figs. 27, 28. *Butomus umbellatus* L. (*Alismatales*). – Figs. 29, 30. *Echinodorus* spec. (*Alismatales*). – *C* carpel, *P* petalum, *S* sepalum, $St_{2,3}$ additional stamens numbered according to their sequence

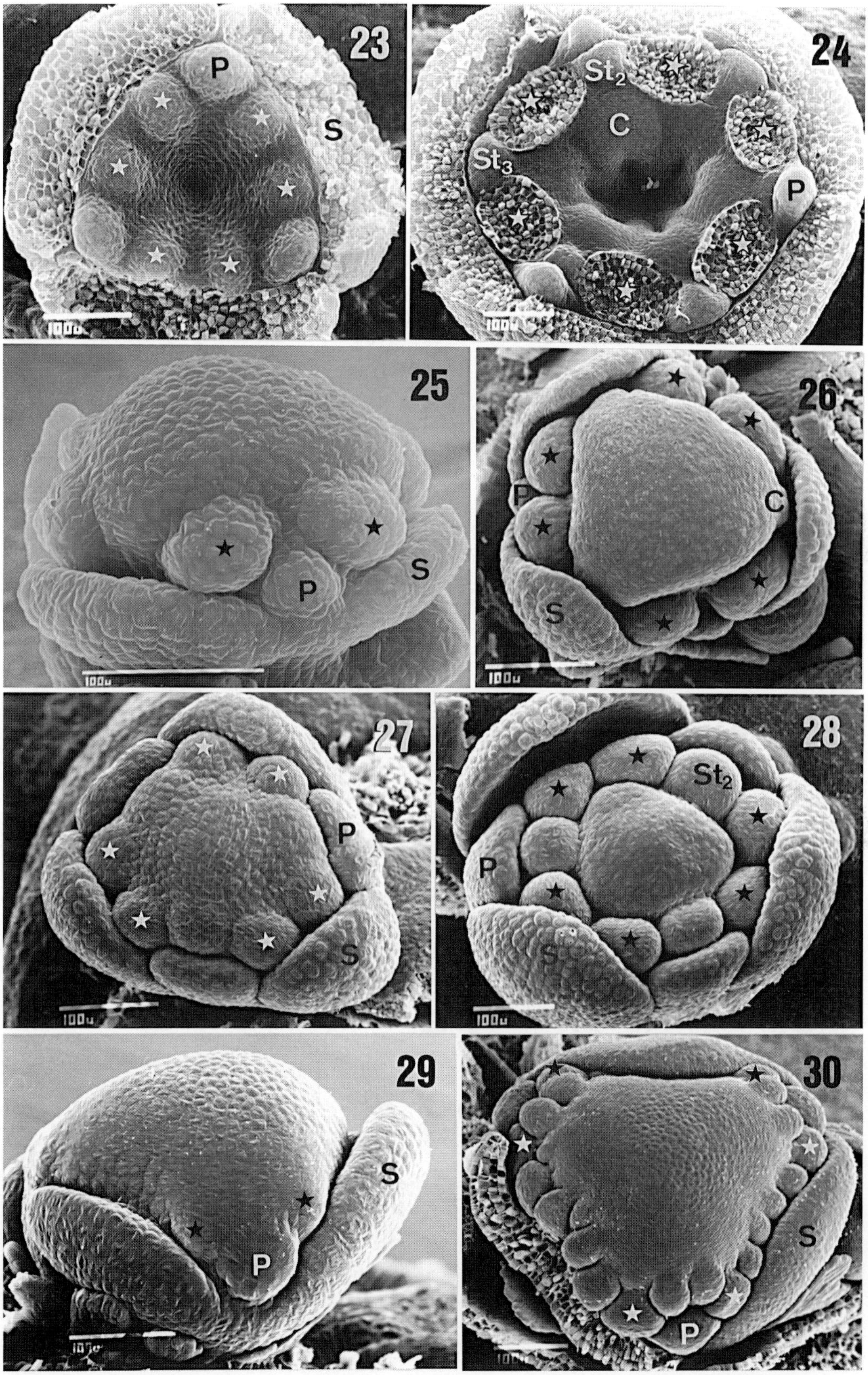
23
P
S
100u
24
St2
C
St3
P
100u
25
P
S
100u
26
P
C
S
100u
27
P
S
100u
28
St2
P
S
100u
29
S
P
30
S
P

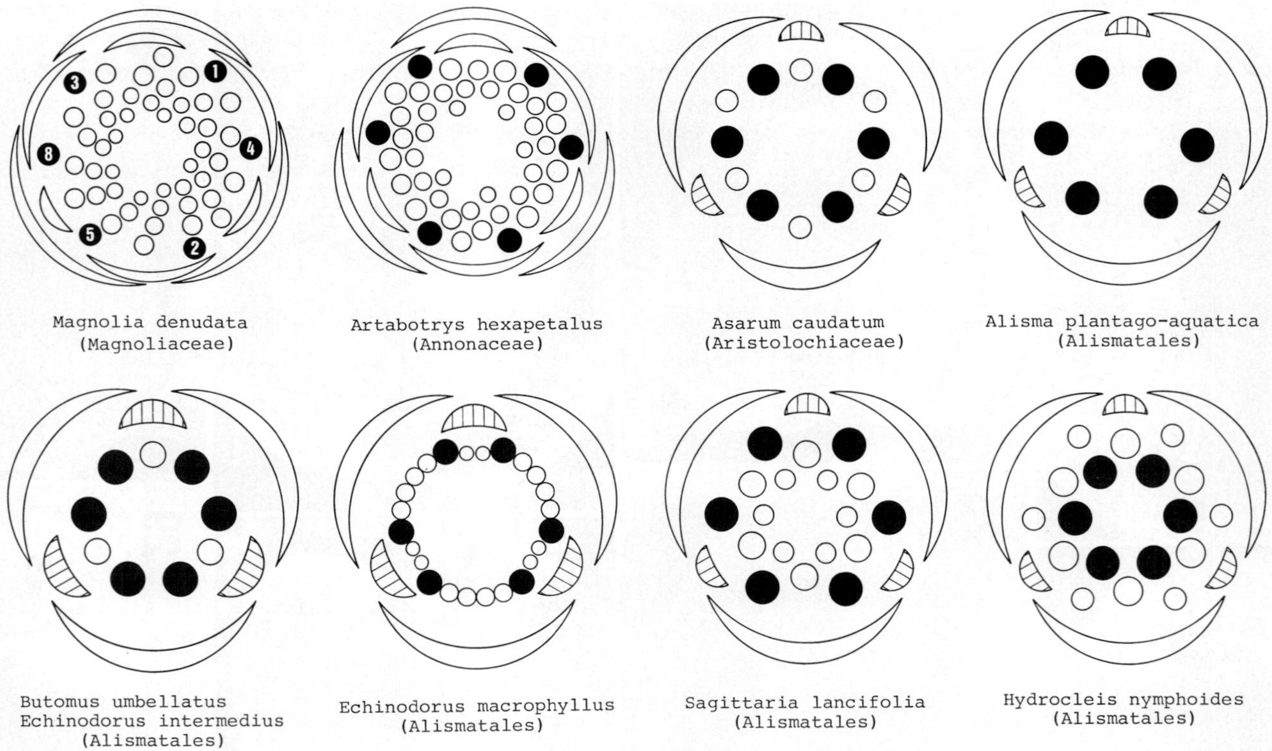


Fig. 31. Floral diagrams from members of *Magnoliales*, *Aristolochiales* and *Alismatales* showing the androecial patterns. The first initiated six stamen primordia are marked black. The diagrams of *Alismatales* are drawn after the results of LEINS & STADLER (1973)

situated pairwise on both sides of the inner perianth members – the same pattern as shown in the *Annonaceae* (Fig. 31). The genera differ, however, in the way in which further stamens are formed. In *Asarum caudatum*, three additional stamens arise, after a relatively long interval, in front of the inner perianth members, rapidly followed by three others in front of the outer perianth members on the outside of the first six stamens (Fig. 24). We can call this a collateral-centrifugal inception of additional stamens (see also LEINS & ERBAR 1985). In *Alismatales*, we often find collateral (*Butomus*: Fig. 28, *Echinodorus*: Fig. 30), centrifugal (*Hydrocleis*) or centripetal (*Sagittaria*) intercalation of additional stamens (cf. LEINS & STADLER 1973). The mode of intercalation within the *Alismatales* seems to depend on the way the floral axis expands, and the number of additional primordia depends on the size of the floral bud (LEINS & STADLER 1973, LEINS 1975). In *Alisma plantago-aquatica*, where there is little enlargement of the floral apex, the androecial development ends with the six stamen primordia differentiated pairwise on both sides of the petals (Figs. 25, 26).

## Comparison of the androecial pattern with other systematic characters

The amazing conformity of the *Annonaceae*, *Aristolochiaceae*, and the monocotyledonous *Alismatales* with regard to the early developmental pattern in the androecium (Fig. 31) is paralleled by similarities in the kind of sieve-tube plastids. *Asarum* and *Saruma* of the *Aristolochiaceae* have sieve-tube plastids identical to the subtype PII of the *Monocotyledoneae* (BEHNKE 1971a, b, 1977, 1988, 1991). They are the only two genera outside the monocotyledons that contain these typical cuneate protein crystalloids. This is remarkable, because there are only two characters that are invariable in all *Monocotyledoneae* studied so far: the single cotyledon[4] and the special kind of sieve-tube plastids. On the other hand, the genus *Aristolochia* and some genera of the *Annonaceae* resemble each other in containing p-type sieve-tube plastids of the PI-subtype with a single polygonal protein crystalloid (BEHNKE 1971a, b, 1977, 1988). Besides the sieve-tube plastids of the p-type, *Annonaceae* and *Aristolochiaceae* have other characteristics in common with the *Monocotyledoneae*: trimerous perianth whorls, sometimes an adaxial prophyll, successive type of microsporogenesis, and monoaperturate pollen grains (within the *Aristolochiaceae* in *Saruma*; ERDTMAN 1966). We believe that there is good evidence that the *Aristolochiaceae* can be linked with the *Annonaceae* on the one hand and the *Monocotyledoneae* on the other hand. Perhaps *Annonaceae*-like dicotyledons were the precursors of the *Aristolochiaceae* and *Alismatales*, which have specialized early in different ways (LEINS & ERBAR 1985, ERBAR 1988).

## The liliaceous androecial pattern

By our detailed ontogenetical studies we tried to overcome the difficulties that arise when we compare the apparently very different androecia of *Magnoliales*, *Aristolochiales* and *Alismatales*. But what about flowers with the typical liliaceous floral diagram, i.e., flowers with fully trimerous whorls (P3 + 3A3 + 3G3)? As an

[4]In contradiction to many literature reports all extant monocotyledons are monocotyledonous without exception (TILLICH 1992).

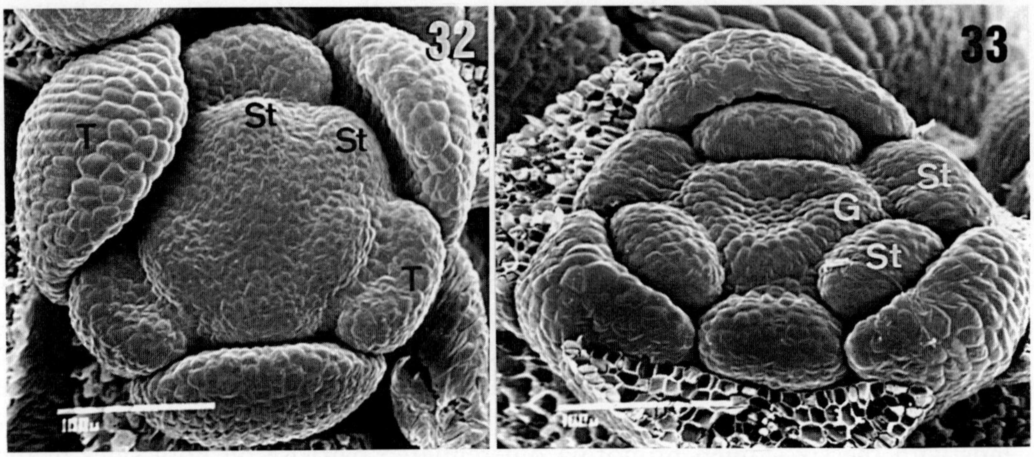


Figs. 32, 33. *Anthercium liliago* L. (*Liliaceae*). Young flower buds with trimerous whorls throughout. – *G* gynoecium, *T* tepalum, *St* stamen

example, young flower buds of *Anthericum* (*Liliaceae*) are shown in Figs. 32 and 33: Two trimerous perianth whorls are followed by six stamens in two whorls[5]. Perhaps the typical trimerous diagram of liliaceous groups can be derived directly from polymerous flowers, or more probably from flowers with paired stamens by a "harmonisation" of the size differences of perianth primordia, stamen primordia and the floral apex (LEINS & ERBAR 1991a, b).

## Other dicotyledonous candidates for presumable precursors of the monocotyledons

We should not disregard that the *Nymphaeales* (clearly dicotyledonous; TILLICH 1990) as well as the *Piperales* have been considered to be closely related to the monocotyledons (e.g., HUBER 1977, 1982; DAHLGREN & BREMER 1985; LOCONTE & STEVENSON 1991; CHASE & al. 1993). The *Nymphaeaceae* (like the *Alismatales*) are typical marsh and water plants and many of their characteristics reflect adaptions to this habitat. It has been suggested that an aquatic or amphibious way of life might have led to the loss of the cambial activity and to the scattered arrangement of closed vascular bundles (TAKHTAJAN 1969). Besides these characters, other ones, like more or less reduced primary root, root structure, and mostly monosulcate pollen grains are in common with the monocotyledons (TAKHTAJAN 1969). With the *Alismatales* (*Butomus*, *Hydrocleis*), the *Nymphaeaceae* share laminar placentation. The nymphaealean genus *Cabomba* exhibits a perianth consisting of two trimerous whorls and six stamina in a position like in *Alisma* (judged from the figures in TAKHTAJAN 1969 and in ROHWEDER & ENDRESS 1983, see also EICHLER 1878). In

[5] Although the members of the two androecial whorls have a different position at the floral apex (the members of the inner one higher up than those of the outer one), we never could find developmental stages, which showed a successive initiation of the two whorls (see also SATTLER 1973 for *Scilla violacea*).

the large polymerous flowers of *Nymphaeaceae* the arrangement of petals and stamens is variable. It may be more or less whorled, rarely spiral (Fibonacci- and Lucas-series) or irregular (WOLF 1991).

The *Piperaceae* show a tendency towards a scattered arrangement of the (in this case open) vascular bundles. The monosulcate pollen and the single prophyll in the genus *Piper* recall monocotyledons. In *Piperaceae* and the neighbouring family *Saururaceae* the flowers seem to be basically trimerous with two whorls of stamens. Detailed floral ontogenetical studies, however, have shown that, e.g., in *Anemopsis californica*, *Piper amalago*, and *Saururus cernuus*, the stamens arise as successive pairs and that the symmetry during stamen initiation is bilateral or dorsiventral (the stamens of a pair arise on opposite sides of the median plane of the flower; TUCKER 1975, 1980, 1981, 1982; LIANG & TUCKER 1989). The order of initiation differs in the members of *Piperaceae* and *Saururaceae* (see Fig. 43 in TUCKER 1982). The initiation of stamens in successive pairs combined with the bilateral symmetry during initiation in *Piperales*, however, differs from the spiral arrangement of stamens in magnolialean flowers, from the paired arrangement of the first six simultaneously formed stamens in *Aristolochiaceae* and *Alismatales*, as well as from the radial trimerous arrangement in liliaceous flowers.

The inclusion of the *Chloranthaceae* in the *Piperales* is debatable (CRONQUIST 1988). ENDRESS (1987b) attributes a key position to the *Chloranthaceae* in that they probably link the *Laurales* and *Piperales* within the *Magnoliidae* as well as the *Magnoliidae* with the *Hamamelididae*. BURGER (1977) speculates that the monocotyledonous (and dicotyledonous) flower can be derived from a cluster of simple flowers similar to those of the *Chloranthaceae*.

The seeds in the *Piperales* have perisperm and endosperm, a construction found elsewhere only in the *Nymphaeales* (CORNER 1976, DAHLGREN & al. 1985). On the other hand, the seeds in the *Alismatales* lack both perisperm and endosperm. Moreover, the sieve-tube plastids in both *Piperales* and *Nymphaeales* completely lack protein crystalloids (BEHNKE 1977, 1988, 1991). The *Nymphaeales* and *Piperales* seem to be sister-groups (DAHLGREN & al. 1985, CHASE & al. 1993), which perhaps have not reached the monocotyledonous state. The two orders may have evolved in parallel to the *Aristolochiales* (LEINS & ERBAR 1985).

## On the systematic value of the early "three-pairs" pattern in the androecium

We hope to have shown convincingly, that this pattern – handled critically – may be of considerable value. KUBITZKI's objection (1993, p. 6) that the paired stamen arrangement has little phylogenetic significance because of its widespread occurrence (examples see in ENDRESS 1987a) is – in our opinion – not well-reasoned. We want to emphasize once more that the taxa which we have chosen for our morphological series (*Illicium*, *Magnolia stellata*, *Liriodendron*, *Magnolia denudata*, *Artabotrys hexapetalus* or *Annona montana*, *Asarum caudatum*) should serve only as models. The progress of cyclisation in the perianth, which occurred repeatedly within the *Magnoliidae*, involved (repeatedly) a paired arrangement of the first stamen primordia. Within the *Monocotyledoneae*, however, paired stamen inception occurs in the *Alismatales* **only**. This fact leads us to the assumption that the *Alismatales*, as an archaic group within the *Monocotyledoneae*, not only in this respect but also because of a number of primitive characters (CRONQUIST 1981), can be linked with *Magnoliidae*. Out of the magnoliid groups with paired stamen arrangement, we must choose those taxa as the closest living relatives that have the largest number of important characters in common with the *Monocotyledoneae*. These are the *Annonaceae* and the *Aristolochiaceae* (LEINS & ERBAR 1985, 1994; ERBAR 1988).

We want to express our sincere thanks to Dr JENS ROHWER for carefully reading and correcting the English text.

**References**

BEHNKE, H.-D., 1971a: Zum Feinbau der Siebröhren-Plastiden von *Aristolochia* und *Asarum* (*Aristolochiaceae*). – Planta **27**: 62–69.

– 1971b: Sieve-tube plastids of *Magnoliidae* and *Ranunculidae* in relation to systematics. – Taxon **20**: 723–730.

– 1977: Transmission electron microscopy and systematics of flowering plants. – Pl. Syst. Evol. [Suppl.] **1**: 155–178.

– 1988: Sieve-element plastids, phloem protein, and evolution of flowering plants: III. *Magnoliidae*. – Taxon **37**: 699–732.

– 1991: Distribution and evolution of forms and types of sieve-tube plastids in the dicotyledons. – Aliso **13**: 167–182.

BURGER, W. C., 1977: The *Piperales* and the monocots. Alternative hypotheses for the origin of monocotyledonous flowers. – Bot. Rev. **43**: 345–393.

CHASE, M. W., SOLTIS, D. E., OLMSTEAD, R. G., MORGAN, D., LES, D. H., MISHLER, B. D., DUVALL, M. R., PRICE, R. A., HILLS, H. G., QIU, Y.-L., KRON, K. A., RETTIG, J. H., CONTI, E., PALMER, J. D., MANHART, J. R., SYTSMA, K. J., MICHAELS, H. J., KRESS, W. J., KAROL, K. G., CLARK, W. D., HEDRÈN, M., GAUT, B. S., JANSEN, R. K., KIM, K.-J., WIMPEE, C. F., SMITH, J. F., FURNIER, G. R., STRAUSS, S. H., XIANG, Q.-Y., PLUNKETT, G. M., SOLTIS, P. S., SWENSEN, S. M., WILLIAMS, S. E., GADEK, P. A., QUINN, C. J., EGUIARTE, L. E., GOLENBERG, E., LEARN, Jr., G. H., GRAHAM, S. W., BARRETT, S. C. H., DAYANANDAN, S., ALBERT, V. A., 1993: Phylogenetics of seed plants: an analysis of nucleotide sequences from the plastid gene *rbc*L. – Ann. Missouri Bot. Gard. **80**: 528–580.

CORNER, E. J. H., 1976: The seeds of dicotyledons I. – Cambridge, London, New York, Melbourne: Cambridge University Press.

CRONQUIST, A., 1981: An integrated system of classification of flowering plants. – New York: Columbia University Press.

– 1988: The evolution and classification of flowering plants. 2nd edn. – Bronx, New York: The New York Botanical Garden.

DAHLGREN, R. M. T., BREMER, K., 1985: Major clades of the angiosperms. – Cladistics **1**: 349–368.

– CLIFFORD, H. T., YEO, P. F., 1985: The families of the monocotyledons. – Berlin, Heidelberg, New York, Tokyo: Springer.

DOYLE, J. A., 1973: The monocotyledons: their evolution and comparative biology. V. Fossil evidence on early evolution of the monocotyledons. – Quart. Rev. Biol. **48**: 399–413.

EICHLER, A. W., 1878: Blüthendiagramme II. – Leipzig: Engelmann.

ENDRESS, P. K., 1987a: Floral phyllotaxis and floral evolution. – Bot. Jahrb. Syst. **108**: 417–438.

1987b: The *Chloranthaceae*: reproductive structures and phylogenetic position. – Bot. Jahrb. Syst. **109**: 153–226.
– 1990: Evolution of reproductive structures and functions in primitive angiosperms (*Magnoliidae*). – Mem. New York Bot. Gard. **55**: 5–34.
ERBAR, C., 1988: Early developmental patterns in flowers and their value for systematics. – In LEINS, P., TUCKER, S. C., ENDRESS, P. K., (Eds): Aspects of floral development, pp. 7–23. – Berlin, Stuttgart: Cramer.
– LEINS, P., 1981: Zur Spirale in Magnolien-Blüten. – Beiter. Biol. Pfl. **56**: 225–241.
– – 1983: Zur Sequenz von Blütenorganen bei einigen Magnoliiden. – Bot. Jahrb. Syst. **103**: 433–449.
ERDTMAN, G., 1966: Pollen morphology and plant taxonomy. – New York, London: Hafner Publ. Co.
HUBER, H., 1977: The treatment of the monocotyledons in an evolutionary system of classification. – Pl. Syst. Evol. [Suppl.] **1**: 285–298.
– 1982: Die zweikeimblättrigen Gehölze im System der Angiospermen. – Mitt. Bot. Staatssamml. München **18**: 59–78.
KUBITZKI, K., 1993: Introduction. – In KUBITZKI, K., ROHWER, J. G., BITTRICH, V., (Eds): The families and genera of vascular plants II., pp. 1–12. – Berlin, Heidelberg, New York: Springer.
LEINS, P., 1975: Die Beziehungen zwischen einfachen und multistaminaten Androeceen. – Bot. Jahrb. Syst. **96**: 231–237.
– ERBAR, C., 1985: Ein Beitrag zur Blütenentwicklung der *Aristolochiaceae*, einer Vermittlergruppe zu den Monokotylen. – Bot. Jahrb. Syst. **107**: 343–368.
– – 1991a: Entwicklungsmuster in Blüten und ihre mutmaßlichen phylogenetischen Zusammenhänge. – Biologie in unserer Zeit **21**: 196–204.
– – 1991b: Fascicled androecia in *Dilleniidae* and some remarks on the *Garcinia* androecium. – Bot. Acta **104**: 336–344.
– – 1994: Early floral developmental studies in *Annonaceae*. – In MORAWETZ, W., (Ed.): *Annonaceae* – evolution and systematics. – Wien Österr. Akad. Wiss. (in press).
– STADLER, P., 1973: Entwicklungsgeschichtliche Untersuchungen am Androeceum der *Alismatales*. – Österr. Bot. Z. **121**: 51–63.
LIANG, H.-X., TUCKER, S. C., 1989: Floral development in *Gymnotheca chinensis* (*Saururaceae*). – Amer. J. Bot. **76**: 806–819.
LOCONTE, H., STEVENSON, D. W., 1991: Cladistics of the *Magnoliidae*. – Cladistics **7**: 267–296.
QIU, Y.-L., CHASE, M. W., LES, D. H., PARKS, C. R., 1993: Molecular phylogenetics of the *Magnoliidae*: cladistic analyses of nucleotide sequences of the plastid gene *rbc*L. – Ann. Missouri Bot. Gard. **80**: 587–606.
ROHWEDER, O., ENDRESS, P. K., 1983: Samenpflanzen: Morphologie und Systematik der Angiospermen und Gymnospermen. – Stuttgart, New York: Thieme.
RONSE DECRAENE, L.-P., SMETS, E., 1990: The floral development of *Popowia whitei* (*Annonaceae*). – Nordic J. Bot. **10**: 411–420.
SATTLER, R., 1973: Organogenesis in flowers. – Toronto, Buffalo: University of Toronto Press.
– SINGH, V., 1973: Floral development of *Hydrocleis nymphoides*. – Canad. J. Bot. **51**: 2455–2458.
– – 1978: Floral organogenesis of *Echinodorus amazonicus* RATAJ and floral constructions of the *Alismatales*. – J. Linn. Soc., Bot. **77**: 141–156.
SINGH, V., SATTLER, R., 1972: Floral development of *Alisma triviale*. – Canad. J. Bot. **50**: 619–627.
– – 1973: Nonspiral androecium and gynoecium of *Sagittaria latifolia*. – Canad. J. Bot. **51**: 1093–1095.

– – 1974: Floral development of *Butomus umbellatus*. – Canad. J. Bot. **52**: 223–230.
– – 1977: Development of the inflorescence and flower of *Sagittaria cuneata*. – Canad. J. Bot. **55**: 1087–1105.
TAKHTAJAN, A., 1969: Flowering plants. Origin and dispersal. – Edinburgh: Oliver & Boyd.
TILLICH, H.-J., 1990: Die Keimpflanze der *Nymphaeaceae* – monocotyl oder dicotyl? – Flora **184**: 169–176.
– 1992: Bauprinzipien und Evolutionslinien bei monocotylen Keimpflanzen. – Bot. Jahrb. Syst. **114**: 91–132.
TUCKER, S. C., 1975: Floral development in *Saururus cernuus* (*Saururaceae*): I. Floral initiations and stamen development. – Amer. J. Bot. **62**: 993–1007.
– 1980: Inflorescence and flower development in the *Piperaceae*. I. *Peperomia*. – Amer. J. Bot. **67**: 686–702.
– 1981: Inflorescence and floral development in *Houttuynia cordata* (*Saururaceae*). – Amer. J. Bot. **68**: 1017–1032.
– 1982: Inflorescence and flower development in the *Piperaceae*. III. Floral ontogeny of *Piper*. – Amer. J. Bot. **69**: 1389–1401.
WOLF, M., 1991: Blütenphyllotaxis von *Nymphaeaceae*: ist das Androecium von *Nymphaea*, *Nuphar* etc. spiralig? – 10. Symp. Morph. Anat. Syst., Göttingen, Abstr., p. 85.

Address of the authors: PD Dr CLAUDIA ERBAR, Prof. Dr PETER LEINS, Institut für Systematische Botanik und Pflanzengeographie der Universität Heidelberg, Im Neuenheimer Feld 345, D-69120 Heidelberg, Federal Republic of Germany.

Accepted January 18, 1994 by P. K. ENDRESS

Pl. Syst. Evol. [Suppl.] 8: 209–218 (1994)

# Flowers in *Magnoliidae* and the origin of flowers in other subclasses of the angiosperms. II. The relationships between flowers of *Magnoliidae, Dilleniidae,* and *Caryophyllidae*

PETER LEINS and CLAUDIA ERBAR

Received November 2, 1993

**Key words:** *Magnoliidae, Dilleniidae, Caryophyllidae, Paeonia.* Floral development, androecial patterns.

**Abstract:** A floral developmental pattern as in *Paeonia* (spiral perianth followed by five spirally arranged **stamen fascicles**) can be linked with that in *Magnolia* (spiral perianth followed by many spirally arranged **individual stamens**). The presumably most archaic flower type in the *Caryophyllidae* is found in the cactaceous genus *Pereskia.* It can be derived from a paeonioid pattern. Both the paeonioid and pereskioid floral patterns repeatedly have given rise to simpler flower constructions in the *Dilleniidae* and *Caryophyllidae*, respectively.

Flowers of the middle and upper evolutionary level of the angiosperms may give informations about archaic flower constructions. Advanced flowers, especially those in which the stamens and the nectar are hidden, e.g., in corolla tubes, often show coloured patterns on their perianth. These coloured patterns, the so-called visual floral guides, function as signals attracting pollinating insects. According to the theory of OSCHE (1979, 1983), the visual floral guides can mostly be interpreted as imitations of the androecium or parts of it. OSCHE argues that these imitations or signal dummies are an expression of conservation of a primary optical signal, the real androecium which is addressed to pollen-eating insects as, e.g., beetles. The insects that have adapted to nectar flowers remain fixed to the primary androecial signal. Under the pressure of competition from open flowers with a visible real androecium, nectar flowers with few and often not visible stamens could only be successful by developing androecial copies. Some examples: At the entrances of many *Iris* flowers, a three-dimensional copy of a whole polymerous androecium can be found (Fig. 1). In the flowers of *Digitalis purpurea*, dark brown dots with whitish margins imitate open anthers (Fig. 2). The same can be observed in flowers of several *Gentiana* species (Fig. 3). The signal of a real androecium consisting of only a few stamens can be intensified by hairy filaments, as in *Verbascum phoeniceum* (Fig. 4). In some *Scrophulariaceae* and related groups, the signal dummies are larger

Figs. 1–6. Flowers with optical signals (imitations of the androecium or parts of it); for explanations see text. – Fig. 1. *Iris* (*Iridaceae*). – Fig. 2. *Digitalis purpurea* L. (*Scrophulariaceae*). – Fig. 3. *Gentiana clusii* PERR. & SONG. (*Gentianaceae*). – Fig. 4. *Verbascum phoeniceum* L. (*Scrophulariaceae*). – Fig. 5. *Antirrhinum majus* L. (*Scrophulariaceae*). – Fig. 6. *Cleome violacea* L. (*Capparaceae*)

than their models and may exert a supernormal stimulus on the pollinators (Fig. 5). We can observe a supernormal anther imitation, e.g., on the lower lip of a snapdragon flower (*Antirrhinum majus*, Fig. 5). Hyperdimensional anther imitations can also intensify the attractiveness of flowers with visible stamens as in the small flowers of *Cleome violacea* (Fig. 6). A simple experiment shows that hyperdimensional visual signals strongly attract many pollinators, e.g., untrained bumblebees which fly directly to artificial dummies (LUNAU 1990, 1991).

## Early flower evolution

According to OSCHE's theory (1979, 1983), we thus may assume that the model to which the flowers of the middle and upper evolutionary level conform with their imitations, is a flower offering pollen as primary food for visiting insects in order to be pollinated. This flower must be hermaphroditic and polyandrous, with a high production of pollen grains compensating the loss by the pollen-eating insects. It is indeed an uneconomical flower, compared to the later evolved nectar-producing flowers, but much more economical than the wind pollinated unisexual flowers

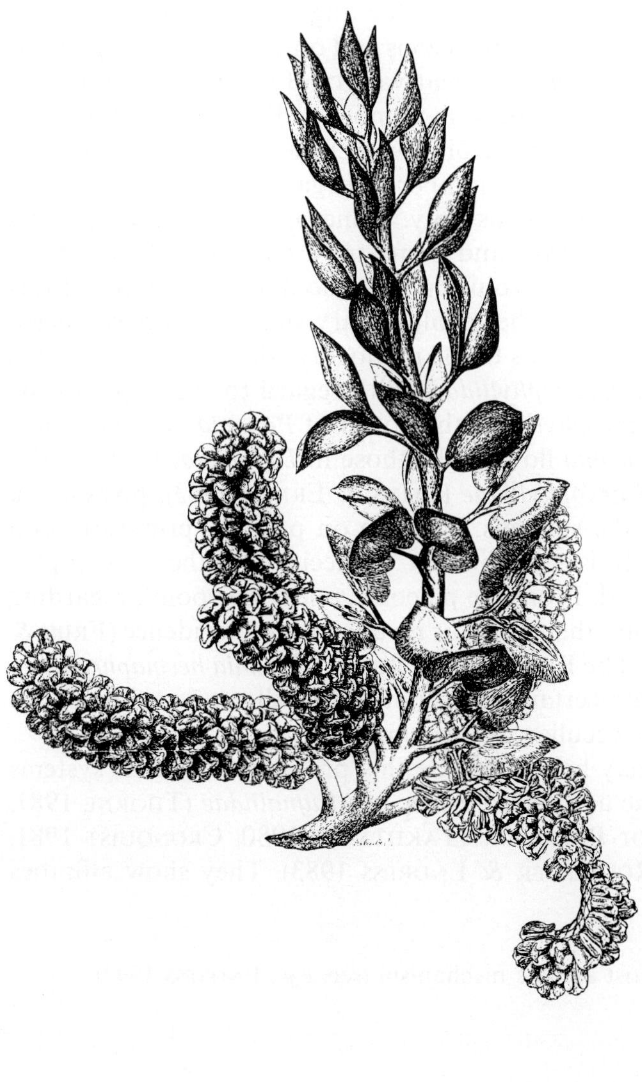

Fig. 7. SCHWEITZER's (1977) reconstruction of *Irania hermaphroditica* from Rhaetic sediments of Northern Iran.

of the gymnospermous ancestors. The hermaphroditic pollen flower can be seen as a response to selection pressure by pollen-eating insects. Thus the parasitic relation changes to a mutualistic one. This had happened repeatedly within the spermatophytes before the angiosperms evolved. We would like to call to mind the Rhaetic (nearly 200 million years old) fossil of a hermaphroditic flower, described as *Irania hermaphroditica* by SCHWEITZER (1977). The flower possesses many microsporangiophores arranged along several androecial axes (Fig. 7). The pollen production must have been high and we can assume that the flower offered part of its pollen as food for visiting insects. The female organs of *Irania hermaphroditica* are very difficult to interpret.

One of the main characters of the angiosperms is the enclosure of the ovules in carpels. We believe that this new condition was an adaption to the destructive behaviour of the first pollinators. Angiospermy is most important for the successful evolution of the flowering plants also in that it makes mechanisms like dichogamy[1] and self-incompatibility possible, avoiding self-fertilisation. Progress in early angiosperm evolution may have involved the emission of floral fragrances and the development of a coloured perianth providing optimal contrast to the androecium. These innovations, however, developed only in a part of the early flowering plants. Recently, the views about the early evolution of the angiosperms have changed. Based on the fossil record from the Lower Cretaceous and on excellent investigations of flower structures in the *Magnoliidae*, the great variety of extant floral forms in this group is thought to reflect a high plasticity in number and arrangement of floral organs at the beginning of angiosperm evolution (ENDRESS 1987a, b, FRIIS & ENDRESS 1990). Large flowers with a high number of organs coexisted very early with small flowers composed of few organs only. Some of such reduced flowers may have returned to wind pollination and some may have attracted flies as pollinators. However that may be, the evolution of some floral forms out of this pool of variety became canalized by further evolutionary steps. Examples of such a canalization are the origin of the flowers of the monocotyledons (ERBAR & LEINS 1994) and of the *Dilleniidae* and *Caryophyllidae*. With regard to the origin of the flowers of the *Dilleniidae* and *Caryophyllidae*, the flower of *Paeonia* occupies a key position. The multistaminate *Paeonia* flower, like those in *Dilleniidae* (LEINS 1983, LEINS & ERBAR 1991a, b) and *Caryophyllidae* (LEINS & ERBAR 1993), possesses a complex androecium, in which the stamens originate on primary primordia in a centrifugal succession (Fig. 9). It is still difficult to decide whether a complex androecium is archaic or advanced. From the paleobotanical viewpoint, regarding the fossils of the Lower Cretaceous, there is so far no unequivocal evidence (FRIIS, & ENDRESS 1990, FRIIS & al. 1993). The Rhaetic fossil flower of *Irania hermaphroditica* would be comparable, but *Irania* certainly can not be the direct ancestor of the angiosperms because of its very peculiar gynoecium.

Systematic considerations may help to tackle this problem. In recent systems of classification, the *Paeoniaceae* are placed either in *Magnoliidae* (THORNE 1981, 1983, 1992; TAKHTAJAN 1987) or *Dilleniidae* (TAKHTAJAN 1980; CRONQUIST 1981, 1988; DAHLGREN 1983, 1989; ROHWEDER & ENDRESS 1983). They show affinities

[1] It seems that protogyny is the most archaic mechanism (see, e.g., ENDRESS 1992).

Table 1. Characters in which *Paeoniaceae* agree with *Magnoliidae* and *Dilleniidae*

| *Magnoliidae* | *Dilleniidae* |
|---|---|
| Choricarpous gynoecium | Choricarpous gynoecium (*Dilleniaceae*) |
| Unfixed number of floral organs | Centrifugal subdivision of primary androecial fascicle primordia |
| Spiral sequence of the flower primordia originating on the floral apex | Persistent calyx |
| Transitions between calyx and corolla leaves | Lack of benzylisoquinoline alkaloids |
| Type of epicuticular wax crystalloids (BARTHLOTT 1991) | |

to both groups in important characters (Table 1). We may conclude that the *Paeoniaceae* can be considered as a link between *Magnoliidae* and *Dilleniidae*, their taxonomic position being a matter of taste (LEINS & ERBAR 1991b). Attention should be paid especially to the fact that in *Paeonia* all floral organ primordia, including the fascicle primordia, are initiated in a spiral sequence. The prevailing archaic characters of the flower in *Paeonia* lead us to the opinion that in a relatively early phase of the evolution of the angiosperms, at least two forms of multistaminate androecia had coexisted, the magnolioid one, in which many stamens originate in a spiral sequence directly on the floral apex (Fig. 8), and the paeonioid one, with a secondary subdivision of only five, still spirally arranged, androecial organs (Fig. 9). Both forms were equally adapted to pollen-eating pollinators (LEINS &

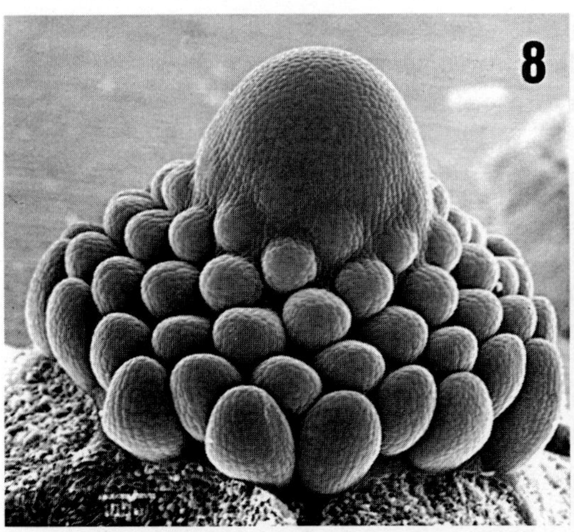


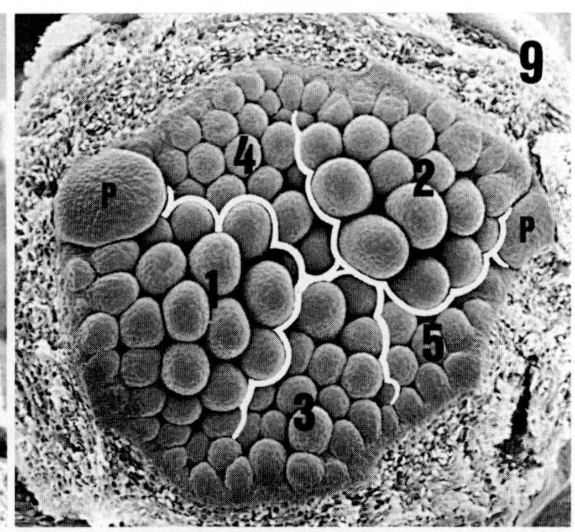


Fig. 8. *Magnolia denudata* DESR. (*Magnoliaceae*). Acropetal and spiral inception of stamen (and carpel) primordia directly on the floral apex. Perianth removed. – Fig. 9. *Paeonia officinalis* L. (*Paeoniaceae*). Spiral inception of five stamen fascicles with a secondary centrifugal subdivision into single stamens. The five stamen fascicles are encircled and numbered according to the spiral sequence. Perianth members removed, except the two youngest (*P*).

Erbar 1991b). It should be noted that the disc in the flower of *Paeonia* is not a nectary disc (Hiepko 1966). It has a postfloral function in stabilizing the fruit, out of which the seeds may be dispersed by birds.

## Floral trends in the *Dilleniidae*

Starting with the paeonioid flower type within the *Dilleniidae* we can envisage trends in different directions and in different ways. Some main steps in phylogeny are: change of the "golden" divergence angle to a 2/5 divergence, cyclisation of the floral parts including the androecial fascicles, formation of an androecial ring primordium, reduction of the androecial primary primordia or the primary ring primordium, and reduction of the stamen number, finally leading to diplostemonous or haplostemonous androecia (see fig. 1 in Leins 1979 and figs. 6, 8 in Leins & Erbar 1991b).

## Floral trends in the *Caryophyllidae*

From the viewpoint of developmental flower morphology also the *Caryophyllidae* can be linked to paeonioid ancestors. In other words, the *Paeoniaceae* presumably occupy a place near the point of divergence of *Dilleniidae* and *Caryophyllidae* from the *Magnoliidae* (see fig. 13.2 in Leins & Erbar 1993). In the *Caryophyllidae* multistaminate androecia basically correspond to those of the *Paeoniaceae* and the *Dilleniidae*. One of the families within the *Caryophyllidae* that is characterized by a multistaminate complex androecium is the *Cactaceae*. In this group we usually find a uniform prominent ring primordium on which the stamens arise in centrifugal succession (Figs. 10–13). Only in the most primitive genus of the *Cactaceae*, in *Pereskia* (Figs. 14, 15), there are some species with nearly separate androecial primary primordia which follow the spiral sequence of the numerous perianth members[2] (Leins & Schwitalla 1986). This androecial pattern of *Pereskia* seems closely related to that of *Paeonia*. In our present state of knowledge we consider the *Pereskia* flower to be most archaic not only within the *Cactaceae* but in the whole subclass *Caryophyllidae*. From ancestors with *Pereskia*-like flowers we can perhaps derive the *Aizoaceae*, in which stamen fascicles or androecial formation centres are usually preserved (Leins & Erbar 1993)[3].

Also in the *Caryophyllidae* there may have been a tendency in the early evolution towards a reduction of the primary primordia or of the common primary ring

[2] Later on, but still before the centrifugal subdivision starts, they become connected; then the spiral sequence can be inferred only from the different degree of subdivision of the primary primordia.

[3] In the species that we investigated the primary primordia of the fascicled androecium are not strictly separate from each other (there is more of a tendency to form a ring wall). When the centrifugal subdivision starts, however, five (or four) formation centres become distinct, alternating with the five (or four) spirally arranged perianth members. The borderlines between the formation centres later on become blurred at least basally (see plate 13.1, 3–6 in Leins & Erbar 1993). In *Aizoaceae* the peripheral members of the androecium develop into petaloid staminodia.

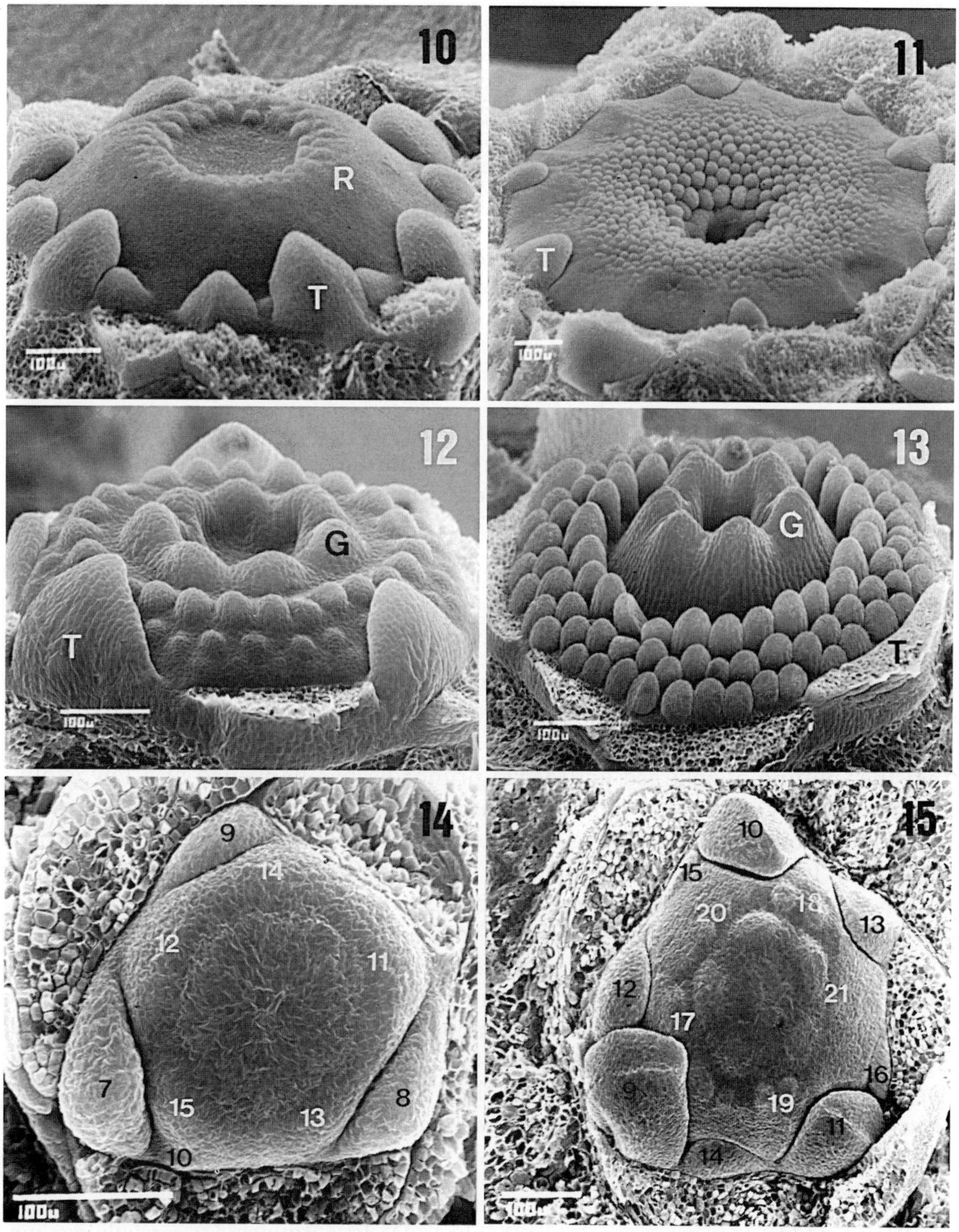


Figs. 10, 11. *Neobuxbaumia polylopha* (DC.) BACKEB. (*Cactaceae*). Centrifugal stamen initiation on an androecial ring primordium (R). – Figs. 12, 13. *Schlumbergera truncata* (HAW.) MORAN (*Cactaceae*). Centrifugal initiation of a lower stamen number on an androecial ring primordium. – Fig. 14. *Pereskia diaz-romeroana* CÁRDENAS (*Cactaceae*). Spiral sequence of floral organs (black numbers: perianth members, white numbers: androecial primary primordia). – Fig. 15. *Pereskia bleo* (KUNTH) DC. (*Cactaceae*). Different degree of subdivision of the androecial primary primordia (white numbers; black numbers: spirally arranged inner perianth members). – *G* gynoecium, *T* tepalum, SEM-photographs by S. SCHWITALLA.

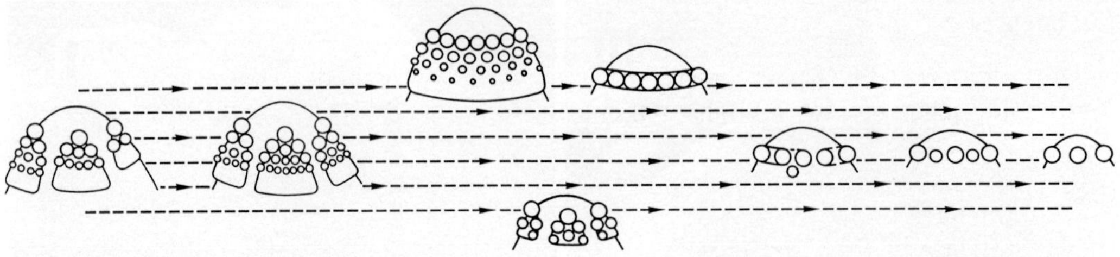

Fig. 16. Transitions from fascicled to diplostemonous (or haplostemonous) androecia via common primary ring primordium and reduction of primary primordia correlated with the reduction of stamen number

primordium, correlated with a reduction of the stamen number. There are transitions between androecia with reduced primary primordia and fixed diplo- or haplostemonous androecia. A tendency towards a diplostemonous androecium is, e.g., obvious in *Phytolaccaceae* (see plate 13.3, 13–17 in Leins & Erbar 1993), which we perhaps can link with the *Aizoaceae*.

## Final remarks

Recently many botanists favour the opinion that the fascicled androecia have evolved repeatedly from haplostemonous (or diplostemonous) ones (e.g., Dahlgren 1977, Ehrendorfer 1991). This may be justified to a certain extent if we assume that the splitting of enlarged androecial primordia has happened at a low evolutionary level. We believe it is more plausible to read the sequence of transitions in the direction from fascicled to diplostemonous androecia (Fig. 16) rather than vice versa. It is difficult to imagine that a fixed diplostemonous androecium can lose its stability and change **stepwise** to a fascicled one[4]. The plausibility is increased by the fact that our starting point, a *Paeonia*-like flower, has rather archaic characters, such as a choricarpous gynoecium, a spiral sequence of the floral organs, and an unfixed organ number. Already in earlier papers (e.g., Erbar 1988; Leins & Erbar 1991a,b) we stressed the importance of the spiral sequence of the primary androecial primordia in *Paeonia*. Therefore, we cannot understand why Ronse Decraene & Smets (1992: 623) object that we "overlook the importance of the fact that the fascicled stamens of, e.g., *Paeonia* arise on primordia, which fit into the phyllotactic sequence of a spiral flower". Just to the contrary – we have used it repeatedly as a key argument.

We would like to express our sincere thanks to Dr Jens Rohwer for carefully reading and correcting the English text.

[4] Even if the whole sequence is reductive (Fig. 16), it is easy to envisage a reversal in some individual steps, e.g., an increase in stamen number when the floral apex enlarges. It is, however, difficult to imagine that regularly organized stamen bundles could be reestablished from cases in which the primary primordia had already disappeared completely. Increasing or decreasing stamen numbers per fascicle are irrelevant in this context.

**References**

BARTHLOTT, W., 1991: Epicuticulare Wachse der Tracheophyten: systematische, mikromorphologische und ökologische Aspekte. – 10. Symp. Morph., Anat., Syst. Göttingen, Abstr. p. 2.

CRONQUIST, A., 1981: An integrated system of classification of flowering plants. – New York: Columbia University Press.

– 1988: The evolution and classification of flowering plants, 2nd edn. – Bronx, New York: The New York Botanical Garden.

DAHLGREN, G., 1989: The last Dahlgrenogram. System of classification of dicotyledons. – In KIT TAN, (Ed.): The Davis and Hedge Festschrift, pp. 249–260. – Edinburgh: University Press.

DAHLGREN, R., 1977: A commentary on a diagrammatic presentation of the angiosperms in relation to the distribution of character states. – Pl. Syst. Evol. [Suppl.] **1**: 253–283.

– 1983: General aspects of angiosperm evolution and macrosystematics. – Nordic J. Bot. **3**: 119–149.

EHRENDORFER, F., 1991: Dritte Abteilung: *Spermatophyta*, Samenpflanzen. – In SITTE, P., ZIEGLER, H., EHRENDORFER, F., BRESINSKY, A., (Eds): Lehrbuch der Botanik für Hochschulen, 33rd edn., pp. 699–828. – Stuttgart, New York: Fischer.

ENDRESS, P. K., 1987a: Floral phyllotaxis and floral evolution. – Bot. Jahrb. Syst. **108**: 417–438.

– 1987b: The early evolution of the angiosperm flower. – Trends Ecol. Evol. **2**: 300–304.

– 1992: Primitive Blüten: Sind Magnolien noch zeitgemäß? – Stapfia **28**: 1–10.

ERBAR, C., 1988: Early developmental patterns in flowers and their value for systematics. – In LEINS, P., TUCKER, S. C., ENDRESS, P. K., (Eds): Aspects of floral development, pp. 7–23. – Berlin, Stuttgart: Cramer.

– LEINS, P., 1994: Flowers in *Magnoliidae* and the origin of flowers in other subclasses of the angiosperms. I. The relationships between flowers of *Magnoliidae* and *Alismatidae*. – Pl. Syst. Evol. [Suppl.] **8**: 193–208.

FRIIS, E. M., ENDRESS, P. K., 1990: Origin and evolution of angiosperm flowers. – Adv. Bot. Res. **17**: 99–162.

– PEDERSEN, K. R., CRANE, P. R., 1993: Early Cretaceous floral structures from Portugal. – XV Internat. Bot. Congress Yokohama, Abstr. 1.7.2–4, p. 24.

HIEPKO, P., 1966: Zur Morphologie, Anatomie und Funktion des Diskus der *Paeoniaceae*. – Ber. Deutsch. Bot. Ges. **79**: 233–245.

LEINS, P., 1979: Der Übergang vom zentrifugalen komplexen zum einfachen Androeceum. – Ber. Deutsch. Bot. Ges. **92**: 717–719.

– 1983: Muster in Blüten. – Bonner Universitätsblätter **1983**: 21–33.

– ERBAR, C., 1991a: Entwicklungsmuster in Blüten und ihre mutmaßlichen phylogenetischen Zusammenhänge. – Biol. unserer Zeit **21**: 196–204.

– ERBAR, C., 1991b: Fascicled androecia in *Dilleniidae* and some remarks on the *Garcinia* androecium. – Bot. Acta **104**: 336–344.

– ERBAR, C., 1993: Putative origin and relationships of the order from the viewpoint of developmental flower morphology. – In BEHNKE, H.-D., MABRY, T. J., (Eds): Evolution and systematics of the *Caryophyllales*, pp. 303–316. – Berlin, Heidelberg, New York: Springer.

– SCHWITALLA, S., 1986: Studien an Cacteen-Blüten. I. Einige Bemerkungen zur Blütenentwicklung von *Pereskia*. – Beitr. Biol. Pflanzen **60**: 313–323.

LUNAU, K., 1990: Colour saturation triggers innate reactions to flower signals: Flower dummy experiments with bumblebees. – J. Comp. Physiol. **A166**: 827–834.

– 1991: Innate flower recognition in bumblebees (*Bombus terrestris*, *B. lucorum*; *Apidae*): Optical signals from stamens as landing reaction releasers. – Ethology **88**: 203–214.

OSCHE, G., 1979: Zur Evolution optischer Signale bei Blütenpflanzen. – Biol. unserer Zeit **9**: 161–170.
– 1983: Optische Signale in der Coevolution von Pflanzen und Tieren. – Ber. Deutsch. Bot. Ges. **96**: 1–27.
ROHWEDER, O., ENDRESS, P. K., 1983: Samenpflanzen: Morphologie und Systematik der Angiospermen und Gymnospermen. – Stuttgart, New York: Thieme.
RONSE DECRAENE, L.-P., SMETS, E. F., 1992: Complex polyandry in the *Magnoliatae*: definition, distribution and systematic value. – Nordic J. Bot. **12**: 621–649.
SCHWEITZER, H.-J., 1977: Die Räto-Jurassischen Floren des Iran und Afghanistans. 4. Die Rätische Zwitterblüte *Irania hermaphroditica* nov. spec. und ihre Bedeutung für die Phylogenie der Angiospermen. – Palaeontographica **B161**: 98–145.
TAKHTAJAN, A., 1980: Outline of the classification of flowering plants (*Magnoliophyta*). – Bot. Rev. (London) **46**: 225–359.
– 1987: Systema Magnoliophytorum. – Leningrad: Nauka. (In Russian.).
THORNE, R. F., 1981: Phytochemistry and angiosperm phylogeny – a summary statement. – In YOUNG, I. A., SEIGLER, D. S., (Eds): Phytochemistry and angiosperm phylogeny, pp. 233–294. – New York: Praeger.
– 1983: Proposed new realignments in the angiosperms. – Nordic J. Bot. **3**: 85–117.
– 1992: Classification and geography of the flowering plants. – Bot. Rev. **58**: 225–348.

Address of the authors: Prof. Dr PETER LEINS and PD Dr CLAUDIA ERBAR, Institut für Systematische Botanik und Pflanzengeographie der Universität Heidelberg, Im Neuenheimer Feld 345, D-69120 Heidelberg, Federal Republic of Germany.

Accepted January 18, 1994 by P. K. ENDRESS

# Taxonomic Index

# Subject Index

M. Hesse, E. Pacini,
M. Willemse (eds.)

## The Tapetum

### Cytology, Function, Biochemistry and Evolution

1993. 98 figures. VII, 152 pages.
Cloth DM 170,–, US$ 100.00 *
ISBN 3-211-82486-3
(Plant Systematics and Evolution / Supplement 7)

The anther tapetum, present in all land plants, is a highly specialized, transient tissue surrounding the (micro-)spores and/or pollen grains during their development. Any tapetum malfunction causes male sterility. The exact knowledge of tapetum form and function therefore is indispensable not only for basic research, but also and especially in plant breeding and plant genetics.
In fourteen contributions by reknown experts, a comprehensive account of the various characters and functions of the tapetum is provided, covering the areas of cytology, cytophysiology, biochemistry, tapetum development and function.

P. F. Yeo

## Secondary Pollen Presentation

### Form, Function and Evolution

1993. 55 figures. VIII, 268 pages.
Cloth DM 220,–, US $ 130.00 *
ISBN 3-211-82448-0
(Plant Systematics and Evolution / Supplement 6)

Secondary pollen presentation is presentation of pollen to vectors by structures other than anthers, either passively or via a specialized protection and delivery system. The main part of the book describes secondary pollen presentation genus-by-genus in 25 families. The process is surveyed in detail on the basis of published and hitherto unpublished observations. Discussions of the families and of the whole topic attempt to explain it and point the way to future work which could fill the numerous gaps in our knowledge. This is a reference work for morphologists, systematists and floral ecologists.

M. Hesse, F. Ehrendorfer (eds.)

## Morphology, Development, and Systematic Relevance of Pollen and Spores

1990. 122 figures. VII, 124 pages.
Cloth DM 138,–, US $ 80.00 *
ISBN 3-211-82182-1
(Plant Systematics and Evolution / Supplement 5)

The contributions in this volume cover the comparative morphology and the systematic/evolutionary significance of pollen/spores in critical taxa, aspects of pollen development, the substructure of sporopollenin, homologies between wall strata of ferns, gymnosperms and angiosperms, and important (but so far underrated) physical aspects of harmomegathy and pollen transport.

T. J. Mabry, G. Wagenitz (eds.)

## Research Advances in the Compositae

1990. 20 figures. V, 124 pages.
Cloth DM 138,–, US $ 80.00*
ISBN 3-211-82174-0
(Plant Systematics and Evolution / Supplement 4)

The volume presents modern research approaches for understanding evolution among members of the family *Compositae* foregrounding chemical and serological techniques, chloroplast DNA restrictive site analyses as well as classical methods for investigating the systematics of the family.

* *10% price reduction for subscribers to "Plant Systematics and Evolution"*

*Prices are subject to change without notice*

**Springer-Verlag Wien New York**

Sachsenplatz 4–6, P.O.Box 89, A-1201 Wien · 175 Fifth Avenue, New York, NY 10010, USA
Heidelberger Platz 3, D-14197 Berlin · 3-13, Hongo 3-chome, Bunkyo-ku, Tokyo 113, Japan

# Plant Gene Research

## Basic Knowledge and Application

Editors: E. S. Dennis, B. Hohn, Th. Hohn, P. J. King, F. Meins, J. Schell, D. P. S. Verma

Recombinant DNA techniques have contributed very much to molecular biology in general and to plant molecular biology in particular.

This experimental capacity combined with the concomitant development of plant tissue culture techniques, somatic embryogenesis, protoplast fusions, anther cultures etc., have already turned the previously neglected plants into very attractive subjects for molecular and cellular biologists. Possibly the strongest influence in making plant gene research a growing field of scientific interest and activity is the expectation that this research will open new and very effective ways for the breeding of agriculturally and biotechnologically important plants.

The present series on Plant Gene Research is therefore very timely and can be expected to play a major role.

Th. Boller, F. Meins (eds.)

**Genes Involved in Plant Defense**

1992. 34 figures. X, 364 pages.
Cloth DM 218,–, US $ 136.00
ISBN 3-211-82312-3

The co-evolution of plants and microbes has led to an elaborate system of genes involved in recognition, attack and defense. This volume deals with these genes and regulation of their expression. The topic is treated broadly ranging from considerations of population genetics to the identification of defense-related genes and their regulation. The book provides a concise, authoritative review of latest developments in a rapidly developing and active field of agronomic importance.

R. G. Herrmann (ed.)

**Cell Organelles**

1992. 55 figures. XII, 467 pages.
Cloth DM 229,–, US $ 143.00
ISBN 3-211-82264-X

*Cell Organelles* is a comprehensive overview of developments in the biology of plastids, mitochondria, glyoxisomes and peroxisomes. Written by well-known scientists in their area of plant research, the twelve articles in this volume cover topics ranging from organelle biogenesis and heredity to evolution, from both molecular and classical points of view. The authors have also included a modern treatment of genetic compartmentalization – one of the fundamental characteristics of eukaryotes.

E. S. Dennis, D. J. Llewellyn (eds.)

**Molecular Approaches to Crop Improvement**

1991. 34 figures. IX, 166 pages.
Cloth DM 112,–, US $ 70.00
ISBN 3-211-82230-5

D. P. S. Verma, R. B. Goldberg (eds.)

**Temporal and Spatial Regulation of Plant Genes**

1988. 55 figures. XIII, 344 pages.
Cloth DM 213,–, US $ 133.00
ISBN 3-211-82046-9

Th. Hohn, J. Schell (eds.)

**Plant DNA Infectious Agents**

1987. 76 figures. XIV, 348 pages.
Cloth DM 198,–, US $ 124.00
ISBN 3-211-81995-9

A. D. Blonstein, P. J. King (eds.)

**A Genetic Approach to Plant Biochemistry**

1986. 30 figures. IX, 291 pages.
Cloth DM 128,–, US $ 80.00
ISBN 3-211-81912-6

B. Hohn, E. S. Dennis (eds.)

**Genetic Flux in Plants**

1985. 40 figures. XII, 253 pages.
Cloth DM 109,–, US $ 68.00
ISBN 3-211-81809-X

D. P. S. Verma, Th. Hohn (eds.)

**Genes Involved in Microbe-Plant Interactions**

1984. 54 figures. XIV, 393 pages.
Cloth DM 169,–, US $ 106.00
ISBN 3-211-81789-1

*Prices are subject to change without notice*

**Springer-Verlag Wien New York**

Sachsenplatz 4–6, P.O.Box 89, A-1201 Wien · 175 Fifth Avenue, New York, NY 10010, USA
Heidelberger Platz 3, D-14197 Berlin · 3-13, Hongo 3-chome, Bunkyo-ku, Tokyo 113, Japan

*Springer-Verlag and the Environment*

WE AT SPRINGER-VERLAG FIRMLY BELIEVE THAT AN international science publisher has a special obligation to the environment, and our corporate policies consistently reflect this conviction.

WE ALSO EXPECT OUR BUSINESS PARTNERS – PRINTERS, paper mills, packaging manufacturers, etc. – to commit themselves to using environmentally friendly materials and production processes.

THE PAPER IN THIS BOOK IS MADE FROM NO-CHLORINE pulp and is acid free, in conformance with international standards for paper permanency.

DATE DUE